BARRON'S

SAT®
SUBJECT TEST
Physics

WITH 5 PRACTICE TESTS

FOURTH EDITION

Robert Jansen, M.A.

Aliso Niguel High School

Aliso Viejo, California

and

Greg Young, M.S.Ed.

San Clemente High School

San Clemente, California

®SAT is a registered trademark of the College Board, which was not involved in the production of, and does not endorse, this product.

About the Authors

Robert Jansen has taught Advanced Placement Physics at Aliso Niguel High School in Aliso Viejo, California, since 1998. He holds a bachelor's degree in psychobiology from the University of California, Los Angeles, and a master's degree in education from Pepperdine University. He gravitated toward teaching physics because of the challenging material and a sustained belief that physics does not need to be mysterious and difficult, but rather can be comprehensible and achievable. The result has been a large and competitive physics program where each year over 220 students participate in the AP Physics 1 and AP Physics C courses. The overarching goal is preparedness and confidence for students who will be studying science during their undergraduate years.

Greg Young has been teaching high school science for more than twenty years. He currently teaches Honors Physics, AP Chemistry, and Chemistry at San Clemente High School in San Clemente, California. He holds a bachelor's degree in biochemistry from the University of California, San Diego and a master's degree in science education from USC. Having always been interested in science and how to make it relevant to others, Greg's interest in teaching lies in being able to create interactive lessons that engage students in their learning and form a relevant context for difficult concepts in physics and chemistry. Science made interesting is science worth learning.

Published by Kaplan, Inc., d/b/a Barron's Educational Series
750 Third Avenue
New York, NY 10017
www.barronseduc.com

ISBN: 978-1-5062-6309-0

10 9 8 7 6 5 4 3 2 1

Kaplan, Inc., d/b/a Barron's Educational Series print books are available at special quantity discounts to use for sales promotions, employee premiums, or educational purposes. For more information or to purchase books, please call the Simon & Schuster special sales department at 866-506-1949.

Contents

Introduction 1

Using This Resource Efficiently 1
The Physics Exam 2
General Examination Strategies 4

Diagnostic Test 7

Diagnostic Test 11
Answer Key 36
Diagnostic Chart 37
Scoring Your Test 37
Answers Explained 39

1 Conventions and Graphing 47

Fundamental and Derived Units 47
Graphing Variables 49
Slope and Area 49
Interpreting Graphs 51
Summary 54
Practice Exercises 55
Answers Explained 57

2 Vectors 59

Coordinate System 59
Scalars 60
Vectors 61
Vector Mathematics 63
Summary 68
Practice Exercises 69
Answers Explained 71

3 Kinematics in One Dimension 73

Kinematic Quantities 73
Identifying Variables 77
Kinematic Equations 79
Kinematic Graphs 82
Summary 84
Practice Exercises 86
Answers Explained 89

4 Kinematics in Two Dimensions 91

Independence of Motion 92
True Velocity and Displacement 93
Relative Velocity 93
Projectile Motion 96
Summary 100
Practice Exercises 102
Answers Explained 105

5   Dynamics...109

Inertia...110
Force..110
Common Forces...111
Force Diagrams..114
Newton's Laws of Motion..115
Solving Force Problems...117
Summary...130
Practice Exercises...131
Answers Explained..136

6   Circular Motion...141

Uniform Circular Motion..141
Angular Displacement...145
Dynamics in Circular Motion...145
Angular Velocity..146
The False Centrifugal Force...147
Summary..150
Practice Exercises...152
Answers Explained..154

7   Energy, Work, and Power..157

Mechanical Energy..157
Work...160
Power...167
Conservation of Energy...168
Summary..175
Practice Exercises..177
Answers Explained..180

8   Momentum and Impulse...183

Momentum..183
Impulse..184
Conservation of Momentum..187
Energy in Collisions..187
Summary..190
Practice Exercises...191
Answers Explained..193

9   Gravity...195

Universal Gravity...196
Gravitational Field...197
Circular Orbits..198
Kepler's Laws...201
Summary...202
Practice Exercises...203
Answers Explained..206

10 Electric Fields..209

Charge...210
Electric Fields...212
Uniform Electric Fields...212
Electric Fields of Point Charges..216

Summary 221
Practice Exercises 222
Answers Explained 225

11 Electric Potential 227

Potential of Uniform Fields 228
Potential of Point Charges 230
Electric Potential Energy 231
Motion of Charges and Potential 233
Capacitors 236
Summary 239
Practice Exercises 241
Answers Explained 244

12 Circuit Elements and DC Circuits 247

Principal Components of a DC Circuit 247
DC Circuits 249
Heat and Power Dissipation 257
Summary 258
Practice Exercises 260
Answers Explained 263

13 Magnetism 265

Permanent or Fixed Magnets 266
Current-Carrying Wires 267
Solenoids and Electromagnets 269
Force on Moving Charges 270
Force on Current-Carrying Wires 274
Electromagnetic Induction 275
Summary 281
Practice Exercises 283
Answers Explained 287

14 Simple Harmonic Motion 289

Terms Related to SHM 289
Oscillations of Springs 291
Oscillations of Pendulums 295
Graphical Representations of SHM 295
Trends in Oscillations 296
Summary 298
Practice Exercises 300
Answers Explained 303

15 Waves 305

Traveling Waves 305
Mechanical Waves 308
Electromagnetic Waves 308
Doppler Effect 311
Superposition and Standing Waves 314
Summary 318
Practice Exercises 319
Answers Explained 322

16 Geometric Optics...323

Ray Model of Light..324
Reflection..324
Refraction...325
Pinhole Camera...328
Thin Lenses...329
Spherical Mirrors...335
Summary...339
Practice Exercises..341
Answers Explained...344

17 Physical Optics..347

Diffraction...347
Interference of Light...348
Polarization of Light...353
Color...356
Summary...357
Practice Exercises..358
Answers Explained.. 361

18 Thermal Properties...363

Thermal Systems...363
Thermal Energy...364
Temperature..365
Thermal Expansion...365
Ideal Gases...366
Heat and Heat Transfer...369
Heating and Cooling...370
Summary...374
Practice Exercises..376
Answers Explained...379

19 Thermodynamics...381

Internal Energy.. 381
Energy Transfer in Thermodynamics.......................................382
Energy Model Summarized...384
First Law of Thermodynamics...384
Heat Engines...386
Entropy...387
Second Law of Thermodynamics...388
Summary...389
Practice Exercises..391
Answers Explained...393

20 Atomic and Quantum Phenomena...395

Development of the Atomic Theory...395
Energy-Level Transitions..399
Ionization Energy/Work Function...403
Photoelectric Effect..403
Summary...407
Practice Exercises..409
Answers Explained.. 412

21 Nuclear Reactions 415

Quarks 415

Nucleons 415

Subatomic Particles 416

Isotopes 418

The Strong Force 419

Mass-Energy Equivalence 419

Radioactive Decay 420

Fission and Fusion 424

Summary 426

Practice Exercises 428

Answers Explained 431

22 Relativity 433

Special Theory of Relativity 433

Time, Length, and Mass 434

Summary 436

Practice Exercises 437

Answers Explained 439

23 Historical Figures and Contemporary Physics 441

Historical Figures 441

Contemporary Physics 444

Practice Exercises 446

Answers Explained 447

Practice Tests 449

Practice Test 1 453

Answer Key 478

Diagnostic Chart 479

Scoring Your Test 479

Answers Explained 480

Practice Test 2 493

Answer Key 515

Diagnostic Chart 516

Scoring Your Test 516

Answers Explained 517

Practice Test 3 531

Answer Key 556

Diagnostic Chart 557

Scoring Your Test 557

Answers Explained 558

Appendix I: Key Equations 569

Mechanics 569

Electricity and Magnetism 570

Simple Harmonic Motion 571

Waves and Optics 571

Thermal Physics/Thermodynamics 572

Atomic and Modern Physics 572

Appendix II: Physical Constants..573

Appendix III: Conversion Factors..575

Metric Conversion Factors 575
Other Conversion Factors 575

Glossary..577

Index..591

Introduction

The SAT Subject Test in Physics is designed to assess the outcome of completing a college-preparatory physics course in high school. Although state and course requirements for physics may vary, all college-preparatory physics courses should address certain core topics and principles. The SAT Subject Test in Physics focuses on this common ground. The goal of this book is to review the main topics and concepts that are likely to appear on the SAT Subject Test in Physics and help you prepare for the exam.

USING THIS RESOURCE EFFICIENTLY

The chapters are organized to maximize the effectiveness of your study time. Each chapter begins with a summary of the topics to be covered, bulleted points of the major topics, and a list of new variables discussed in the chapter. The body of the chapter includes a discussion of the topics along with relevant example questions. Each chapter also includes a unique "What's the Trick?" approach to help you solve the questions quickly and effectively. The margins contain tips called "If You See . . ." that point out some of the major insights into critical topics and difficult concepts. The end of the chapter contains a bulleted summary along with a table grouping the critical "If You See . . ." elements as a concentrated review. Each chapter is followed by multiple-choice practice questions with answers and explanations.

> To see the Table of Contents for Video Problems, go to *online.barronsbooks.com*

In addition to the chapters that review the exam content, the book includes four complete practice exams. The first practice exam is a diagnostic test to assess your current level of understanding of the subject matter and to establish a baseline score to improve upon. Ideally, you should take this first diagnostic examination using the same guidelines as an actual SAT Subject Test in Physics:

- Time limit of 1 hour for 75 multiple-choice questions.
- NO calculators allowed.
- No physics formula sheet is allowed, and none will be provided.
- Correct answers receive 1 point.
- Subtract ¼ point for each incorrect answer.
- Answers left blank receive 0 points.

A complete list of test-taking parameters and how to find your approximate raw score is provided near the end of this introduction. You should take the other three examinations after you have completed all or portions of your review.

Different students will approach this review in a variety of ways. Some may choose to work methodically through each chapter, which will require starting well before the actual exam date and setting aside adequate review time. Students with limited time may decide to read the important "If You See . . ." tips in the margins and attempt the end-of-chapter questions to determine if they should study a particular chapter in depth. Keep in mind that each chapter builds on the material from previous chapters. Skimming the material too quickly, especially in chapters containing key foundational material, can result in errors throughout the entire exam. Remember these helpful tips as you use this review book:

- Start reviewing the material well before the exam date. Set aside an hour or two each day to read through the chapters. Trying to cram in all the information at once is not as effective as reviewing smaller portions over time.
- Solve the practice problems as though they are an actual exam. Merely reading the solutions without actually attempting to solve the problems will not help you to understand the material.
- Being able to visualize the events described in an exam question is a valuable skill in physics. Students who construct diagrams to represent the situations described in physics problems tend to earn better scores on the exam.
- Some questions require you to recall facts, and others require you to understand concepts and principles. Many involve the use or understanding of formulas without complicated arithmetical calculations. Calculators are *not* allowed on the examination, and a list of formulas is *not* provided. Therefore, memorizing key physics formulas and having a working knowledge of how to manipulate variables are crucial for success.

THE PHYSICS EXAM

A complete outline of the contents of the SAT Subject Test in Physics can be obtained from the College Board's website at *http://sat.collegeboard.org*. The College Board, which writes and administers the examination, does not publish copies of former examinations. However, they do offer sample questions on their website.

All questions are multiple choice and have five answer choices. The practice tests and sample questions in this book reflect both the content and the question formats found on the SAT Subject Test in Physics. The exam tests students' knowledge in six topics.

The content and approximate percentage of the test devoted to that content is as follows:

Mechanics—Approximately 40% (~30 questions)

- **Kinematics:** may include velocity, acceleration, motion in one dimension, projectile motion, and graphical analysis.
- **Dynamics:** may include force, Newton's laws, static equilibrium, vectors, circular motion, centripetal force, universal gravitation, Kepler's laws, and simple harmonic motion, such as pendulums and mass on a spring.
- **Energy and momentum:** may include potential and kinetic energy, work, power, impulse, momentum, conservation of energy, and conservation of momentum.

Electricity and Magnetism—Approximately 20% (~15 questions)

- **Electrostatics:** may include Coulomb's law, induced charge, electric fields, electric potential, electric potential difference, electric potential energy, and parallel plate capacitors.
- **Circuits:** may include solving for series and/or parallel circuits involving resistors and lightbulbs, Ohm's law, and Joule's law.

- **Magnetism:** may include permanent magnets, Faraday's law, Lenz's law, magnetic fields created by moving charges, currents created by changing magnetic fields, forces on charges in magnetic fields, and the right-hand rule.

Waves and Optics—Approximately 20% (~15 questions)

- **General aspects of waves:** may include wave speed, frequency, wavelength, amplitude, the effect of the medium on wave properties, superposition, standing waves, and Doppler effect.
- **Ray optics:** may include reflection, refraction, Snell's law, ray tracing as it pertains to pinholes, mirrors, and lenses.
- **Physical optics:** may include single-slit diffraction, double-slit interference, polarization, and color.

Heat and Thermodynamics—Approximately 8% (~6 questions)

- **Thermal properties:** may include temperature, heat, heat transfer, specific and latent heats of fusion and vaporization, changes in state, and thermal expansion.
- **Laws of thermodynamics:** may include first and second laws of thermodynamics, entropy, internal energy, heat engines, and efficiency.

Modern Physics—Approximately 8% (~6 questions)

- **Quantum and atomic phenomena:** may include Rutherford and Bohr models of the atom, energy levels, atomic spectra, photons, and the photoelectric effect.
- **Nuclear physics:** may include fundamental particles, radioactivity, nuclear reactions, half-life, fission, and fusion.
- **Relativity:** may include length contraction, speed of light, time dilation, and mass-energy equivalence.

Miscellaneous—Approximately 4% (~3 questions)

- **General:** may include the history of physics and important persons in the development of physics.
- **Analytical skills:** may include graphical analysis, measurement, and math skills as related to the topics covered.
- **Contemporary physics:** may include astronomy, superconductivity, and current events in the world of physics.

Format of the SAT Subject Test in Physics

The following list describes the overall format of the SAT Subject Test in Physics:

- The test is 1 hour and consists of 75 multiple-choice questions.
- No calculators are allowed on the test.
- A list of physics formulas is *not* provided on the examination.
- To simplify calculations, $g = 10 \text{ m/s}^2$ is used in all such problems.
- The total score for the test is reported on a 200-to-800 point scale.
- There is a ¼-point deduction for any incorrect answers marked.

Raw Score and Approximate Scaled Score

After you have taken one of the practice tests included in this book, you will want to determine your raw score. To do so, use the following formula:

$$\text{Raw score} = \text{\# Correct} - (\text{\# Incorrect} \times \tfrac{1}{4}) = \underline{\hspace{2cm}}$$

Multiply by ¼ the number of questions answered incorrectly. This is known as the "guessing penalty." Do not deduct points for unanswered, blank questions. Questions that are left blank receive 0 points.

Scaled scores vary from test session to test session, so there is no accurate way to predict what raw score will produce a particular scaled score. However, raw scores between 65 and 75 will typically qualify for a scaled score of around 800, and a raw score of 45 will usually qualify for a scaled score of around 700.

Although colleges do not publish their SAT Subject Test admission data, it is fairly safe to assume that a score of 700 or better on any SAT Subject Test is considered to be an excellent score. Admission to any university is a complicated process and encompasses many factors, one of which can be the SAT Subject Test scores.

GENERAL EXAMINATION STRATEGIES

Multiple-choice exam questions in physics often involve many elements simultaneously. They require students to know definitions, concepts, and how variables are mathematically related. In addition, the answer choices include well-thought-out distracters. Use the following exam strategies to help overcome these challenges.

Write on the Exam

You are allowed to write on the exam booklet. Use this to your advantage.

1. **WHEN READING A DIFFICULT PROBLEM, UNDERLINE OR CIRCLE WHAT THE QUESTION IS ASKING FOR.** When you choose an answer, make sure it answers what the question is looking for and is not a partial answer. Example: Students are often tricked into choosing an answer that describes velocity, when the question actually asks for the trend in acceleration.

2. **DRAWING A SKETCH IS EXTREMELY ADVANTAGEOUS.** When in doubt, making a quick sketch of the problem always improves the odds of arriving at a correct answer.

3. **IN COMPLEX PROBLEMS, MAKE A LIST OF VARIABLES.** Doing this turns a word problem into a math problem. Be aware of hidden variables. These are often zero quantities hidden in the language of the problem. Example: The phrase "constant velocity" is a way of indicating that acceleration is zero.

4. **WRITE DOWN FORMULAS DURING THE EXAM.** Doing this will help you avoid making silly errors when solving mathematical problems. It also helps you determine more easily the relationships among variables in conceptual problems.

Make Educated Guesses

Should you guess? This is a good question that is best answered while completing the practice tests. There are ways you can improve the odds of choosing a correct answer, and there are ways to figure out if guessing is wise or not.

1. **IF YOU ARE ABLE TO IDENTIFY OBVIOUS INCORRECT ANSWERS, CROSS THEM OUT IN THE EXAM BOOKLET.** If a correct answer does not immediately present itself, then eliminate obviously wrong answers. Simply cross them out.

2. **USUALLY YOUR FIRST IMPULSE IS CORRECT.** When you cannot decide between two answers, your first choice is most often correct. If you reconsider your first choice and change the answer, you may likely change a correct answer into a wrong answer. Statistically, it is safer to keep the original answer when undecided between two possible answers. However, if you revisit a problem and are certain that you answered it incorrectly, changing an answer is a must.

3. **THERE IS A WAY TO DETERMINE IF ANSWERING QUESTIONABLE PROBLEMS IS WISE.** While taking each practice exam, circle any answers on the answer sheet that required you to guess. At the end of the exam, score these problems separately, giving 1 point for each correct answer and subtracting ¼ of a point for each incorrect answer. If the outcome is a positive score, then the strategies you are using to guess are paying off. However, if the outcome is negative, you should not guess.

4. **BEFORE TAKING THE NEXT PRACTICE EXAM REVIEW THE PROBLEMS WITH INCORRECT ANSWERS OR THOSE THAT REQUIRED GUESSING.** It may be wise to record all the corrections for missed and guessed questions in one place. Index cards and Cornell notes are ideal methods of recording a difficult question followed by the essential knowledge that leads to the correct solution. Study and review this information before taking the next practice exam and before taking the actual subject test.

Physics questions can be very challenging. For many students, improvement is a process that gets easier with each practice exam. Attempting problems and developing awareness of your own strengths and weaknesses are the key to future success. Making mistakes is not a problem as long as you make a determined effort to learn from them.

Diagnostic Test

The intent of this exam is to assess your current strengths and weaknesses. To be meaningful, take it under the same conditions as the actual exam.

- Remove the answer sheet on the following page, or number a sheet of paper from 1 to 75.
- Find a quiet place to take the test that will be free of interruptions.
- Work for no more than 1 hour on the exam.
- *Do not* use a calculator or any other resources.

Part A

In the first portion of the exam, two or more questions are grouped together. One set of answer choices is given and must be used to answer all questions in the group. Each answer choice may be used once, more than once, or not at all.

Part B

This portion of the exam is very traditional. Questions may be grouped if they use the same introductory information, diagrams, and/or graphs. However, in this portion of the exam, each question has its own unique list of answer choices.

ANSWER SHEET
Diagnostic Test

1. Ⓐ Ⓑ Ⓒ Ⓓ Ⓔ	21. Ⓐ Ⓑ Ⓒ Ⓓ Ⓔ	41. Ⓐ Ⓑ Ⓒ Ⓓ Ⓔ	61. Ⓐ Ⓑ Ⓒ Ⓓ Ⓔ
2. Ⓐ Ⓑ Ⓒ Ⓓ Ⓔ	22. Ⓐ Ⓑ Ⓒ Ⓓ Ⓔ	42. Ⓐ Ⓑ Ⓒ Ⓓ Ⓔ	62. Ⓐ Ⓑ Ⓒ Ⓓ Ⓔ
3. Ⓐ Ⓑ Ⓒ Ⓓ Ⓔ	23. Ⓐ Ⓑ Ⓒ Ⓓ Ⓔ	43. Ⓐ Ⓑ Ⓒ Ⓓ Ⓔ	63. Ⓐ Ⓑ Ⓒ Ⓓ Ⓔ
4. Ⓐ Ⓑ Ⓒ Ⓓ Ⓔ	24. Ⓐ Ⓑ Ⓒ Ⓓ Ⓔ	44. Ⓐ Ⓑ Ⓒ Ⓓ Ⓔ	64. Ⓐ Ⓑ Ⓒ Ⓓ Ⓔ
5. Ⓐ Ⓑ Ⓒ Ⓓ Ⓔ	25. Ⓐ Ⓑ Ⓒ Ⓓ Ⓔ	45. Ⓐ Ⓑ Ⓒ Ⓓ Ⓔ	65. Ⓐ Ⓑ Ⓒ Ⓓ Ⓔ
6. Ⓐ Ⓑ Ⓒ Ⓓ Ⓔ	26. Ⓐ Ⓑ Ⓒ Ⓓ Ⓔ	46. Ⓐ Ⓑ Ⓒ Ⓓ Ⓔ	66. Ⓐ Ⓑ Ⓒ Ⓓ Ⓔ
7. Ⓐ Ⓑ Ⓒ Ⓓ Ⓔ	27. Ⓐ Ⓑ Ⓒ Ⓓ Ⓔ	47. Ⓐ Ⓑ Ⓒ Ⓓ Ⓔ	67. Ⓐ Ⓑ Ⓒ Ⓓ Ⓔ
8. Ⓐ Ⓑ Ⓒ Ⓓ Ⓔ	28. Ⓐ Ⓑ Ⓒ Ⓓ Ⓔ	48. Ⓐ Ⓑ Ⓒ Ⓓ Ⓔ	68. Ⓐ Ⓑ Ⓒ Ⓓ Ⓔ
9. Ⓐ Ⓑ Ⓒ Ⓓ Ⓔ	29. Ⓐ Ⓑ Ⓒ Ⓓ Ⓔ	49. Ⓐ Ⓑ Ⓒ Ⓓ Ⓔ	69. Ⓐ Ⓑ Ⓒ Ⓓ Ⓔ
10. Ⓐ Ⓑ Ⓒ Ⓓ Ⓔ	30. Ⓐ Ⓑ Ⓒ Ⓓ Ⓔ	50. Ⓐ Ⓑ Ⓒ Ⓓ Ⓔ	70. Ⓐ Ⓑ Ⓒ Ⓓ Ⓔ
11. Ⓐ Ⓑ Ⓒ Ⓓ Ⓔ	31. Ⓐ Ⓑ Ⓒ Ⓓ Ⓔ	51. Ⓐ Ⓑ Ⓒ Ⓓ Ⓔ	71. Ⓐ Ⓑ Ⓒ Ⓓ Ⓔ
12. Ⓐ Ⓑ Ⓒ Ⓓ Ⓔ	32. Ⓐ Ⓑ Ⓒ Ⓓ Ⓔ	52. Ⓐ Ⓑ Ⓒ Ⓓ Ⓔ	72. Ⓐ Ⓑ Ⓒ Ⓓ Ⓔ
13. Ⓐ Ⓑ Ⓒ Ⓓ Ⓔ	33. Ⓐ Ⓑ Ⓒ Ⓓ Ⓔ	53. Ⓐ Ⓑ Ⓒ Ⓓ Ⓔ	73. Ⓐ Ⓑ Ⓒ Ⓓ Ⓔ
14. Ⓐ Ⓑ Ⓒ Ⓓ Ⓔ	34. Ⓐ Ⓑ Ⓒ Ⓓ Ⓔ	54. Ⓐ Ⓑ Ⓒ Ⓓ Ⓔ	74. Ⓐ Ⓑ Ⓒ Ⓓ Ⓔ
15. Ⓐ Ⓑ Ⓒ Ⓓ Ⓔ	35. Ⓐ Ⓑ Ⓒ Ⓓ Ⓔ	55. Ⓐ Ⓑ Ⓒ Ⓓ Ⓔ	75. Ⓐ Ⓑ Ⓒ Ⓓ Ⓔ
16. Ⓐ Ⓑ Ⓒ Ⓓ Ⓔ	36. Ⓐ Ⓑ Ⓒ Ⓓ Ⓔ	56. Ⓐ Ⓑ Ⓒ Ⓓ Ⓔ	
17. Ⓐ Ⓑ Ⓒ Ⓓ Ⓔ	37. Ⓐ Ⓑ Ⓒ Ⓓ Ⓔ	57. Ⓐ Ⓑ Ⓒ Ⓓ Ⓔ	
18. Ⓐ Ⓑ Ⓒ Ⓓ Ⓔ	38. Ⓐ Ⓑ Ⓒ Ⓓ Ⓔ	58. Ⓐ Ⓑ Ⓒ Ⓓ Ⓔ	
19. Ⓐ Ⓑ Ⓒ Ⓓ Ⓔ	39. Ⓐ Ⓑ Ⓒ Ⓓ Ⓔ	59. Ⓐ Ⓑ Ⓒ Ⓓ Ⓔ	
20. Ⓐ Ⓑ Ⓒ Ⓓ Ⓔ	40. Ⓐ Ⓑ Ⓒ Ⓓ Ⓔ	60. Ⓐ Ⓑ Ⓒ Ⓓ Ⓔ	

Diagnostic Test

Do not use a calculator. To simplify numerical calculations, use $g = 10 \text{ m/s}^2$.

PART A

Directions: In this section of the exam, the same lettered choices are used to answer several questions. Each group of questions is preceded by five lettered choices. When answering questions in each group, select the best answer from the available choices and fill in the corresponding bubble on the answer sheet. Each possible answer may be used once, more than once, or not at all.

Questions 1–3

    (A) Amplitude
    (B) Frequency
    (C) Resonance
    (D) Wave speed
    (E) Oscillation

1. The brightness of light and the volume of sound are associated with which wave characteristic?

2. Which wave property is controlled by the medium that the wave propagated through?

3. Which wave property remains constant when light waves enter a medium that has a greater optical density?

Questions 4–6 refer to the following field diagrams.

(A)

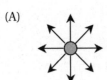

(B)

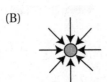

(C)

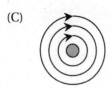

(D)

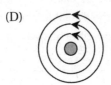

(E)

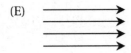

4. Which diagram correctly depicts a uniform magnetic field?

5. Which diagram correctly depicts the magnetic field of a wire carrying a current into and perpendicular to the page?

6. Which diagram correctly depicts the electric field surrounding an electron?

Questions 7–8

    (A) Albert Einstein
    (B) Albert Michelson
    (C) James Maxwell
    (D) Ernest Rutherford
    (E) J. J. Thomson

7. Which scientist suggested that light could be thought of as packets of energy and that the exact quantity of energy could be determined by the equation $E = hf$?

8. Which scientist determined that the atom consisted mostly of empty space with a small, dense, positive nucleus?

Questions 9–11

    (A) Coulomb's law
    (B) Faraday's law
    (C) First law of thermodynamics
    (D) Lenz's law
    (E) Second law of thermodynamics

9. The change in internal energy of a system is equal to the energy transferred into or out of the system by work and/or heat.

10. This law describes the direction an induced current must flow so that the induced magnetic field opposes the change in flux of the original magnetic field.

11. The entropy of a system always increases until the system reaches equilibrium.

Directions: This section of the exam consists of questions or incomplete statements followed by five possible answers or completions. Select the best answer or completion, and fill in the corresponding bubble on the answer sheet.

Questions 12–14

The motion of an object is depicted in the following speed-time graph.

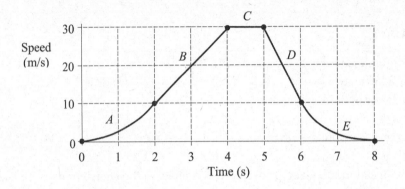

12. Determine the magnitude of acceleration during interval *B*, from 2 seconds to 4 seconds.

    (A) 0 m/s^2
    (B) 10 m/s^2
    (C) 15 m/s^2
    (D) 20 m/s^2
    (E) 40 m/s^2

13. During which interval(s) is the object moving at a constant velocity?

    (A) *A* only
    (B) *B* only
    (C) *C* only
    (D) Both *A* and *E*
    (E) Both *B* and *D*

14. During which interval did the object travel the farthest?

    (A) *A*
    (B) *B*
    (C) *C*
    (D) *D*
    (E) *E*

15. An object is accelerating. Which of the following is NOT possible?

    (A) The speed of the object may be constant.
    (B) The magnitude of the object's velocity may be constant.
    (C) The velocity of the object may be constant.
    (D) The object may be turning.
    (E) The magnitude of the force acting on the object is constant.

Questions 16–17

The motion of an object is depicted in the following position-time graph.

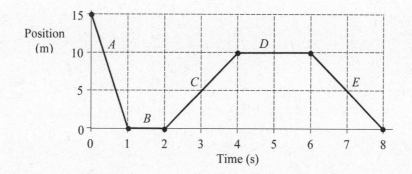

16. During which interval(s) is the magnitude of the object's velocity decreasing?

    (A) *A* only
    (B) *B* only
    (C) *E* only
    (D) *A* and *E* only
    (E) None of these

17. During which interval does the object have the greatest speed?

    (A) *A*
    (B) *B*
    (C) *C*
    (D) *D*
    (E) *E*

---

18. An object initially at rest uniformly accelerates for t seconds and moves distance x. An identical object that has twice the force applied to it during the same time, t, will move a distance

    (A) $\frac{1}{2}x$
    (B) x
    (C) $\sqrt{2}(x)$
    (D) $2x$
    (E) $4x$

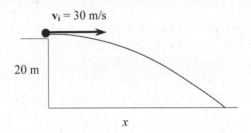

19. A ball is thrown horizontally at 30 meters per second from the top of a 20-meter-tall platform, as shown above. Determine the horizontal distance traveled by the ball.

(A) 10 m
(B) 20 m
(C) 40 m
(D) 50 m
(E) 60 m

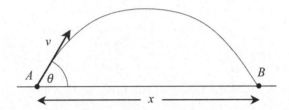

20. The diagram above depicts a projectile launched from point A with a speed v at an angle of θ above the horizontal. The projectile hits the ground at point B, achieving a final range of x. The total time of flight from point A to point B is t seconds. Determine the speed of the projectile at point B.

(A) zero
(B) v
(C) $\frac{1}{2}v$
(D) $v \cos \theta$
(E) $v \sin \theta$

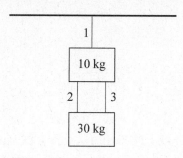

21. Two masses, 10 kilograms and 30 kilograms, are suspended by massless ropes from the ceiling, as shown in the diagram above. Determine the tension in rope 3.

    (A) 100 N
    (B) 150 N
    (C) 200 N
    (D) 300 N
    (E) 400 N

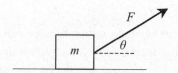

22. As shown in the figure above, mass m is pulled along a rough horizontal surface by force F, acting at an angle θ measured from the surface. The resulting motion is constant velocity. Which statement below is true?

    (A) The weight, W, of the object is equal to the normal force, N.
    (B) The weight, W, of the object is less than the normal force, N.
    (C) The force of friction, f, is equal to the applied force, F.
    (D) The force of friction, f, is less than the applied force, F.
    (E) The force of friction, f, is greater than the applied force, F.

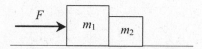

23. A force, $F = 12$ newtons, pushes two masses, $m_1 = 3$ kilograms and $m_2 = 1$ kilogram, horizontally along a frictionless surface, as shown in the diagram above. Determine the acceleration of mass m_2.

    (A) 1 m/s^2
    (B) 2 m/s^2
    (C) 3 m/s^2
    (D) 4 m/s^2
    (E) 12 m/s^2

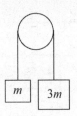

24. Masses m and $3m$ are connected by a string, which is draped over a pulley, as shown in the diagram above. The masses are released from rest. Determine the magnitude of acceleration of mass m.

(A) $\dfrac{g}{4}$

(B) $\dfrac{g}{3}$

(C) $\dfrac{g}{2}$

(D) g

(E) $2g$

25. At $t = 0$ seconds, a force, $F_1 = 10$ newtons, acting in the $+x$-direction is applied to a 5-kilogram mass that is initially at rest. At $t = 2$ seconds, a new force is added to the first force. The new force, $F_2 = 10$ newtons, acts in the $-x$-direction. Determine the acceleration of the object at $t = 5$ seconds while both forces continue to be applied.

(A) zero

(B) 1 m/s^2

(C) 2 m/s^2

(D) 4 m/s^2

(E) 5 m/s^2

26. A 50-kilogram person stands on a scale in an elevator that is accelerating upward at 1 meter per second squared. What is the apparent weight of the person?

(A) zero

(B) 50 N

(C) 450 N

(D) 500 N

(E) 550 N

27. A mass remains at rest on an incline, as shown above. Which free-body diagram is correct?

(A)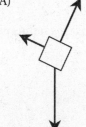

(B)

(C)

(D)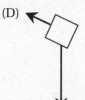

(E)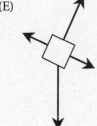

28. An object in uniform circular motion with a radius of 20 meters has a frequency of 0.10 hertz. Determine the speed of the object.

    (A) π m/s
    (B) 2π m/s
    (C) 4π m/s
    (D) 8π m/s
    (E) 16π m/s

29. What is the maximum speed possible that a car can turn on a road with a radius of 5 meters and a coefficient of friction of 0.5 without slipping?

    (A) 1 m/s
    (B) 5 m/s
    (C) 10 m/s
    (D) 25 m/s
    (E) 30 m/s

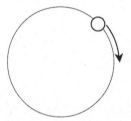

30. The object in the diagram above is in uniform circular motion. Which vectors show the direction of tangential velocity and centripetal acceleration for the object at the instant diagrammed?

(A)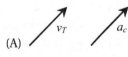

(B)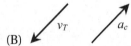

(C)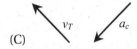

(D)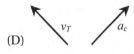

(E)

31. A roller coaster needs to complete a vertical loop that has a radius of 14.4 meters. What must the coaster's minimum speed be at the top of the loop?

    (A) 5 m/s
    (B) 7 m/s
    (C) 10 m/s
    (D) 12 m/s
    (E) 14 m/s

Questions 32–33

In the figure below, a 10-newton force, F, is applied at a 37° angle with respect to the horizontal to a mass, m. The mass is pulled horizontally to the right at constant velocity along a rough surface. Force F and its components are shown in the diagram.

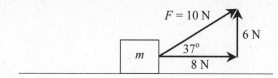

32. Determine the work done by force F as it moves the mass a distance of 5.0 meters horizontally.

    (A) zero
    (B) 30 J
    (C) 40 J
    (D) 50 J
    (E) 100 J

33. Determine the net work done on the object during the 5.0-meter motion.

    (A) zero
    (B) 30 J
    (C) 40 J
    (D) 50 J
    (E) 100 J

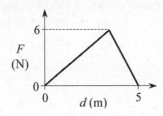

34. A variable force acts on a 2.0-kilogram mass, displacing the mass 5.0 meters. The force and displacement are graphed above. Determine the work done on the mass by the variable force.

(A) 5 J
(B) 10 J
(C) 15 J
(D) 20 J
(E) 30 J

Questions 35–36

A 50-kilogram roller coaster car is initially at rest at the top of a 25-meter-high hill. When it is released, the car rolls down the hill and passes through a loop that has a radius of 10 meters.

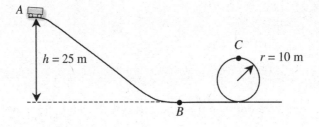

35. Determine the speed of the roller coaster when it reaches point B at the bottom of the hill.

(A) $2\sqrt{5}$ m/s
(B) $10\sqrt{5}$ m/s
(C) $30\sqrt{5}$ m/s
(D) $40\sqrt{5}$ m/s
(E) $50\sqrt{5}$ m/s

36. What is the change in potential energy as the roller coaster car moves from point B to point C?

(A) 1,000 J
(B) 2,500 J
(C) 5,000 J
(D) 10,000 J
(E) 20,000 J

---

37. Stretching a spring a distance of x requires a force of F. In the process, potential energy, U, is stored in the spring. If that same spring is stretched so that it stores $4U$ of potential energy, how far is the spring stretched?

(A) $2x$
(B) $4x$
(C) x^2
(D) x^4
(E) $2x^2$

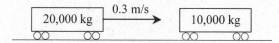

38. As shown in the diagram above, a 20,000-kilogram railroad freight car is moving at 0.3 meter per second when it strikes and couples with a 10,000-kilogram freight car that is initially at rest. What is the resulting speed of the railroad freight cars after the collision?

(A) 0.1 m/s
(B) 0.2 m/s
(C) 0.3 m/s
(D) 0.6 m/s
(E) 0.9 m/s

39. Which of the following quantities is conserved in a perfectly elastic collision?

(A) Total velocity
(B) Total linear momentum
(C) Total kinetic energy
(D) Both A and C
(E) Both B and C

Two planets are observed orbiting a star. The star has a mass of M. The smaller planet has a mass of m and is orbiting at a radius of r. The larger planet has twice the mass, $2m$, of the smaller planet and is orbiting at twice the distance, $2r$, as measured from the center of the star.

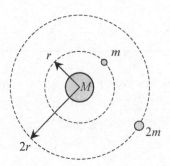

40. What is the ratio of the force of gravity acting between the central star and planet m compared with the force of gravity acting between the central star and planet $2m$?

    (A) $\frac{1}{4}$

    (B) $\frac{1}{3}$

    (C) $\frac{1}{1}$

    (D) $\frac{2}{1}$

    (E) $\frac{4}{1}$

41. Which statement is true?

    (A) Planet m has a faster tangential speed and a longer orbital period than planet $2m$.
    (B) Planet m has a faster tangential speed and a shorter orbital period than planet $2m$.
    (C) Planet m has a slower tangential speed and a longer orbital period than planet $2m$.
    (D) Planet m has a slower tangential speed and a shorter orbital period than planet $2m$.
    (E) Both planets have the same orbital period.

42. Two identical conducting spheres are initially separated. The left sphere has a negative 4-coulomb charge, and the right sphere has a positive 8-coulomb charge. The spheres are allowed to touch each other briefly, and then they are separated. Determine the charge on the left sphere.

    (A) −4 C
    (B) −2 C
    (C) 0 C
    (D) +2 C
    (E) +4 C

43. A proton and an electron are released from rest in the same uniform electric field. Assume the proton and electron do not interact with one another. How do the force and acceleration of the electron compare with those of the proton?

	Magnitude of Force on the Electron	Direction of Force on the Electron	Acceleration of the Electron
(A)	Less	Same	Less
(B)	Less	Opposite	Less
(C)	Same	Same	Greater
(D)	Same	Opposite	Less
(E)	Same	Opposite	Greater

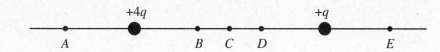

44. In the diagram above, two point charges, $+4q$ and $+q$, are held stationary. Determine the approximate location where the electric field is zero.

(A) A
(B) B
(C) C
(D) D
(E) E

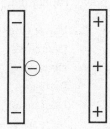

45. Two charged plates have a potential difference, ΔV, as shown in the diagram above. An electron with mass m and charge e is initially at the negative plate. The electron is accelerated through the potential difference and reaches a speed of v. The potential difference between the plates is doubled to $2(\Delta V)$. An electron accelerated through this potential difference will have a speed of

(A) $\frac{1}{2} v$
(B) v
(C) $\sqrt{2}\, v$
(D) $2v$
(E) $4v$

Questions 46–47

A set of parallel plates are charged to 0.30 coulombs and are separated by a distance of 10 centimeters. The plates have a potential difference of 6.0 volts.

46. Determine the magnitude of the electric field between the plates.

   (A) 0.050 V/m
   (B) 0.60 V/m
   (C) 1.8 V/m
   (D) 20 V/m
   (E) 60 V/m

47. What is the capacitance of the charged plates?

   (A) 0.050 F
   (B) 0.60 F
   (C) 1.8 F
   (D) 20 F
   (E) 60 F

_____

48. A simple circuit consists of a battery and a single resistor. An additional resistor is added to the circuit and is wired in series with the original resistor. How does the addition of this new resistor affect the total resistance of the circuit, the total current leaving the battery, and the total power consumed by the circuit?

	Resistance	Current	Power Consumed
(A)	Decreases	Decreases	Decreases
(B)	Decreases	Increases	Decreases
(C)	Decreases	Increases	Increases
(D)	Increases	Decreases	Decreases
(E)	Increases	Decreases	Increases

The following diagram depicts three resistors connected to an ideal battery. The switch is initially open as shown.

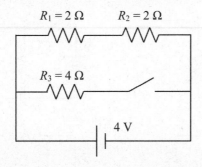

49. What is the current flowing through the battery initially, when the switch is open?

    (A) 0.5 A
    (B) 1.0 A
    (C) 2.0 A
    (D) 4.0 A
    (E) 16.0 A

50. What is the total equivalent resistance of the circuit when the switch is closed?

    (A) 1/2 Ω
    (B) 3/4 Ω
    (C) 4/3 Ω
    (D) 2 Ω
    (E) 4 Ω

51. The resistors in the circuit above are actually lightbulbs. When the switch is closed, how is the brightness of lightbulb 1 (R_1) affected?

    (A) The brightness is halved.
    (B) The brightness doubles.
    (C) The brightness is four times greater.
    (D) The brightness is eight times greater.
    (E) The brightness remains the same.

Questions 52–53

The circuits shown below all contain the same three identical resistors, each with resistance *R*, and the same identical battery with potential *V*.

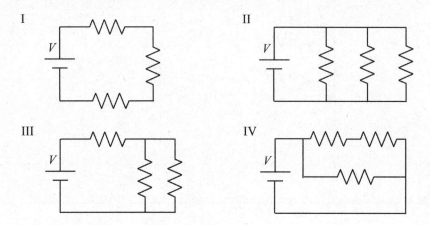

52. Which of the circuits will dissipate the most power?

   (A) I only
   (B) II only
   (C) III only
   (D) IV only
   (E) They will each dissipate the same amount of power.

53. In which circuit will the voltage drop across each resistor be identical to the voltage of the battery?

   (A) I only
   (B) II only
   (C) III only
   (D) I and II only
   (E) III and IV only

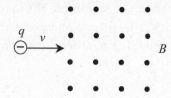
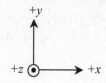

54. As shown in the diagram above, an electron with a charge of 1.6×10^{-19} coulombs is moving 1.0×10^5 meters per second in the +x-direction. The electron enters a 2.0-tesla uniform magnetic field that is oriented in the +z-direction. What are the magnitude and direction of the force that acts on the electron at the instant it enters the magnetic field?

(A) 0.8×10^{-14} N, +y-direction
(B) 0.8×10^{-14} N, −y-direction
(C) 3.2×10^{-14} N, +x-direction
(D) 3.2×10^{-14} N, +y-direction
(E) 3.2×10^{-14} N, −y-direction

55. A loop of wire and a bar magnet are moving relative to one another. Which motion in the diagrams shown below will NOT induce a current in the loop?

(A)

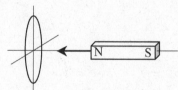

(B)

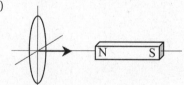

(C)

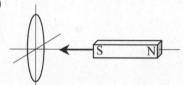

(D)

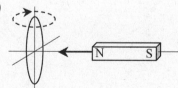

(E)

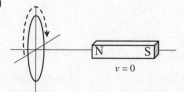

56. A mass m is attached to a spring and displaced from equilibrium. The mass is released, and the system begins to oscillate at frequency f. The mass is replaced with a new mass of $2m$ and is again displaced by the same amount. The new frequency of oscillation will be

(A) $\dfrac{\sqrt{2}}{2}f$

(B) f

(C) $\sqrt{2}f$

(D) $2f$

(E) $4f$

57. Which graph correctly depicts the potential energy, U, of a spring-mass system during an oscillation of amplitude A?

(A)

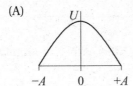

(B)

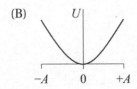

(C)

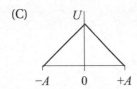

(D)

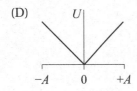

(E)

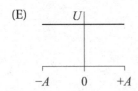

58. When light enters an optically denser medium, its

    (A) speed decreases and its wavelength decreases
    (B) speed decreases and its wavelength increases
    (C) speed decreases and its wavelength remains contstant
    (D) speed increases and its wavelength decreases
    (E) speed increases and its wavelength increases

59. During each cycle of a wave traveling through a medium, the individual oscillating
    particles move through a total distance equal to

    (A) 2 amplitudes, while the wave itself travels a $\frac{1}{2}$ wavelength and completes 1 period.

    (B) 4 amplitudes, while the wave itself travels a $\frac{1}{2}$ wavelength and completes 1 period.

    (C) 2 amplitudes, while the wave itself travels 1 wavelength and completes 1 period.
    (D) 4 amplitudes, while the wave itself travels 1 wavelength and completes 1 period.

    (E) 2 amplitudes, while the wave itself travels 2 wavelengths and completes $\frac{1}{2}$ of
        a period.

60. A sound source is moving away from an observer. As compared with the actual
    wavelength and frequency of the waves, how would the observer describe the waves?

    (A) They have shorter wavelengths and a lower frequency.
    (B) They have shorter wavelengths and a higher frequency.
    (C) They have shorter wavelengths and the same frequency.
    (D) They have longer wavelengths and a lower frequency.
    (E) They have longer wavelengths and a higher frequency.

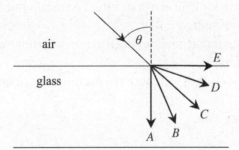

61. In the diagram above, light moving in air enters a piece of glass at an angle of θ as
    measured from a normal drawn perpendicular to the surface of the glass. Which ray
    shows the path of the light in the glass?

    (A) *A*
    (B) *B*
    (C) *C*
    (D) *D*
    (E) *E*

62. The image of a large, distant object viewed by a pinhole camera is

(A) upright, smaller than the object, and real
(B) upright, larger than the object, and real
(C) inverted, smaller than the object, and real
(D) inverted, smaller than the object, and virtual
(E) inverted, larger than the object, and real

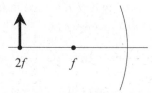

63. An object is initially placed at a distance of $2f$ from the focal point of a concave mirror, as shown in the diagram above. If the object is moved away from the mirror, how is the image affected?

(A) The image size remains constant but moves away from the mirror.
(B) The image increases in size and moves away from the mirror.
(C) The image increases in size and moves toward the mirror.
(D) The image decreases in size and moves away from the mirror.
(E) The image decreases in size and moves toward the mirror.

64. Monochromatic light passes through two narrow slits and is projected onto a screen, creating a double-slit interference pattern. Which of the following is true?

(A) The double-slit interference pattern is evidence that light has a wave characteristic.
(B) The path difference for light arriving at the first maximum from two different slits is equal to one wavelength.
(C) Increasing the separation between the two slits will compress the observed interference pattern.
(D) Increasing the wavelength will cause the maximums displayed on the screen to spread out.
(E) All of the above are true.

65. The bending of light caused by the change in the light wave's speed as it enters a new optical medium at an angle is called

(A) refraction
(B) reflection
(C) diffraction
(D) interference
(E) polarization

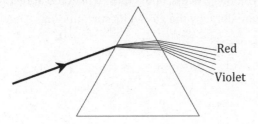

66. Why does a prism disperse white light into the colors of the spectrum in the pattern shown in the diagram above?

   (A) Violet light has more energy and therefore bends at a larger angle.
   (B) The amplitude of violet light is greater, causing greater refraction.
   (C) Each wavelength of light has a slightly different index of refraction.
   (D) The red light travels a shorter distance and bends less.
   (E) The violet light travels a longer distance and has more time to bend.

Questions 67–68 refer to the heating and cooling curve shown below.

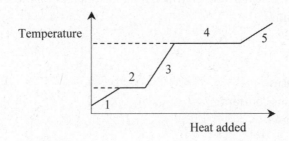

67. Which statement is true regarding process 3 in the diagram above?

   (A) The rate of temperature change is dependent on the specific heat capacity of the substance that is being heated.
   (B) The rate of temperature change is dependent on the latent heat of fusion of the substance that is being heated.
   (C) The rate of temperature change is dependent on the latent heat of vaporization of the substance that is being heated.
   (D) The substance is in the solid phase and is expanding.
   (E) The substance is in the gas phase and is expanding.

68. Which statement is true regarding process 4 in the diagram above?

   (A) This occurs at the boiling point of the substance.
   (B) Process 4 is dependent on the latent heat of vaporization.
   (C) The temperature cannot rise until the phase change is completed.
   (D) The process involves a liquid/gas phase change.
   (E) All of the above are correct.

69. During a thermodynamic process, 400 joules of heat are added to a gas, while 300 joules of work are done by the gas on its surroundings. Determine the change in internal energy.

    (A) zero
    (B) 100 J
    (C) 300 J
    (D) 400 J
    (E) 700 J

70. An engine operates between 127°C and 227°C. Determine its maximum theoretical efficiency.

    (A) 10%
    (B) 20%
    (C) 30%
    (D) 40%
    (E) 50%

71. The entropy of an isolated system, which undergoes a natural process and evolves toward equilibrium,

    (A) is zero
    (B) is one
    (C) never decreases
    (D) always remains constant
    (E) always increases

Questions 72–73

The energy-level diagram below is for a large sample of atoms that are all identical. These atoms all contain one electron that is initially in the ground state. The sample is radiated with photons that all have 10-electron volts of energy.

```
0 eV  – – – – – – – – –
−1 eV ————————————  n = 4
−2 eV ————————————  n = 3


−6 eV ————————————  n = 2



−12 eV ———————————  n = 1
```

72. Determine the energy of the excited electrons resulting from the absorption of the 10-electron volt photons.

(A) −1 eV
(B) −2 eV
(C) −6 eV
(D) −10 eV
(E) −12 eV

73. Shortly after the absorption, the atoms begin to emit photons spontaneously. What are all the possible energies of the emitted photons?

(A) 4 eV only
(B) 6 eV only
(C) 4 eV, 6 eV only
(D) 4 eV, 6 eV, and 10 eV only
(E) 4 eV, 6 eV, 10 eV, and 12 eV

---

74. A radioactive sample with a half-life of 10 days is discovered to have 1/16 of its radioactive material remaining. How many days has the sample been experiencing radioactive decay?

(A) 20 days
(B) 30 days
(C) 40 days
(D) 60 days
(E) 120 days

75. Two spaceships approach each other. Spaceship A has a speed of $0.8c$ (80% of the speed of light). Spaceship B has a speed of $0.6c$ (60% of the speed of light). A passenger on spaceship A aims a laser at spaceship B. How fast does the laser light appear to be moving as observed by a passenger on spaceship B?

(A) $0.2c$
(B) $0.6c$
(C) $0.8c$
(D) c
(E) $1.4c$

ANSWER KEY
Diagnostic Test

1.	A	26.	E	51.	E
2.	D	27.	A	52.	B
3.	B	28.	C	53.	B
4.	E	29.	B	54.	D
5.	C	30.	E	55.	E
6.	B	31.	D	56.	A
7.	A	32.	C	57.	B
8.	D	33.	A	58.	A
9.	C	34.	C	59.	D
10.	D	35.	B	60.	D
11.	E	36.	D	61.	B
12.	B	37.	A	62.	C
13.	C	38.	B	63.	E
14.	B	39.	E	64.	E
15.	C	40.	D	65.	A
16.	E	41.	B	66.	C
17.	A	42.	D	67.	A
18.	D	43.	E	68.	E
19.	E	44.	D	69.	B
20.	B	45.	C	70.	B
21.	B	46.	E	71.	E
22.	D	47.	A	72.	B
23.	C	48.	D	73.	D
24.	C	49.	B	74.	C
25.	A	50.	D	75.	D

For more practice, go to
online.barronsbooks.com

DIAGNOSTIC CHART

Subject Area	Question Numbers	Questions Incorrect	Chapter(s) to Study
Mechanics	12, 13, 14, 15, 16, 17, 18, 19, 20, 21, 22, 23, 24, 25, 26, 27, 28, 29, 30, 31, 32, 33, 34, 35, 36, 37, 38, 39, 40, 41, 56, 57		1–9
Electricity and Magnetism	4, 5, 6, 10, 42, 43, 44, 45, 46, 47, 48, 49, 50, 51, 52, 53, 54, 55		10–13
Waves and Optics	1, 2, 3, 58, 59, 60, 61, 62, 63, 64, 65, 66		14–17
Heat and Thermodynamics	9, 11, 67, 68, 69, 70, 71		18–19
Modern Physics	72, 73, 74, 75		20–22
Miscellaneous	7, 8		23

SCORING YOUR TEST

How to Determine Your Raw Score

Your raw score is the number of correctly answered questions minus the incorrectly answered questions multiplied by ¼. An incorrectly answered question is one that you bubbled in but was incorrect. If you leave the answer blank, it does not count as an incorrect answer.

Number of correctly answered questions: _____

Number of incorrectly answered questions: _____ $\times \dfrac{1}{4}$ = _____

$$\underline{\hspace{3cm}} - \underline{\hspace{3cm}} = \underline{\hspace{3cm}}$$

Number Correct   Number Incorrect $\times \dfrac{1}{4}$ Raw Score

How to Determine Your Scaled Score

The SAT Subject Tests are routinely revised and rewritten, resulting in varying point distributions for each exam. Some exams are harder, and some are easier. Statistics are applied to adjust the scaled scores on each exam so the test results can be compared with one another. In other words, the exams are curved differently. Students taking a difficult exam may need a raw score of only 58 to receive a scaled score of 800, while a student taking an easier exam may need a raw score of 65 to receive a scaled score of 800. As a result, predicting the exact scaled score for a sample test is impossible unless it is given to a very large group of students and scored by the College Board. Therefore, the values in the table below are given as approximate ranges.

Raw Score	Scaled Score	Raw Score	Scaled Score
60–75	800	20–24	520–550
55–59	770–790	15–19	490–510
50–54	730–760	10–14	450–480
45–49	700–720	5–9	420–440
40–44	660–690	0–4	380–410
35–39	630–650	−5 to −1	350–370
30–34	590–620	−10 to −6	310–340
25–29	560–580	−11 and lower	300 and less

Improving Your Score

Remember that your score on the Diagnostic Test is an approximate baseline. Do not read too much into the score itself. It is more important to identify errors and to correct them. You should also look at the questions you guessed on, even if you answered them correctly. Concentrate your efforts on correcting misconceptions to avoid making the same errors repeatedly.

Review the answer explanations. Then turn to the chapters that follow for detailed explanations of key concepts and terminology to help you make these corrections. Each chapter also contains a variety of examples and practice problems that highlight common tricks and misconceptions. You should record tough concepts, forgotten formulas, and missed questions in a format (index cards, Cornell notes, etc.) that will enable you to review them quickly. When taking the practice exams at the end of the book, treat them as though they are an actual subject test in physics. Add any errors you make to your review notes. Preparing for an exam that covers a year of study is a process, and any effort you make will lead to a better score. Students who actually attempt the practice problems, search out the information to correct their misconceptions, and review their previous mistakes will experience the greatest improvement in test scores.

ANSWERS EXPLAINED

1. **(A)**	Amplitude is volume for sound and brightness for light.
2. **(D)**	Wave speed is dependent upon the medium through which a wave travels.
3. **(B)**	Frequency remains constant when a wave enters a new medium. Only wave speed and wavelength are affected by a change in medium.
4. **(E)**	A uniform magnetic field has the same magnitude and direction at every point within a region of space. The field lines are parallel to each other.
5. **(C)**	With the thumb of your right hand pointing into the page, your fingers will curl in the clockwise direction and indicate the direction of the magnetic field created by the current. Choice D is the magnetic field created by a current coming out of the page.
6. **(B)**	Electric field lines point toward negative particles and away from positive particles. Choice A is a positive particle.
7. **(A)**	This statement describes Albert Einstein and the photoelectric effect.
8. **(D)**	This statement describes Ernest Rutherford and the gold foil experiment.
9. **(C)**	The first law of thermodynamics states $\Delta U = Q + W$.
10. **(D)**	Lenz's law is a restatement of the law of conservation of energy as it applies to induced electrical currents and their subsequently induced magnetic fields.
11. **(E)**	The second law of thermodynamics states that the entropy of an isolated system always increases until equilibrium is reached.
12. **(B)**	The slope of a speed-time graph is acceleration. The slope of the line during interval B is 10 m/s².
13. **(C)**	During interval C, the object moved at 30 m/s constantly.
14. **(B)**	The area under the curve of a speed-time graph is displacement. The greatest distance will occur in the interval that has the greatest area under the graphed interval. The area under B is 40 m.
15. **(C)**	Acceleration is the rate of change in velocity. It is also a vector quantity consisting of both magnitude and direction. Therefore, only choice C is NOT possible. Choice A is possible because the speed, which is the magnitude of velocity, of a turning object may remain constant even while direction is changing. A constant force, choice E, produces a constant acceleration and is therefore also possible.
16. **(E)**	The magnitude of velocity is the absolute value of the slope of a position-time graph. All slopes in this graph are constant; therefore, they all represent magnitudes of velocity that remain constant. In intervals A and E, the magnitude of velocity will be negative, but the value is not changing (not decreasing).
17. **(A)**	The slope of a position-time graph is speed. Interval A has the steepest slope and therefore the greatest speed.

18. **(D)**	No mention was made of mass, so it is assumed to remain constant. Under the conditions described, force, F, and acceleration, a, are directly proportional, $F = ma$. Doubling the force must then also double the acceleration.
	$$2F = m(2a)$$
	For an object initially at rest, the relationship between displacement and acceleration is:
	$$x = \frac{1}{2}at^2$$
	Displacement, x, is directly proportional to acceleration. Doubling acceleration will double the displacement during the same time interval.
	$$2x = \frac{1}{2}(2a)t^2$$
19. **(E)**	Time is dependent on y-direction variables. For a horizontal launch, the initial velocity in the y-direction is zero. Here, acceleration is g.
	$$y = \frac{1}{2}gt^2 \qquad t = \sqrt{\frac{2y}{g}} = \sqrt{\frac{2(20 \text{ m})}{(10 \text{ m/s}^2)}} = 2 \text{ s}$$
	The motion in the x-direction has constant velocity. So the horizontal distance traveled by the ball is:
	$$\Delta x = v_{0x} t = (30 \text{ m/s})(2 \text{ s}) = 60 \text{ m}$$
20. **(B)**	Air resistance is assumed to be negligible unless specifically stated otherwise. Under these conditions, the speed (magnitude of true velocity) of a projectile at a specific height will be the same on the way upward and downward at that same height. This projectile lands at the same height as it was launched, and its speed when landing is the same as when it was launched.
21. **(B)**	Rope 2 and rope 3 are equally sharing the 300 N weight of the 30 kg mass. Therefore, each has a tension of 150 N. The 10 kg mass suspended by rope 1 acts as another "ceiling" for the 30 kg mass. Its mass is not relevant to solving the tension caused by the 30 kg mass suspended beneath it.
22. **(D)**	The force of friction will always be equal to, but opposite of, the component of force acting in the direction of motion. Since the applied force, F, is at an angle to the direction of motion, the component of force actually acting in the direction of motion will be $F\cos\theta$. This is less than the applied force, F.
23. **(C)**	The masses are connected and act as if they were a single mass. Sum the forces for the entire system.
	$$\Sigma F_{\text{sys}} = (m_1 + m_2)a \qquad a = \frac{\Sigma F}{m_1 + m_2} = \frac{12 \text{ N}}{3 \text{ kg} + 1 \text{ kg}} = 3 \text{ m/s}^2$$

24. (C)	The masses are connected by a string and act as if they were a single mass of $4m$. The $3m$ mass is being pulled in one direction by the force of gravity. The m mass is being pulled in the other direction, also by the force of gravity. Sum the forces for the entire system all at once. $$\Sigma F_{\text{sys}} = F_{g_{3m}} - F_{g_m}$$ $$(3m + m)a_{\text{sys}} = (3m)g - (m)g$$ $$a_{\text{sys}} = \frac{1}{2}g$$
25. (A)	Since the two forces have equal magnitude but act in opposite directions, their net sum will be zero. The magnitude of acceleration is directly proportional to the magnitude of the net force. If the magnitude of the net force is zero, then acceleration has a magnitude of zero. An acceleration of zero is consistent with constant velocity.
26. (E)	Spring scales measure apparent weight, which is the normal force acting on an object. The normal force acts upward, while the force of gravity acts downward. Sum the forces, and solve for the normal force. $$\Sigma F = N - F_g$$ $$N = F_g + \Sigma F = mg + ma = (50 \text{ kg})(10 \text{ m/s}^2) + (50 \text{ kg})(1 \text{ m/s}^2) = 550 \text{ N}$$
27. (A)	The sum of the force vectors added tip to tail must result in a zero sum for the object to remain stationary or move at a constant, nonzero velocity. Choice A is the only diagram with a zero sum for the vectors. The others all show a resulting force vector and therefore both a net force and a net acceleration.
28. (C)	Determine the period, which is the time to complete one cycle. $$T = \frac{1}{f} = \frac{1}{0.10 \text{ Hz}} = 10 \text{ s}$$ Solve for the speed in circular motion. $$v = \frac{2\pi r}{T} = \frac{2\pi(20 \text{ m})}{10 \text{ s}} = 4\pi \text{ m/s}$$
29. (B)	When a car makes a turn, the net force resulting in the circular motion, F_C, is equal to the friction force holding the car in the turn. $$F_C = f$$ $$m\frac{v^2}{r} = \mu N \quad \text{(where } N = F_g = mg) \quad m\frac{v^2}{r} = \mu mg$$ $$v = \sqrt{\mu g r} = \sqrt{(0.5)(10 \text{ m/s}^2)(5 \text{ m})} = 5 \text{ m/s}$$
30. (E)	Tangential velocity is tangent to the circular path at any given point along the path. Centripetal acceleration always points toward the center of the circle from any given point along a circular path.

31. **(D)**	In order to complete a loop with a minimum speed at the top, the centripetal force must be a minimum. At the top of the loop gravity and the normal force of the track hold the roller coaster in the loop. Although gravity cannot be decreased, the normal force can be reduced to zero. The normal force is zero for only an instant at the very top of the loop. $$F_C = F_g + N$$ $$m\frac{v^2}{r} = mg + 0$$ $$v = \sqrt{gr} = \sqrt{\left(10 \text{ m/s}^2\right)\left(14.4 \text{ m}\right)} = 12 \text{ m/s}$$
32. **(C)**	Work is equal to the component of force parallel to the motion of an object multiplied by the distance traveled by the object. $$W = F_{\text{parallel}}\, d = (8 \text{ N})(5 \text{ m}) = 40 \text{ J}$$
33. **(A)**	There are two ways to arrive at the answer. The first uses force, and the second uses energy. The resisting force of friction acts opposite to the 8 N component of force applied in the direction of motion. This results in a net force of zero. The net work is directly proportional to the net force. $$W_{\text{net}} = F_{\text{net}}\, d = \Sigma F d = (0 \text{ N})(5 \text{ m}) = 0 \text{ J}$$ The net work is also equal to the change in kinetic energy (work–kinetic energy theorem). For an object moving at constant velocity, the change in kinetic energy and the net work are both zero.
34. **(C)**	The area under the curve of a force-displacement graph is the work done in N • m (Joules). This curve is a triangle. As such, the area can be determined by $\frac{1}{2}$ base × height. The magnitude of the mass does not affect the answer.
35. **(B)**	This question is about conservation of energy. The potential energy at the top of the hill is converted to (and equal to) the kinetic energy at the bottom of the hill. $$mgh = \frac{1}{2}\, mv^2$$ $$v = \sqrt{2gh} = \sqrt{2\left(10 \text{ m/s}^2\right)\left(25 \text{ m}\right)} = 5\sqrt{20} \text{ m/s}$$
36. **(D)**	When an object changes height, its gravitational potential energy changes. $$\Delta U = mg\Delta h = (50 \text{ kg})(10 \text{ m/s}^2)(20 \text{ m}) = 10{,}000 \text{ J}$$
37. **(A)**	Potential energy stored in a spring can be determined by: $$U_s = \frac{1}{2}\, kx^2 \qquad (4U_s) = \frac{1}{2}\, k(2x)^2$$ If the potential energy is quadrupled, it would require a doubling of the stretch of the spring, x.

38. **(B)**	Momentum is conserved in the inelastic collision between the two cars. $$m_1 v_1 + m_2 v_2 = (m_1 + m_2) v$$ $$(20{,}000 \text{ kg})(0.3 \text{ m/s}) + (10{,}000 \text{ kg})(0 \text{ m/s}) = (30{,}000 \text{ kg}) v$$ $$v = 0.2 \text{ m/s}$$
39. **(E)**	Linear momentum and kinetic energy are both conserved in perfectly elastic collisions. However, velocity is *not* conserved during collisions.
40. **(D)**	The force of gravity between the small planet and the star is: $$F_g = G \frac{mM}{r^2}$$ The force of gravity between the large planet and the star is: $$F_g = G \frac{2mM}{(2r)^2}$$ Dividing the force acting on the smaller planet by the force acting on the larger planet will give the ratio of these forces. This cancels all the variables, represented by letters, and leaves only the numerical coefficients. $$\frac{(1)}{\left(\frac{2}{2^2}\right)} = \frac{2}{1}$$
41. **(B)**	The closer a planet is to the central star, the faster its tangential speed and the shorter its period of orbit.
42. **(D)**	When the two spheres are brought into contact, their charges combine. The negative 4-coulomb charge of the left sphere and the positive 8-coulomb charge of the right sphere combine to make a total of a positive 4-coulomb charge. Upon separation of the spheres, these charges separate equally, leaving both the left and right spheres with a charge of positive 2 coulombs each.
43. **(E)**	The magnitude of force is determined by the charge multiplied by the magnitude of the electric field, $F_E = qE$. The magnitude of force on both the electron and proton are the same because the particles have equal charge. However, the direction of force is opposite for electrons and protons. Protons move with electric fields, while electrons move against electric fields. Although the magnitude of force for the two particles is equal, the acceleration is not equal. The electron has very little mass compared with protons. The same force will give the electron a much greater acceleration.
44. **(D)**	In order for the electric field to cancel, the two charges must produce electric fields that are opposite in direction and equal in magnitude. At positions *B, C,* and *D*, the electric field due to point charge $+4q$ is pointed toward the right and the electric field due to point charge $+q$ is pointed toward the left. The magnitude of the electric field is influenced by the size of the charges. The larger charge $+4q$ creates a larger overall electric field. However, the electric field also diminishes with distance. Point *D* is closer to the small charge $+q$. So at point *D* the smaller charge can generate an electric field that is equal in magnitude to the larger-far away charge.

45. **(C)**	This question involves conservation of energy. The potential energy of the electron, $U_E = qV$, accelerates the electron. When the electron reaches the positive plate, the potential energy has been converted entirely to kinetic energy. $$qV = \frac{1}{2}mv^2$$ By doubling the voltage to $2V$, this will increase the speed to $\sqrt{2}\,v$. $$q(2V) = \frac{1}{2}m(\sqrt{2}\,v)^2$$
46. **(E)**	Voltage equals the magnitude of the electric field multiplied by the distance between the plates. Convert 10 centimeters to meters. $$V = Ed \qquad E = \frac{V}{d} = \frac{6.0 \text{ V}}{0.10 \text{ m}} = 60 \text{ V/m}$$
47. **(A)**	Capacitance equals the amount of charge divided by the voltage. $$C = \frac{Q}{V} = \frac{0.30 \text{ C}}{6.0 \text{ V}} = 0.050 \text{ F}$$
48. **(D)**	Adding resistors in series will increase the resistance to the flow of current; therefore, the current will decrease. With less current flowing, less power is consumed.
49. **(B)**	The current flows only through series resistors R_1 and R_2. $$R_S = R_1 + R_2 = 2 \text{ }\Omega + 2 \text{ }\Omega = 4 \text{ }\Omega$$ Ohm's law, $V = IR$, can be used to solve for the current. $$I = \frac{V}{R} = \frac{4 \text{ V}}{4 \Omega} = 1 \text{ A}$$
50. **(D)**	When the switch is closed, the 4 Ω resistor is added in parallel into the circuit. $$\frac{1}{R_P} = \frac{1}{(R_1 + R_2)} + \frac{1}{R_3} = \frac{1}{(2\Omega + 2\Omega)} + \frac{1}{4\Omega} = \frac{1}{2\Omega}$$ Remember to invert the value above to find the resistance in parallel. $$R = 2 \text{ }\Omega$$
51. **(E)**	Closing the switch creates a new parallel path through R_3 where voltage can be applied and current can flow through. This parallel path does not affect the current flowing through or the voltage across resistors R_1 and R_2 as they are on a separate parallel path. So the brightness of the lightbulb R_1 remains the same.
52. **(B)**	Resistors in parallel will dissipate the most power.
53. **(B)**	Resistors in parallel to the battery receive the same voltage. Resistors in circuit A also have the same voltage; however, it is $\frac{1}{3}V$ for each of those resistors.
54. **(D)**	Determine the magnitude of the magnetic force on the moving charge. $$F_B = qvB = (1.6 \times 10^{-19} \text{ C})(1.0 \times 10^5 \text{ m/s})(2.0 \text{ T}) = 3.2 \times 10^{-14} \text{ N}$$ The direction of the field can be found with the right-hand rule. However, the moving charge is negative. The right hand can still be used, but then the answer must be reversed. Another way to determine the direction of negative charges is to use the left hand. Point the thumb of the left hand in the direction

	the charge is moving. Extend the fingers so they point in the direction of the magnetic field (out of the page in this case). The direction in which the palm of the hand pushes is the direction of force (toward the top of the page if using the left hand), which is the $+y$-direction.
55. **(E)**	Inducing a current requires a change in flux. Flux is the amount of magnetic field passing through the loop of wire. Only choice E demonstrates no change in the amount of magnetic field passing through the loop. All others show either an increase or a decrease in flux.
56. **(A)**	Period of oscillation for a spring system is $$T = 2\pi\sqrt{\frac{m}{k}}$$ Doubling the mass will increase the period, T, by $\sqrt{2}$. $$(\sqrt{2})T = 2\pi\sqrt{\frac{(2m)}{k}}$$ Frequency is the reciprocal of period, $f = 1/T$. So the inverse of $\sqrt{2}$ will be applied to the frequency. $$\frac{1}{\sqrt{2}}f = \frac{\sqrt{2}}{2}f$$
57. **(B)**	The potential energy is at its maximum when the spring-mass is at maximum displacement. The potential energy is zero as the spring-mass passes through the equilibrium position, where displacement is zero. Kinetic energy of the system is depicted in choice A, and total energy is depicted in choice E.
58. **(A)**	Wave speed is affected by the medium through which a wave travels. In a more dense optical medium, light speed decreases. Frequency is unaffected by the medium, and so the wavelength must adjust according to the equation $v = f\lambda$.
59. **(D)**	There is one wavelength and one period in one complete cycle. However, there are four amplitudes. Answer C is a frequently chosen distracter. Amplitude is measured from the equilibrium position to the maximum displacement, and oscillators pass through four amplitudes in one cycle.
60. **(D)**	According to the Doppler effect, a source moving away has longer wavelengths and lower frequency.
61. **(B)**	When light enters a more optically dense medium, the speed of light decreases and its wavelength shortens. Snell's law puts refraction into mathematical terms. $$n_1 \sin\theta_1 = n_2 \sin\theta_2$$ When light enters glass, the index of refraction in glass, n_2, is greater than that in air, n_1. In order to maintain the equality in Snell's law, the angle of refraction in glass, θ_2, must be smaller than the angle of incidence in air, θ_1.
62. **(C)**	Light rays passing through the pinhole will converge at the opening and invert as they pass through the opening. This creates a real, inverted image at the back of the camera. Since the image distance is small compared to the object distance, the image will be small.
63. **(E)**	As the object moves farther away from the focal point, the image continues to get closer to the mirror with the maximum distance of the image being one focal length away from the mirror. The image size, however, continues to decrease as the object moves farther and farther away from the focal point.

64. **(E)**	Each of the results listed will occur as monochromatic light is projected onto a screen and creates a double-slit interference pattern.
65. **(A)**	This is the definition of refraction.
66. **(C)**	Each wavelength of light does have a slightly different index of refraction in an optical medium. Choice A is close but is not the best answer. Color is a function of frequency, and frequency is not affected by the medium.
67. **(A)**	Process 3 represents the rate of temperature change as the substance is heated from its liquid phase to its boiling point. Latent heats occur only during processes 2 and 4, when there is a phase change and therefore no temperature change.
68. **(E)**	All of the statements are correct about process 4.
69. **(B)**	This question is about the first law of thermodynamics. Heat is being added to the system, so Q is positive. If work is being done by the system, the gas must be expanding to move the piston. In doing so, the gas loses energy. So work, W, done by the gas is negative. $$\Delta U = Q + W = (400 \text{ J}) + (-300 \text{ J}) = 100 \text{ J}$$
70. **(B)**	When temperatures are given, maximum efficiency is calculated using $$e = \frac{T_H - T_C}{T_H} = \frac{\left(227°\text{C} + 273 \text{ K}\right) - \left(127°\text{C} + 273 \text{ K}\right)}{\left(227°\text{C} + 273 \text{ K}\right)} = 0.20 = 20\%$$ When formulas contain temperature, T, degrees Kelvin are required. Degrees Celsius can only be used when a formula contains a change in temperature, ΔT. When in doubt, use degrees Kelvin.
71. **(E)**	This is the definition of entropy, the measured amount of disorder of isolated systems.
72. **(B)**	Only discreet states above the ground state, $n = 1$, exist for a particular atom. Only photons with energies matching the difference between electron energy levels can be absorbed. When the 10 eV photon is added to electrons in the ground state, the electrons acquire an energy of –2 eV and move to the $n = 3$ energy level.
73. **(D)**	The electrons will lose energy in order to return to the ground state. In some atoms, the electrons may drop from energy level $n = 3$ to energy level $n = 2$, losing 4 eV. Then these electrons will drop from energy level $n = 2$ to energy level $n = 1$, losing 6 eV. However, in other atoms the electrons may drop from $n = 3$ all the way to energy level $n = 1$, losing 10 eV. Each of these three energy-level drops produces a distinct photon with a matching energy.
74. **(C)**	During one half-life, the sample is reduced to half. In order for only $\frac{1}{16}$ of the sample to be remaining, four half-lives must have transpired. $$\frac{1}{16} = \left(\frac{1}{2}\right)\left(\frac{1}{2}\right)\left(\frac{1}{2}\right)\left(\frac{1}{2}\right) = \left(\frac{1}{2}\right)^4$$ If each half-life lasts 10 days, then four half-lives would last 40 days.
75. **(D)**	Light is always measured by all observers to be at the constant speed of c.

Conventions and Graphing

1

→ **FUNDAMENTAL AND DERIVED UNITS**

→ **GRAPHING VARIABLES**

→ **SLOPE AND AREA**

→ **INTERPRETING GRAPHS**

Understanding the measurement and unit conventions used in the SAT Subject Test in Physics and being able to identify and interpret graphs quickly will help solve many of the questions posed on the examination. The majority of the measurements and units on the examination will employ the metric system. Graphs will be used to establish relationships between dependent and independent variables as well as solve for quantities based on either the slope of a line or area under a curve. This chapter will do the following:

- Review the fundamental metric units (SI units) and some of the derived metric units (SI units) used in physics.
- Determine the dependent and independent variables of a graph.
- Explain the importance of slope and area to a graph.
- Interpret the graphical representation of common physics equations.

FUNDAMENTAL AND DERIVED UNITS

The **fundamental metric units** (SI units) in physics cover the basic quantities measured, such as length, mass, and time. The units measure a quantity and are given a unit name and symbol. Table 1.1 lists the fundamental quantities along with the unit names and symbols.

Table 1.1 Fundamental Quantities and Units

Quantity (Symbol)	Unit Name	Symbol
Length (l)	Meter	m
Mass (m)	Kilogram	kg
Time (t)	Second	s
Electric current (I)	Ampere	A
Temperature (T)	Kelvin	K
Amount of substance (n)	Mole	mol

Derived units are combinations of one or more of the fundamental units. Table 1.2 lists common derived units used in physics.

Table 1.2 Derived Units

Quantity (Symbol)	Unit Name	Unit Symbol	Fundamental Units
Area (A)	Area	m^2	m^2
Volume (V)	Volume	m^3	m^3
Density (ρ)	Density	kg/m^3	kg/m^3
Frequency (f)	Hertz	Hz	$1/s = s^{-1}$
Force (F)	Newton	N	$kg \cdot m/s^2$
Energy (E)	Joule	J	$N \cdot m = kg \cdot m^2/s^2$
Power (P)	Watt	W	$J/s = kg \cdot m^2/s^3$
Pressure (P)	Pascal	Pa	$N/m^2 = kg/m \cdot s^2$
Electric charge (q)	Coulomb	C	$A \cdot s$
Electric potential (V)	Volt	V	$J/C = J/A \cdot s = kg \cdot m^2/A \cdot s^3$

Some questions on the SAT Subject Test in Physics may ask which units correctly belong to a specific quantity. An easy way to do this is to write out the principal formula for the quantity and then replace each variable on the right side of the equation with its unit symbol. There may be more than one correct answer including the unit symbol, other derived units, and fundamental units. For example, all of the following are correct ways to express units of energy: J, $N \cdot m$, and $kg \cdot m^2/s^2$.

EXAMPLE 1.1

Derived Units

The unit of force is the newton. What are the fundamental units that make up the newton?

(WHAT'S THE TRICK?)

Write down the foundational formula for force.

$$\vec{F} = m\vec{a}$$

Replace the variable symbols with their matching units. Force is measured in newtons, N. Mass is measured in kilograms, kg. Acceleration is measured in meters per second squared, m/s^2.

$$N = kg \cdot m/s^2$$

GRAPHING VARIABLES

The graphing techniques of mathematics are used in science to compare dependent and independent variables. In mathematics, you are familiar with the traditional x- and y-coordinate axes. In science, the x-axis represents the independent variable and the y-axis represents the dependent variable. The value of the dependent variable depends upon the independent variable.

Graphs are always titled so that the dependent variable is listed first, and the independent variable is listed second. As an example, a position versus time graph would have position (dependent variable) plotted on the y-axis and time (independent variable) plotted on the x-axis.

IF YOU SEE
a graph title

Plot the first variable listed on the y-axis.

This is the dependent variable.

SLOPE AND AREA

Slope

Slopes are very important and are often the key to answering many of the graphing questions on the SAT Subject Test in Physics. Slope is determined by dividing the rise (y-axis value) by the run (x-axis value). The trick is to look at the units written on the axes of the graph. If you divide these units, you can easily identify the significance of the slope.

EXAMPLE 1.2

Slope of a Graphed Function

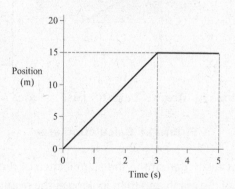

(A) What is the value and significance of the slope in the time interval from 0 to 3 seconds?

WHAT'S THE TRICK?

Determining the slope is simply a matter of dividing the rise (y-axis values) by the run (x-axis values). The significance of the slope is determined by examining the resulting units.

$$\text{slope} = \frac{\text{rise}}{\text{run}} = \frac{15m - 0m}{3s - 0s} = 5 \text{ m/s}$$

The resulting units, meters per second (m/s), are the units of velocity. Therefore, the slope of the position versus time graph is equal to velocity. During the first 3 seconds, the object has a velocity of 5 m/s.

(B) What is the value and significance of the slope in the time interval from 3 to 5 seconds?

WHAT'S THE TRICK?

The slope in the interval between 3 and 5 seconds is zero.

$$\text{slope} = \frac{\text{rise}}{\text{run}} = \frac{15m - 15m}{5s - 3s} = 0 \text{ m/s}$$

During this time interval, the object has a velocity of zero and the y-axis value (position) is not changing. The object's position remains constant at a location 15 m from the origin.

IF YOU SEE
a graph

Slope or
area may be
important.

Include units in calculations of slope and area to verify if they match the quantity you are solving for.

Area

The **area** formed by the boundary between the x-axis and the line of a graph is also very useful. Areas are calculated by multiplying the height (y-axis value) by the base (x-axis value). In problems where the area forms a triangle, the area is found with $\frac{1}{2}$ height $\times$ base. In cases where the line of the graph is below the x-axis, the area is negative. See Figure 1.1.

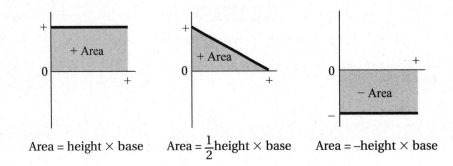

Area = height × base   Area = $\frac{1}{2}$height × base   Area = –height × base

Figure 1.1 Calculating area

As with slope, you can easily determine the significance of the area. By multiplying the units written on the axes of the graph and then looking at the resulting units, you can quickly determine the significance of the area.

EXAMPLE 1.3

Area of a Graphed Function

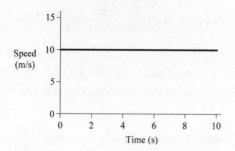

What is the value and significance of the area of the graph during the time interval between 0 and 10 seconds?

WHAT'S THE TRICK?

Determine the area, and examine the resulting units.

$$\text{area} = \text{height} \times \text{base} = (10 \text{ m/s})(10 \text{ s}) = 100 \text{ m}$$

Meters (m) are the units of displacement. The area under a speed versus time graph is therefore the displacement of the object during that time interval. The object graphed above traveled 100 m in 10 seconds.

INTERPRETING GRAPHS

Consider the graph of velocity versus time in Figure 1.2.

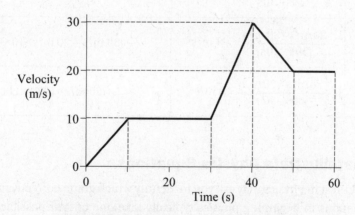

Figure 1.2 Velocity versus time graph

The graph tells the story of an object, such as a car, as it moves over a 60-second period of time. At time zero, the object has a velocity of 0 meters per second and is therefore starting from rest. The y-intercept of a speed versus time graph is the initial velocity of the object, v_0.

What the object is doing during the 60 seconds can be determined by analyzing the slope and area during the separate time intervals. Determine the significance of the slope by dividing the rise units (y-axis values) by the run units (x-axis values).

$$\text{slope units} = \frac{\text{rise units}}{\text{run units}} = \frac{\text{m/s}}{\text{s}} = \text{m/s}^2$$

The slope units, meters per second squared (m/s^2), are the units of acceleration. Thus, the slope of speed versus time is acceleration. Determine the significance of the area between the graphed function and the x-axis by multiplying the units of the y-axis by the units of the x-axis.

$$\text{area units} = \text{height units} \times \text{base units} = \frac{\text{m}}{\text{s}} \times \text{s} = \text{m}$$

Meters (m) are the units of displacement. The area of a velocity versus time graph is displacement.

To analyze the motion mathematically, divide the graph into a series of line segments and evaluate each section. The following chart shows the acceleration and displacement for the time intervals corresponding to the graphed line segments.

Time (s)	Slope (Acceleration)	Area (Displacement)
0 to 10	$\frac{10\,\text{m/s}-0\,\text{m/s}}{10\,\text{s}-0\,\text{s}} = 1\,\text{m/s}^2$	$\frac{1}{2}(10\,\text{m/s})(10\,\text{s}) = 50\,\text{m}$
10 to 30	$\frac{10\,\text{m/s}-10\,\text{m/s}}{30\,\text{s}-10\,\text{s}} = 0\,\text{m/s}^2$	$(10\,\text{m/s})(20\,\text{s}) = 200\,\text{m}$
30 to 40	$\frac{30\,\text{m/s}-10\,\text{m/s}}{40\,\text{s}-30\,\text{s}} = 2\,\text{m/s}^2$	$\frac{1}{2}(10\,\text{m/s} + 30\,\text{m/s})(10\,\text{s}) = 200\,\text{m}$
40 to 50	$\frac{20\,\text{m/s}-30\,\text{m/s}}{50\,\text{s}-40\,\text{s}} = -1\,\text{m/s}^2$	$\frac{1}{2}(30\,\text{m/s} + 20\,\text{m/s})(10\,\text{s}) = 250\,\text{m}$
50 to 60	$\frac{20\,\text{m/s}-20\,\text{m/s}}{60\,\text{s}-50\,\text{s}} = 0\,\text{m/s}^2$	$(20\,\text{m/s})(10\,\text{s}) = 200\,\text{m}$

Graphs That Illustrate Physics Equations

The SAT Subject Test in Physics may ask you to identify which graph correctly matches a given equation. Equations in beginning physics typically take one of four possible forms: linear, quadratic, square root, and inverse. You can quickly deduce the shape of a graph by looking at the relationship between the dependent and independent variables as shown in Table 1.3.

Table 1.3 Identifying Graphs

Desired Graph	What's the Trick?	Graph
How would $x = vt$ appear on an x versus t graph?	Solve for x (dependent variable). $$x = vt$$ Examine t (independent variable). t is *not* squared, *not* under a square root, and *not* inversely related to x. The graph must be linear (directly proportional).	
How would $x = \frac{1}{2}at^2$ appear on an x versus t graph?	Solve for x (dependent variable). $$x = \frac{1}{2}at^2$$ Examine t (independent variable). t is *squared*. The graph is quadratic (a parabola).	
How would $T = 2\pi\sqrt{L/g}$ appear on a T versus L graph?	Solve for T (dependent variable). $$T = 2\pi\sqrt{L/g}$$ Examine L (independent variable). L is under a *square root*.	
How would $P = nRT$ appear on a P versus V graph?	Solve for P (dependent variable). $$P = nRT\frac{1}{V}$$ Examine V (independent variable). V has an *inverse relationship* with P. The graph is a hyperbola (inversely proportional).	

SUMMARY

1. **FUNDAMENTAL AND DERIVED UNITS.** The fundamental metric units are for the basic units of measure, such as length, mass, and time. As formulas are created with these smaller units, derived units will result. You should know the fundamental unit components of derived units. You can do this by replacing quantity symbols in a physics formula with their fundamental units to determine the derived unit.

2. **HOW A GRAPH IS PLOTTED AND TITLED.** The dependent variable is plotted on the y-axis, and the independent variable is plotted on the x-axis. The title of a graph always lists the dependent variable first and the independent variable second.

3. **THE IMPORTANCE OF SLOPE.** The slope of a line is determined by dividing the rise (y-axis value) by the run (x-axis value). Whenever you see a graph with units listed on its axes, you should immediately divide the units and see if their quotient is a unit with some significance. For example, the slope of a position (meters) versus time (seconds) graph will have units in meters per second. The slope is therefore the velocity.

4. **THE IMPORTANCE OF AREA.** The area under a line segment can be determined by multiplying the rise (y-axis value) by the run (x-axis value). The areas encountered in beginning physics will be zero, rectangular, triangular, and/or trapezoidal. Areas below the x-axis are negative. Multiplying the units listed on the axes will indicate their significance. For example, the area of a velocity (m/s) versus time (s) graph will have units in meters. This area represents displacement.

5. **USING A GRAPH TO PREDICT THE FUNCTION THAT CREATED IT.** Plotting a function based on a physics equation will produce one of four likely curves: linear, quadratic, square root, and inverse. You should become familiar with the basic shapes of these four likely curves and then be able to associate them with physics equations.

If You See	Try	Keep in Mind
Derived units	Replace quantity symbols with fundamental units.	Memorizing basic formulas will help. Replace quantity symbols with fundamental unit symbols to show derived units.
A graph title	Plot the first variable listed on the y-axis.	This first variable is the dependent variable.
A graph	Determine the slope or area; they may be important.	Include units when calculating slope and area to verify if they match the quantity you are solving for.

PRACTICE EXERCISES

Questions 1–2

(A) $kg \cdot m/s^2$

(B) $kg \cdot m^2/s^2$

(C) $kg \cdot m^2/s^3$

(D) $kg/m \cdot s^2$

(E) $A \cdot s$

Select the combination of fundamental units above that should be used to answer each of the following questions about derived units.

1. The formula for finding pressure is $P = F/A$. The derived unit for pressure is the pascal. Which combination of fundamental units is equivalent to a pascal?

2. The formula for finding kinetic energy (K) is $K = \frac{1}{2} mv^2$. The derived unit for energy is the joule. Which combination of fundamental units is equivalent to a joule?

Questions 3–4 refer to the following velocity-time graph.

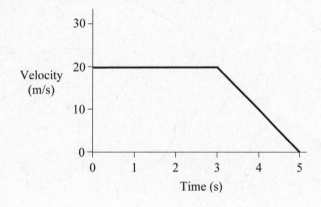

3. Determine the magnitude of acceleration of the object when time $t = 4$ seconds.

(A) $0 \ m/s^2$

(B) $2 \ m/s^2$

(C) $5 \ m/s^2$

(D) $10 \ m/s^2$

(E) $20 \ m/s^2$

4. Determine the displacement (change in position), Δx, of the object during the 5-second time interval.

(A)   40 m

(B)   60 m

(C)   80 m

(D) 100 m

(E) 120 m

5. Which graph best represents the equation, $P = I^2R$?

(A)
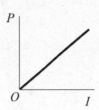

(B)
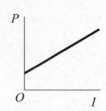

(C)
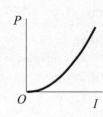

(D)
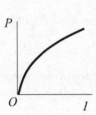

(E)
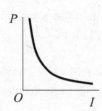

6. Which graph best represents the equation $p = h/\lambda$?

(A)
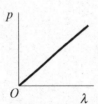

(B)
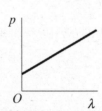

(C)
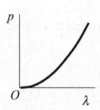

(D)
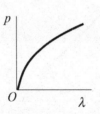

(E)
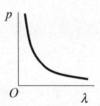

7. Which graph best represents the equation $F = kx$?

(A)
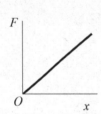

(B)
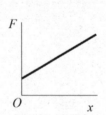

(C)
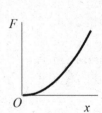

(D)
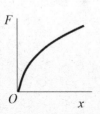

(E)
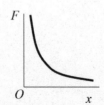

ANSWERS EXPLAINED

	Key Words	Needed for Solution	Now Solve It
1. **(D)**	Derived unit for pressure	$P = F/A$	Formulas are not provided on the examination. The formula was provided in this question for convenience. However, you should memorize formulas for such purposes as derived-unit analysis. Replacing the symbols for force, F, and area, A, with their units would produce: $$\text{pressure} = \text{N/m}^2$$ A newton, N, is also a kg • m/s^2, so the combined fundamental units for pressure would be kg/ms^2.
2. **(B)**	Derived unit for energy	$K = \frac{1}{2}mv^2$	Formulas are not provided on the examination. The formula was provided in this question for convenience. However, you should memorize formulas for such purposes as derived-unit analysis. Replacing symbols for mass, m, and velocity, v, with their units would produce: $$\text{kinetic energy} = \frac{1}{2}(\text{kg})(\text{m/s})^2$$ The resulting combination would be kg • m^2/s^2. The one-half has no effect on the unit combination.
3. **(D)**	Velocity-time graph; acceleration	Knowledge/definitions or Test if units of slope or area match the units of acceleration	The units of slope match the units needed in the answer. The slope of velocity versus time is acceleration. The time given, $t = 4$ s, lies on a constantly sloping line between $t = 3$ s and $t = 5$ s. $$\text{acceleration} = \text{slope} = \frac{\text{rise}}{\text{run}}$$ $$= \frac{0\,\text{m} - 20\,\text{m}}{5\,\text{s} - 3\,\text{s}} = -10 \text{ m/s}^2$$ The question requires the magnitude (numerical value without direction) of acceleration, which is the absolute value of acceleration.
4. **(C)**	Velocity-time graph; displacement	Knowledge/definitions or Test if units of slope or area match the units of displacement	The units of area match the units needed in the answer. The area of velocity versus time is displacement. Add the areas of the rectangle and triangle, bounded by the function and the x-axis. $$\text{displacement} = \text{area} = \text{height} \times \text{base}$$ $$\text{displacement} = \text{area of rectangle}$$ $$+ \text{ area for triangle}$$ $$(20 \text{ m/s} \times 3 \text{ s}) + \frac{1}{2}(20 \text{ m/s} \times 2 \text{ s})$$ $$= 80 \text{ m}$$

	Key Words	Needed for Solution	Now Solve It
5. **(C)**	Which graph; $P = I^2R$	Are the graphed variables squared, under a square root, or inverted?	The independent variable, I, is squared. A squared value indicates a quadratic function, which will graph as a parabola. The independent variable, P, will increase by the square of the dependent variable, I. This creates a parabola consistent with answer C.
6. **(E)**	Which graph; $p = h/\lambda$	Are the graphed variables squared, under a square root, or inverted?	The independent variable, λ, is inverted. The dependent and independent variables are inversely proportional, and the resulting graph is a hyperbola. Only one hyperbola is shown in the answers.
7. **(A)**	Which graph; $F = kx$	Are the graphed variables squared, under a square root, or inverted?	The equation contains no squared values, no square roots, and no inverted values. This is a linear equation in the following form: $$y = mx + b$$ There are two possible answers: A and B. Answer B includes a y-intercept. However, the equation $F = kx$ is missing the addition of a constant (such as $F = kx + b$). The y-intercept must therefore be zero. So the answer is A.

Vectors

2

→ **COORDINATE SYSTEM**

→ **SCALARS**

→ **VECTORS**

→ **VECTOR MATHEMATICS**

Understanding vector quantities and mastering basic vector mathematics are essential skills in physics. After reading this chapter, you will be able to

- Identify a mathematical coordinate system that will provide a common frame of reference to orient direction in physics problems.
- Understand the differences and similarities among scalar and vector quantities.
- Resolve vectors into components and add vector quantities.

COORDINATE SYSTEM

Problems in physics often involve the motion of objects. Position, displacement, velocity, and acceleration are key numerical quantities needed to describe the motion of an object. Position involves a specific location, while velocity and acceleration act in specific directions. Using the mathematical **coordinate system** is ideal to visualize both position and direction. The coordinate system provides a common frame of reference in which the quantities describing motion can be easily and consistently compared with one another.

We can place an axis anywhere, and we can orient the axis in any direction of our choosing. If a problem does not specify a starting location or direction, then position the origin at the object's starting location. In Figure 2.1, a problem involving the motion of a car can be visualized as starting at the origin and moving horizontally along the positive x-axis.

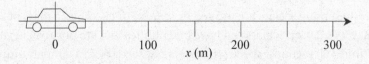

Figure 2.1 Horizontal motion

In more complex problems, some quantities cannot be oriented along a common axis. In these problems, direction must be specified in degrees measured counterclockwise (ccw) from the positive x-axis, as shown in Figure 2.2.

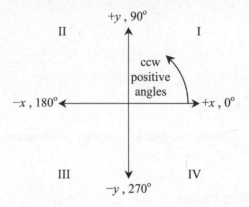

Figure 2.2 Coordinate system

A coordinate system is a valuable tool that provides a frame of reference when position and direction are critical factors.

SCALARS

A **scalar** is a quantity having only a numerical value. No direction is associated with a scalar. The numerical value describing a scalar is known as its magnitude. Some examples of commonly used scalars are listed in Table 2.1.

Table 2.1 Commonly Used Scalars

Quantities Involving	Examples of Common Scalars
Time	Time—t, period—T, and frequency—f
Motion	Distance—d and speed—v
Energy	Kinetic energy—K or KE and potential energy—U or PE
Mass	Mass—m and density—ρ
Gases	Pressure—P, volume—V, and temperature—T
Electricity	Charge—q or Q and potential (voltage)—V
Circuits	Current—I, resistance—R, and capacitance—C

The symbols representing scalars are printed in *italics*. For example, a mass of 2.0 kilograms will be written as $m = 2.0$ kg. Scalars can have magnitudes that are positive, negative, or zero. For example, time = 60 seconds, speed = 0 meters per second, and temperature = $-10°$ C.

VECTORS

Although scalars possess only magnitude, **vectors** possess both magnitude and a specific direction. Examples of commonly encountered vectors are listed in Table 2.2.

Table 2.2 Commonly Used Vectors

Vector Quantity	Vector Symbol	Component Symbol
Displacement	$\vec{s}$ or $\vec{r}$	x or y
Velocity	$\vec{v}$	v_x or v_y
Acceleration	$\vec{a}$	a_x or a_y
Force	$\vec{F}$	F_x or F_y
Momentum	$\vec{p}$	p_x or p_y
Electric field	$\vec{E}$	E_x or E_y
Magnetic field	$\vec{B}$	B_x or B_y

Formal vector variables are usually written in italics with a small arrow drawn over the letter, as shown in the middle column in Table 2.2. The College Board converted to this style in its Advanced Placement courses for the 2015 exam. Before that change, the College Board indicated vector quantities using bold standard print. Why mention this? The SAT Subject Exam in Physics is separate from the Advanced Placement Exams. As a result, there may be some differences in the conventions that these two separate College Board organizations use to express vector quantities.

You may encounter vector quantities, such as force, in any one of these forms: $\vec{F}$, **F**, F_x, F_y, and F. The first, $\vec{F}$, is the most accepted and distinctly indicates a vector quantity. It is the format used in this book and in the SAT Subject Test in Physics. The second, **F**, is an alternate way to indicate a vector quantity. The next two, F_x and F_y, signify vector components that lie along the specific axis indicated by their subscripts. The last, F, appears to be the convention to indicate a scalar quantity. This is actually used very frequently in textbooks and on formal exams. It is typically used when only the magnitude of the vector is needed and the direction is understood.

Distinguishing between vectors and scalars by simply looking at an equation can be confusing. How, then, do we tell scalars and vectors apart? Physics problems may contain clues in the text of the problem to help distinguish vectors from scalars. The mention of a specific direction definitely indicates a vector quantity. However, it is up to the student to learn which quantities are vectors and when the use of vector components is necessary. Counting on the use of a specific set of symbol conventions may not be wise.

Vectors do follow certain mathematical conventions that are worth noting. Vector magnitudes can be only positive or zero. However, vectors can have negative direction. Consider the acceleration of gravity, a vector quantity acting in the negative y-direction. The gravity vector includes both magnitude and direction ($\vec{g}$ = 10 m/s², $-y$). Substituting this exact expression, including the negative y-direction, into an equation is not really workable. Instead the value -10 m/s² may be substituted into equations. The negative sign in front

IF YOU SEE
a specific
direction

Vector
Quantity

Direction is important, and it influences vector mathematics.

of the magnitude indicates the negative y-direction. This can be done only if all the vector quantities used in an equation lie along the same axis and it is understood that the signs on all vector quantities represent direction along that axis. This essentially transforms the vector quantities into scalar quantities, allowing normal mathematical operations. As a result, the variable may be shown as a scalar in italics ($g = -10$ m/s^2) rather than in bold print. When a negative sign is associated with a vector quantity, it technically specifies the vector's direction and assists with proper vector addition.

Vectors are represented graphically as arrows. For displacement vectors, the tail of the arrow is the initial position of the object, x_i, and the tip of the arrow is the final position of the object, x_f. The length of the arrow represents the vector's magnitude, and its orientation on the coordinate axis indicates direction. This may give some insight into the reason that some vector quantities are displayed in italics.

Figure 2.3 shows a car moving 200 meters and its associated vector.

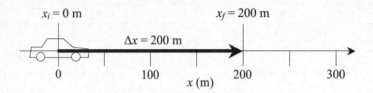

Figure 2.3 Horizontal displacement

The magnitude of the displacement vector, Δx, is the absolute value of the difference between the final position, x_f, and the initial position, x_i. Direction can be seen in the diagram.

$$\Delta x = x_f - x_i = 200 - 0 = 200 \text{ m, to the right } (+x)$$

For other vectors, such as velocity and force, the quantity described by the vector occurs at the tail of the arrow. The tail of the arrow shows the actual location of the object being acted upon by the vector quantity. The tip of the arrow points in the direction the vector is acting. The length of the arrow represents the magnitude of the vector quantity. The magnitude and direction described by these types of vectors may be instantaneous values capable of changing as the object moves. In addition, the object may not reach the location specified by the tip of the arrow.

These types of vectors are readily seen in projectile motion. In Figure 2.4, a projectile is launched with a speed of 50 meters per second at an angle of 37° above the horizontal.

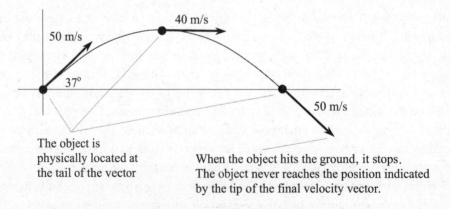

Figure 2.4 Projectile motion

Although only three key velocity vectors are shown in the diagram, they clearly demonstrate how the magnitude and direction of velocity change throughout the flight. During the motion depicted in the diagram, no two instantaneous velocity vectors are completely alike.

You will encounter a variety of vector quantities in the chapters ahead. Knowing how to recognize vectors quantities like displacement, velocity, acceleration, and force will improve your problem-solving skills. The importance of vector direction cannot be overstated. Including the correct sign representing a vector's direction is often the key to arriving at the correct solution. The next sections will demonstrate the importance of vector direction as we review basic vector mathematics.

VECTOR MATHEMATICS

Components

Vectors aligned to the x- and y-axes are mathematically advantageous. However, some problems involve diagonal vector quantities. Diagonal vectors act simultaneously in both the x- and y-directions, and they are difficult to manipulate mathematically. Fortunately, diagonal vectors can be resolved into x- and y-component vectors. The x- and y-component vectors form the adjacent and opposite sides of a right triangle where the diagonal vector is its hypotenuse. Aligning the component vectors along the x- and y-axes simplifies vector addition.

The magnitudes of component vectors are determined using right-triangle trigonometry. In Figure 2.5, vector **A** is a diagonal vector. It has a magnitude of A and a direction of θ.

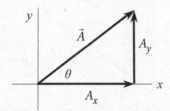

Figure 2.5 Magnitudes of vectors

Vector A_x is the x-component of $\vec{A}$ and is adjacent to angle θ. Vector A_y is the y-component of $\vec{A}$ and is opposite angle θ. Normally, the magnitude of the components of vector $\vec{A}$ would be determined using the following right-triangle trigonometry.

$$A_x = A \cos \theta, +x\text{-direction}$$
$$A_y = A \sin \theta, +y\text{-direction}$$

The SAT Subject Test in Physics does not allow the use of a calculator. If component vectors are needed to solve a problem, there has to be an easy way to avoid using trigonometry. This can be accomplished if questions involve only three well-known, memorized right triangles.

IF YOU SEE
a diagonal vector, with direction given in degrees

Find the x- and y- component vectors.

x- and y- vectors act independently.

Important Right Triangles

On exams excluding calculator use, determining component vectors will be restricted to the key right triangles shown in Table 2.3.

Table 2.3 Important Right Triangles

30°-60°-90°	3-4-5 (37°-53°-90°)	45°-45°-90°
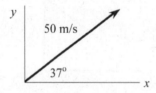		
hypotenuse = 2 × side opposite 30°	hypotenuse = 5/3 side opposite 37° hypotenuse = 5/4 side adjacent 37°	hypotenuse = $\sqrt{2}$ × side

---

EXAMPLE 2.1

Determining Component Vectors

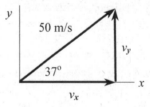

A projectile is launched with an initial velocity of 50 meters per second at an angle of 37° above the horizontal. Determine the x- and y-component vectors of the velocity.

WHAT'S THE TRICK?

Draw the component vectors and identify the adjacent and opposite sides.

The 37° angle indicates a 3-4-5 triangle. Determining the magnitudes of each component requires multiplying the hypotenuse by the correct fraction. The direction of each component can be determined by looking at the diagram.

$$v_x = 4/5 \text{ hyp} = 4/5 \ (50) = 40 \text{ m/s, } +x\text{-direction}$$
$$v_x = 3/5 \text{ hyp} = 3/5 \ (50) = 30 \text{ m/s, } +y\text{-direction}$$

In some problems, the component vectors are known or given and you must determine the vector they describe. Pythagorean theorem and inverse tangent are used to calculate the magnitude and direction of the diagonal vector described by the component vectors.

$$|\vec{A}| = \sqrt{A_x^2 + A_y^2} \quad \text{and} \quad \tan\theta = \frac{A_y}{A_x}$$

Taking an exam without using a calculator will limit your ability to perform these calculations. Problems will be limited to those easily solved by the Pythagorean theorem, or they will involve the three key right triangles described on page 64.

IF YOU SEE
components
and need the
resultant

Pythagorean
theorem
or the
memorized
key right
triangles

Adding Vectors

One important aspect of working with vectors is the ability to add two or more vectors together. Only vectors with the same units for magnitude can be added to each other. The result of adding vectors together is known as the vector sum, or resultant.

You can use two visual methods to add vectors. The first is the tip-to-tail method, and the second is the parallelogram method. In some problems, the resultant is known or given and you must determine the magnitude and direction of one of the vectors contributing to the vector sum. The sections below detail examples of each of these scenarios.

TIP-TO-TAIL METHOD

Adding vectors **tip to tail** is advantageous when a vector diagram is not given. Begin by sketching a coordinate axis. Vectors can be added in any order. However, drawing x-direction vectors first, followed by y-direction vectors, is best. Choose the first vector and draw it starting from the origin and pointing in the correct direction. Start drawing the tail of the next vector at the tip of the previous vector. Keep the orientation of the second vector the same as it was given in the question. Continue this process, adding any remaining vectors to the tip of each subsequent vector. Finally, draw the resultant vector from the origin (tail of the first vector) pointing to the tip of the last vector. Vector addition on the SAT Subject Test in Physics will most likely be limited to the following simple cases:

- Vectors pointing in the same direction
- Vectors pointing in opposite directions
- Vectors that are 90° apart

The following examples demonstrate tip-to-tail vector addition for these three common scenarios.

EXAMPLE 2.2

Adding Vectors Pointing in the Same Direction
A person walks 40 meters in the positive x-direction, pauses, and then walks an additional 30 meters in the positive x-direction. Determine the magnitude and direction of the person's displacement.

(WHAT'S THE TRICK?)

When vectors point in the same direction, simply add them together. Sketch or visualize the vectors tip to tail. The resultant is equal to the total length of both vectors added together.

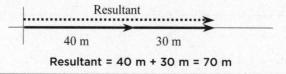

Resultant = 40 m + 30 m = 70 m

EXAMPLE 2.3

Adding Vectors Pointing in Opposite Directions

A person walks 40 meters in the positive *x*-direction, pauses, and then walks an additional 30 meters in the negative *x*-direction. Determine the magnitude and direction of the person's displacement.

WHAT'S THE TRICK?

When a vector points in the opposite (negative) direction, you can insert a minus sign in front of the magnitude. Technically, vectors cannot have negative magnitudes. The minus sign actually indicates the vector's direction, and it represents a vector turned around 180°. Again, sketching or visualizing the vectors tip to tail will help you arrive at the correct resultant. The resultant is drawn from the origin to the tip of the last vector added.

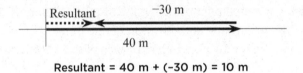

Resultant = 40 m + (–30 m) = 10 m

EXAMPLE 2.4

Adding Vectors That Are 90° Apart

An object moves 100 meters in the positive *x*-direction and then moves 100 meters in the positive *y*-direction. Determine the magnitude and direction of the object's displacement.

WHAT'S THE TRICK?

Start at the origin and draw the *x*-direction vector first. Then add the tail of the *y*-direction vector to the tip of the first vector. Finally, draw the resultant from the origin pointing toward the tip of the final vector added.

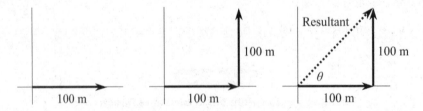

The components and resultant form a 45°-45°-90° triangle. The magnitude of the hypotenuse can be obtained by multiplying a side by the square root of two.

$$\text{hypotenuse} = (\sqrt{2})(\text{side}) = (\sqrt{2})(100), \ \theta = 45°$$

Without a calculator, $100\sqrt{2}$ is the mathematically simplified answer.

PARALLELOGRAM METHOD

In some exam questions, a vector diagram may be provided that shows the vectors in a tail-to-tail configuration. You can add these vectors by constructing a **parallelogram**, as shown in the example below.

Adding Vectors Using the Parallelogram Method

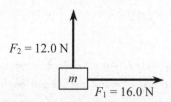

A mass, *m*, is acted upon by two force vectors, $\vec{F}_1$ = 16.0 N in the +*x*-direction and $\vec{F}_2$ = 12.0 N in the +*y*-direction, as shown in the diagram above. Determine the magnitude and direction of the resultant force acting on mass *m*.

WHAT'S THE TRICK?

Construct a parallelogram. The diagram below on the left shows a dashed line drawn from the tip of $\vec{F}_1$ parallel to $\vec{F}_2$ and a second dashed line drawn from the tip of $\vec{F}_2$ parallel to $\vec{F}_1$. In the diagram below on the right, the resultant is drawn with its tail starting at the origin and the tip extending to the intersection of the dashed lines. The resultant is the sum of the force vectors ($\Sigma\vec{F}$).

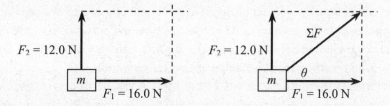

The dashed lines have the same length as the given vectors. Adding the resultant to the diagram creates two right triangles. Look carefully at the ratio of the sides. Two 3-4-5 triangles have been formed.

$$\Sigma\vec{F} = 20.0 \text{ N at } 37°$$

Finding a Missing Vector

In some problems, the resultant is known and the problem requires you to find the magnitude and direction of a missing vector. This frequently occurs when clues in the problem lead you to the conclusion that the resultant vector has a magnitude equal to zero. In order for two vectors to add up to zero, the vectors must have equal magnitudes and point in opposite directions.

EXAMPLE 2.6

Deducing the Existence of a Missing Vector

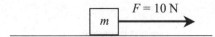

A mass, *m*, is initially at rest on a horizontal surface. A 10-newton force acting in the positive *x*-direction is applied to mass *m*. The mass remains at rest. Why?

WHAT'S THE TRICK?

A force is either a push or a pull. When an object remains stationary, all the pushing forces acting on the object must cancel out each other. Therefore, the sum of all the force vectors is zero. You must conclude that a second force is acting on the mass to cancel the force given in the problem. The only force capable of canceling the given force is a 10-newton force acting in the opposite direction.

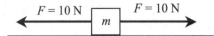

SUMMARY

1. **DIRECTION IS AN IMPORTANT AND SIGNIFICANT ASPECT OF MANY QUANTITIES IN PHYSICS.** At the start of a problem dealing with vector quantities, visualize the subject of the problem (object or mass) at the origin of a coordinate-axis system. Draw vector quantities acting on the object as pointing outward from the object. This allows the coordinate-axis system to act as a frame of reference to display, compare, and mathematically manipulate vector quantities.

2. **QUANTITIES IN PHYSICS ARE EITHER SCALARS OR VECTORS.** Scalars and vectors share one common characteristic: they both have a numerical size or strength known as a magnitude. Vector quantities have an additional characteristic—they point in a specific direction. Vectors are represented in diagrams as arrows, where the length of the arrow is proportional to the vector's magnitude and the direction of the arrow is consistent with the direction of the vector. The magnitude of vectors can be only positive or zero. A negative sign associated with a vector quantity indicates that the vector points in a negative (opposite) direction.

3. **VECTOR MATHEMATICS INCLUDES RESOLVING VECTORS INTO COMPONENTS AND ADDING THE VECTORS.** Working with vectors that do not lie along a principal axis is difficult. Fortunately, vectors can be mathematically broken down into component vectors that lie along the *x*- and *y*-axes. A vector and its component vectors form a right triangle. You can add vectors and vector components together using either the tip-to-tail method or the parallelogram method of vector addition. Only vectors measuring the same quantities, with the same units, can be added together. Knowing the side relationships for the 30°-60°-90°, 3-4-5, and 45°-45°-90° right triangles is essential when solving vector components and when adding perpendicular vectors.

If You See	Try	Keep in Mind
A specific direction associated with a given quantity	Remember that you are working with a vector quantity.	Vector direction is very important and can influence vector mathematics.
A diagonal vector quantity with a direction specified in degrees	Find the x- and y-component vectors.	Without a calculator, this involves memorizing the side relationships for three key right triangles: 30°-60°-90° 3-4-5 45°-45°-90°
Component vectors and you need the resultant vector	Use the Pythagorean theorem $$\text{hyp} = \sqrt{(\text{adj})^2 + (\text{opp})^2}$$ and/or the memorized sides of the three key right triangles.	

PRACTICE EXERCISES

1. Which of the following statements does NOT describe a vector quantity?

   (A) An object has a speed of 20 m/s in the positive x-direction.
   (B) A 30-newton force acts at an angle of 30° above the horizontal.
   (C) A car travels a distance of 2.0 kilometers.
   (D) The acceleration of gravity, g, is directed downward.
   (E) A mass is displaced 5.0 meters horizontally.

2. A projectile is launched with an initial velocity of 50 meters per second at an angle of 30° above the horizontal. Determine the y-component of the projectile's initial velocity, v_y.

   (A) 25 m/s
   (B) 30 m/s
   (C) $25\sqrt{2}$ m/s
   (D) 40 m/s
   (E) $25\sqrt{3}$ m/s

3. Two forces act on a 5-kilogram mass. A 16-newton force pushes the mass in the positive x-direction, and a 12-newton force pushes it in the negative x-direction. Determine the resultant net force acting on the mass.

(A) 4 N, $-x$ direction
(B) 4 N, $+x$ direction
(C) 14 N, $+x$ direction
(D) 28 N, $-x$ direction
(E) 28 N, $+x$ direction

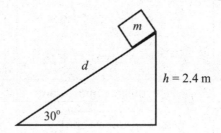

4. A mass is displaced upward along a 30° incline as shown in the diagram above. When it reaches the top of the incline, the mass has a vertical height, h, of 2.4 meters. What is the displacement, Δd, of the mass as measured along the incline?

(A) $1.2\sqrt{2}$ m
(B) $1.2\sqrt{3}$ m
(C) 3.0 m
(D) $2.4\sqrt{3}$ m
(E) 4.8 m

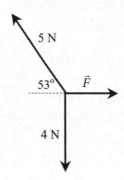

5. Three forces act on a mass as shown above. Determine the magnitude of force $\vec{F}$ so that the resultant force acting on the mass is zero.

(A) 3 N
(B) 4 N
(C) $4\sqrt{2}$ N
(D) $5\sqrt{2}$ N
(E) $5\sqrt{3}$ N

ANSWERS EXPLAINED

	Key Words	Needed for Solution	Now Solve It
1. **(C)**	NOT; vector quantity	Knowledge/definitions	Answer C reports only the magnitude. There is no mention of a specific direction. This is a scalar and *not* a vector.
2. **(A)**	Initial velocity; 30°; y-component; v_y	Side relationships for a 30°-60°-90° triangle	Visualize or sketch the velocity vector and its components. v_y is opposite the 30° angle, so its magnitude is half of the hypotenuse. $$v_y = (1/2)(50) = 25 \text{ m/s}$$
3. **(B)**	Two forces; resultant net force	Tip-to-tail vector addition	Start at the origin and add the vectors tip to tail. The resultant is drawn from the origin to the tip of the second vector. $$\Sigma \vec{F} = 16 + (-12) = +4 = 4 \text{ N},$$ $$+x \text{ direction}$$
4. **(E)**	Incline; vertical height; displacement; along the incline	Side relationships for a 30°-60°-90° triangle	The displacement along the incline is the hypotenuse of a 30°-60°-90° triangle. The hypotenuse is double the side opposite the 30° angle, which in this case is double the height. $$\Delta d = (2)(2.4 \text{ m}) = 4.8 \text{ m}$$

	Key Words	Needed for Solution	Now Solve It
5. **(A)**	Three forces; determine the magnitude and direction of force $\vec{F}$; resultant force acting on the mass is zero	Side relationships for a 3-4-5 triangle	Split the diagonal 5-newton force into components. The upward 4-newton component vector is canceled by the downward 4-newton vector. The leftward 3-newton component vector must also be canceled. $\vec{F} = 3$ newtons in the $+x$ direction

Kinematics in One Dimension

3

→ **KINEMATIC QUANTITIES**

→ **IDENTIFYING VARIABLES**

→ **KINEMATIC EQUATIONS**

→ **KINEMATIC GRAPHS**

*K*inema is a Greek word meaning "movement." **Kinematics** is a mathematical way to describe the motion of objects without investigating the cause of their motion. This chapter focuses on the following objectives:

- Discuss and compare the kinematic quantities.
- Identify kinematic variables in word problems.
- Apply the correct kinematic equation to solve problems.
- Interpret graphical representations of kinematic equations.

Table 3.1 lists the variables and their units that are used in a study of kinematics.

Table 3.1 Variables and Units Used in Kinematics

Variables Used in Kinematics	Units
$\vec{x}$ = Displacement (distance) (Also: Δx, y, Δy, h, Δh, or d)	m (meters)
$\vec{v}_i$ = Initial velocity (speed)	m/s (meters per second)
$\vec{v}_f$ = Final velocity (speed)	m/s (meters per second)
$\vec{a}$ = Acceleration	m/s^2 (meters per second squared)
t = Elapsed time	s (seconds)

KINEMATIC QUANTITIES

Kinematics involves the mathematical relationship among key quantities describing the motion of an object. These quantities include displacement—$\vec{x}$, velocity—$\vec{v}$, and acceleration—$\vec{a}$. You should also note the relationships between displacement and distance and between velocity and speed.

Displacement and Distance

Displacement, $\vec{x}$, is a vector extending from the initial position of an object to its final position. The variable x is typically used for horizontal motion, while y and h (height) are used for vertical motion.

Displacement differs slightly from distance, *d*. **Distance** is a scalar quantity representing the actual path followed by the object. When an object travels in a straight line and does not reverse its direction, then distance and the magnitude of displacement are interchangeable.

EXAMPLE 3.1

Distinguishing Between Distance and Displacement

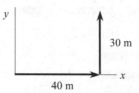

A person runs 40 meters in the positive *x*-direction and then 30 meters in the positive *y*-direction. Determine both the distance traveled by the runner and the magnitude of the displacement.

WHAT'S THE TRICK?

Distance is a scalar using the actual path followed.

$$d = 40 \text{ m} + 30 \text{ m} = 70 \text{ m}$$

Displacement is a vector extending from the initial position to the final position. The vector drawn from the initial point to the final point becomes the hypotenuse of a 3-4-5 right triangle.

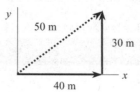

The resulting displacement is 50 m. The variable used to represent this displacement can vary. Although the 50 m displacement does not lie along the *x*-axis, some authors will report this displacement as $x = 50$ m or $\Delta x = 50$ m. Others may choose a letter that does not rely on an axis, such as $d = 50$ m. However, *d* used this way can be confused with distance. You may encounter a variety of conventions, and you need to be flexible. Regardless of the variable chosen, focus on what is being solved (distance or displacement).

Velocity and Speed

Velocity and speed are kinematic quantities measuring the rate of change in displacement and distance. A rate is a mathematical relationship showing how one variable changes compared with another. When the word *rate* appears in a problem, simply divide the quantity mentioned by time. **Velocity**, $\vec{v}$, is a vector describing the rate of displacement, $\Delta \vec{x}$. The equation for velocity

$$\vec{v}_{\text{avg}} = \frac{\Delta \vec{x}}{t}$$

solves for the average velocity during a time interval, t. Additional information is needed to determine if velocity is constant or is changing during the time interval.

If velocity is changing, then it has different values at different moments in time. However, instantaneous velocity is the velocity at a specific time, t. If you report that you are driving north at 65 mph, you have given an instantaneous velocity. This is a snapshot, freezing the problem at a specific instant. In kinematics, you will encounter two specific instantaneous velocities. Initial velocity, v_i, is at the start of a problem. Final velocity, v_f, is at the end of a problem.

Velocity is a vector quantity, so it includes a specific magnitude and a direction. When the magnitude and direction of velocity are both constant, we say that the object is moving at constant velocity. However, when direction is changing, the term *speed* may be used. **Speed** is a scalar quantity that calculates the rate of distance, as opposed to displacement. If an object travels in a straight line, then the terms *speed* and *velocity* are interchangeable.

EXAMPLE 3.2

Distinguishing Between Speed and Velocity

Consider a car moving from point A to point B and then returning to point A.

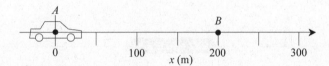

This motion takes place in 20 seconds. Determine the speed and average velocity during the entire motion.

WHAT'S THE TRICK?

The object reversed direction and returned to its starting point. It traveled a specific round-trip distance. However, at the end of the motion, there was no overall displacement. Speed is a scalar that depends on actual distance moved by the object.

$$v = \frac{d}{t} = \frac{200\,\text{m} + 200\,\text{m}}{20\,\text{s}} = 20 \text{ m/s}$$

Average velocity is the rate of the displacement (straight-line distance from starting point to ending point).

$$v = \frac{\Delta x}{t} = \frac{0\,\text{m}}{20\,\text{s}} = 0 \text{ m/s}$$

Acceleration

Acceleration is the rate of change in velocity.

$$\vec{a} = \frac{\Delta \vec{v}}{t}$$

The above equation actually solves for average acceleration. However, acceleration on the SAT Subject Test will usually be uniform acceleration. When acceleration is uniform, its magnitude remains constant. The magnitude of acceleration indicates how quickly velocity is changing. In other words, acceleration is the rate of a rate, which is why students new to physics often have difficulty comprehending it. The effect of acceleration on velocity depends on the orientation of the vector quantities relative to one another, as shown in Table 3.2.

Table 3.2 The Effect of Acceleration on Velocity

Same Direction	Opposite Direction	Perpendicular Direction
Acceleration in the same direction as initial velocity causes an increase in speed.	Acceleration in the opposite direction as initial velocity causes a decrease in speed.	Acceleration perpendicular to initial velocity causes a change in direction.

IF YOU SEE

change in velocity

Acceleration

Objects could be speeding up, slowing down, or changing direction.

When answering conceptual questions, the possibility of changing direction is often overlooked. A car moving around a circular track at a constant speed is said to have uniform acceleration as opposed to constant acceleration. The phrase "uniform acceleration" indicates that the magnitude of acceleration will remain the same while the direction may be changing. The phrase "constant acceleration," however, indicates that both magnitude and direction are the same.

EXAMPLE 3.3

Conceptual Problem

Can an object be accelerating if it has constant speed?

WHAT'S THE TRICK?

Speed is a scalar. Constant speed implies only constant magnitude; it gives no information about direction. An object at constant speed may be changing direction. For example, a car moving around a circular track at a constant speed must continually change direction. Changing direction involves a change in the velocity vector, and a change in velocity during a time interval is acceleration. Therefore, an object with a constant speed may be accelerating. Had the problem specified constant velocity, the answer would be quite different. An object undergoing constant velocity must have both constant magnitude (constant speed) and constant direction, resulting in no acceleration.

Acceleration of Gravity

All objects on Earth are subject to the acceleration of gravity. This acceleration has a known value at Earth's surface. It is so prevalent in physics problems that it receives its own variable, g. The acceleration of gravity acts downward and has a value of 9.8 m/s^2. For the SAT Subject Test in Physics, the value is rounded to 10 m/s^2.

IDENTIFYING VARIABLES

Correctly identifying the variables in a problem statement will assist you in the problem-solving process. When presented with a word problem, begin by writing down all of the variables given in the question. Typically, you will find that you are given three variables and you must solve for a fourth. However, only one or two numerical values may appear in the word problem. In these situations, you may have overlooked variables hidden in the language of the problem.

Hidden Variables

Throughout physics, hidden variables appear in either of two main forms:

1. Known constants, such as gravity, do not need to be specifically mentioned in the text of a problem even though they are important in solving the problem.
2. When quantities have a zero value, they are often indicated by key phrases in the text of the problem.

The common hidden-zero quantities used in kinematics are listed in Table 3.3. The wording used in the question will help you find these variables.

Table 3.3 Common Quantities with a Zero Value

Key Phrase	Hidden Variable Quantity
Initially at rest	$\vec{v}_i = 0$ m/s
Stops	$\vec{v}_f = 0$ m/s
Constant velocity	$\vec{a} = 0$ m/s^2
Returns to its starting point	$\vec{x} = 0$ m
Dropped, thrown, or falls	$\vec{g} = -10$ m/s^2
Dropped	$\vec{v}_i = 0$ m/s
Maximum height	$\vec{v}_f = 0$ m/s

Sign Conventions

The kinematic quantities—displacement, velocity, and acceleration—are all vectors. The magnitude of a vector is always positive. However, vectors can point in either a positive or a negative direction. When vectors point in a negative direction, a negative sign is added to the magnitude of the vector for calculation purposes.

The coordinate-axis system, discussed in Chapter 2, is the best tool to use when determining the correct sign on vector quantities. Picture the object at the origin of the coordinate axes at the start of the problem ($x_i = 0$ and $y_i = 0$). The default positive directions are right and upward. Any vector quantities pointing to the left or downward will include a negative sign in calculations. Table 3.4 summarizes the sign conventions for kinematic variables.

Table 3.4 Signs Used for Kinematic Variables

Variable and Sign	Horizontal Motion	Vertical Motion
$+x$	Object finishes right of starting point	Object finishes above starting point
$-x$	Object finishes left of starting point	Object finishes below starting point
$+v$	Moving right	Moving upward
$-v$	Moving left	Moving downward
$+a$	Positive acceleration increases the speed of objects that have a positive velocity. However, when an object has a negative velocity a positive acceleration acts to decrease speed.	
$-a$	Negative acceleration decreases the speed of objects that have a positive velocity. However, when an object has a negative velocity a negative acceleration acts to increase speed.	In free-fall problems, acceleration is equal to g and always acts downward

As seen in the table above, the sign on acceleration can have opposing effects depending on the sign of velocity. If acceleration and velocity vectors have the same direction, then speed increases. When acceleration and velocity vectors oppose each other, speed decreases. When solving horizontal-motion problems, the initial motion can always be set as rightward ($+v_{xi}$). However, vertical motion is subject to the constant downward acceleration of gravity and is more complicated. Upwardly launched objects always solve traditionally. The initial upward vertical velocity is positive ($+v_{yi}$), and the acceleration of gravity is negative ($-g$). The opposing acceleration acts to slow objects as they ascend. After reaching maximum height, objects reverse direction ($-v_y$). Now velocity and acceleration have the same direction, causing speed to increase as objects descend. When objects are dropped or thrown downward, the signs on all nonzero kinematic variables are negative. Since all the negative signs cancel, they are often omitted when solving these problems. The solutions appear as though the initial downward motion was set as the positive, resulting in positive downward velocity ($+vy$), displacement ($+\Delta y$), and gravity ($+g$). Setting the direction of initial motion as positive simplifies the sign on acceleration. If this is done, then positive acceleration increases speed, and negative acceleration decreases speed.

The most important aspect to determining vector direction is consistency. Which direction you set as positive does not really matter. However, you must consistently apply this decision to every vector throughout the entire problem. When you pick a positive and a negative direction, do not change it during the problem. Make sure every vector uses the sign convention you have chosen.

EXAMPLE 3.4

Identifying Variables

A ball is thrown upward with an initial velocity of 15 meters per second. Write a complete list of kinematic variables for the instant the ball reaches its maximum height.

(WHAT'S THE TRICK?)

Only one numerical value is mentioned. However, there are two hidden values in the word problem.

1. Unless told otherwise, all problems take place on Earth ($g = -10$ m/s^2).
2. The phrase "maximum height" implies the ball has an instantaneous vertical velocity of zero ($v_f = 0$ m/s).

Adding these hidden variable results in the following variable list:

Δx = not mentioned	v_i = 15 m/s	v_f = 0 m/s	g = -10 m/s^2	t = not mentioned

---

KINEMATIC EQUATIONS

The kinematic equations relate the kinematic variables in a manner that solves for a variety of situations.

$$\Delta x = v_i t + \frac{1}{2} at^2 \qquad v_f = v_i + at \qquad v_f^2 = v_i^2 + 2a\Delta x$$

The SAT Subject Test in Physics will not provide you with a list of equations. However, equations are essential when solving numerical problems. They also assist you in analyzing conceptual problems. Therefore, it is in your best interest to memorize the equations used throughout this book. Different resources vary, and you may be familiar with different versions of the equations presented in this text. To simplify the memorization process and to review information specific to the Subject Test in Physics, the most streamlined versions of all equations have been used in this book.

Choosing the Correct Equation

Choosing the correct equation depends on the variables mentioned in each problem. In addition, when an object is initially at rest ($v_i = 0$), the equations simplify into frequently tested, easier versions of the kinematic equations. Table 3.5 will help you identify which equation you should use based on what is given and what is requested in a particular question. It will also help you identify shortened variations of those equations for objects that are initially at rest.

IF YOU SEE
displacement,
velocity,
acceleration,
and/or time

Kinematic equations

Memorize all three kinematic equations.

Table 3.5 Choosing Which Kinematic Equation to Use

IF YOU SEE

constant

velocity

a = 0

x = vt

Derived by
substituting
$a = 0$ into

$x = v_i t + \frac{1}{2} at^2$.

Problem Includes	Equation	Initially at Rest ($v_i = 0$)
Time is *not* given and is *not* requested	$v_f^2 = v_i^2 + 2ax$	$v_f = \sqrt{2ax}$
Time is given and velocity is requested	$v_f = v_i + at$	$v_f = at$
Time is given and displacement is requested	$x = v_i t + \frac{1}{2} at^2$	$x = \frac{1}{2} at^2$
Constant velocity ($a = 0$)	$v = \dfrac{x}{t}$ or $x = vt$	Not applicable

Note: The constant velocity ($a = 0$) formula can be derived from $x = v_i t + \frac{1}{2} at^2$ by substituting zero for acceleration: $x = v_i t$. If velocity is constant, the initial velocity, v_i, is the same as the velocity, v, at any instant.

IF YOU SEE

kinematic

variables but

time is not

mentioned

$v_f^2 = v_i^2 + 2ax$

EXAMPLE 3.5

Problem Never Mentions Time

Determine the maximum height reached by a ball thrown upward at 20 meters per second.

WHAT'S THE TRICK?

Complete a variable list, including known constants and hidden values. In vertical-motion problems, y and h are often used in place of displacement, x. In addition, the acceleration of gravity, g, replaces the general acceleration, a. When objects reach "maximum height," they come to an instantaneous stop ($v_f = 0$ m/s).

$y = ?$	$v_i = 20$ m/s	$v_f = 0$ m/s	$g = -10$ m/s²	$t =$ not mentioned

Time is not given, and you are not asked to solve for it.

$$v_f^2 = v_i^2 + 2gy$$

$$(0)^2 = (20 \text{ m/s})^2 + (2)(-10 \text{ m/s}^2)\, y$$

$$y = 20 \text{ m}$$

EXAMPLE 3.6

Problem Involves Time and Velocity

A car traveling at 30 meters per second undergoes an acceleration of 5.0 meters per second squared for 3.0 seconds. Determine the final velocity of the car.

WHAT'S THE TRICK?

Complete a variable list. If variables seem to be missing, read the problem again and look for key phrases signaling hidden variables. The problem did not state how the acceleration was affecting the car, so you must assume the simplest scenario. Unless the problem specifies a decrease in speed, assume acceleration is positive and that it acts to increase speed.

x = not mentioned	v_i = 30 m/s	v_f = determine	a = 5.0 m/s²	t = 3.0 s

Time is not given, and the problem involves velocity.

$$v_f = v_i + at$$
$$v_f = (30 \text{ m/s}) + (5 \text{ m/s}^2)(3.0 \text{ s})$$
$$v_f = 45 \text{ m/s}$$

IF YOU SEE
time,
acceleration,
and velocity

$v_f = v_i + at$

EXAMPLE 3.7

Problem Involves Time and Displacement

A ball is dropped from a 45-meter-tall structure. Determine the time the ball takes to hit the ground.

WHAT'S THE TRICK?

Complete a variable list, including known constants and hidden values. A "dropped" object has an initial velocity of zero (v_i = 0 m/s). The structure is 45 m tall, and the ball is moving downward toward the ground (y = –45 m). The acceleration is due to gravity, which also acts downward (g = –10 m/s²).

y = –45 m	v_i = 0 m/s	v_f = not mentioned	g = –10 m/s²	t = determine

You need an equation relating displacement and time.

$$y = v_i t + \frac{1}{2}gt^2$$

Since v_i = 0 m/s, you can simplify the equation.

$$y = \frac{1}{2}gt^2$$
$$(-45 \text{ m}) = \frac{1}{2}(-10 \text{ m/s}^2)t^2$$
$$t = 3.0 \text{ s}$$

IF YOU SEE
time,
acceleration,
and
displacement

$x = v_i t + \frac{1}{2}at^2$

KINEMATIC GRAPHS

Interpreting graphs and determining their significance is discussed at length in Chapter 1, "Conventions and Graphing." The key values to assess are slope, area, and intercepts. To determine if slope or area is important, remember to include units in your calculations. In addition, it may also be important to determine if values are constant or changing. Table 3.6 describes frequently used graphs involving the kinematic formulas and variables.

Table 3.6 Graphs and Kinematics

If You See	Slope	Area	y-intercept	x-intercept
Position, x (or distance, d) versus time	Velocity		Initial position	$x = 0$
Velocity (or speed) versus time	Acceleration	Displacement (change in position)	Initial velocity	$v = 0$
Acceleration versus time		Change in velocity	Initial acceleration	$a = 0$

The velocity versus time graph described in Table 3.6 contains the most information, making it the most valuable and most frequently tested kinematic graph.

EXAMPLE 3.8

Analyzing Velocity versus Time Graphs

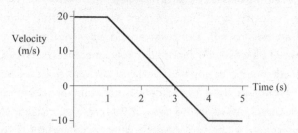

The motion of an object is shown in the velocity versus time graph above.

(A) Determine the initial velocity of the object.

WHAT'S THE TRICK?

Initial conditions occur at zero time. In graphs with time along the *x*-axis, initial values are the *y*-intercept. The initial velocity is 20 m/s.

(B) Describe the motion during the first second.

WHAT'S THE TRICK?

The horizontal line indicates that the independent variable, velocity, remains constant. The motion during the first second is constant velocity.

(C) Determine the displacement during the first second.

WHAT'S THE TRICK?

Displacement is the area under the velocity versus time graph.

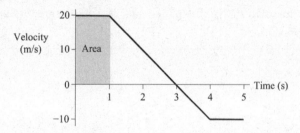

displacement = height × base = (20 m/s × 1 s) = 20 m

(D) Determine the acceleration in the time interval between 1 and 4 seconds.

WHAT'S THE TRICK?

Acceleration is the slope of the velocity versus time graph.

$$\text{slope} = \frac{\text{rise}}{\text{run}} = \frac{(-10\text{m})-(20\text{m})}{(4\text{s})-(1\text{s})} = -10 \text{ m/s}^2$$

(E) Determine the final speed of the object.

WHAT'S THE TRICK?

Simply read the graph. At the end of 5 seconds, the object has a velocity of –10 m/s. However, the question asks for speed. Speed is a scalar representing the magnitude of velocity. Speed is the absolute value of the velocity, 10 m/s.

SUMMARY

1. **RECOGNIZE THE SUBTLE DIFFERENCES IN KINEMATIC QUANTITIES.** Displacement, velocity, and acceleration are all vector quantities, whereas distance and speed are not. Constant speed does not necessarily mean zero acceleration (as in the case of a car driving at a constant speed on a circular track). However, objects moving at constant velocity always have zero acceleration. The term *uniform motion* is used to indicate a constant acceleration in magnitude only but not in direction (as in the case of the car on a circular track). The term *constant acceleration* is used to indicate when acceleration is truly constant in both magnitude as well as in direction.

2. **IDENTIFY THE KINEMATIC VARIABLES.** This is a key step in solving problems. A variable list is a great tool to use when identifying the kinematic quantities, including known constants and hidden variables.

3. **VISUALIZE MOTION SUPERIMPOSED ON A COORDINATE AXIS.** Doing this gives the problem a consistent frame of reference. Many kinematic quantities are vectors. Using a coordinate system assists you in selecting the correct sign for displacement, velocity, and acceleration.

4. **USE THE VARIABLE LIST TO CHOOSE THE CORRECT EQUATION.** Time is a critical variable. Without it, you can use only one possible kinematic equation. When time is given, you can use two possible equations. One of these involves velocity, and the other involves displacement. When an object starts from rest or moves at constant velocity, the kinematic equations simplify greatly.

5. **KINEMATIC GRAPHS ARE FREQUENTLY USED.** Graphs often involve slope, area, or intercepts. Including units in slope and area calculations will help you assess if either of these methods matches the answers given in the problem. When slope and area fail to solve the problem, consider if using intercepts and data points might be important.

If You See	Try	Keep in Mind
A change in velocity	Acceleration	Objects could be speeding up, slowing down, or changing direction.
A combination of the kinematic variables: displacement, velocity, acceleration, and/or time	Kinematics	Constructing a variable list will help you decide which equation to use.
Kinematic variables, but time is never mentioned	$v_f^2 = v_i^2 + 2ax$	For vertical motion, x can be replaced with either y or h. Acceleration, a, is usually replaced with the acceleration of gravity, g.
Kinematic variables, and time is mentioned	$v_f = v_i + at$ or $x = v_i t + \frac{1}{2}at^2$	
Objects initially at rest $v_i = 0$ m/s	The kinematic equations simplify: $v_f = \sqrt{2ax}$ $v_f = at$ $x = \frac{1}{2}at^2$	
Constant velocity	$x = v_i t + \frac{1}{2}at^2$ simplifies to $x = vt$	At constant velocity, acceleration is zero. $a = 0$

PRACTICE EXERCISES

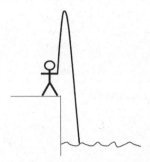

1. A ball tossed straight up from a bridge just misses the bridge on the way down and lands in the water below, as shown in the diagram above. The bridge is 24 meters above the surface of the water. The elapsed time for this motion is 3 seconds. Determine the initial speed of the upward throw.

   (A) 3 m/s
   (B) 5 m/s
   (C) 7 m/s
   (D) 9 m/s
   (E) 11 m/s

2. The driver of a car moving at 20 meters per second notices the light has changed to red. When the driver applies the brakes, the car experiences a deceleration of 10 meters per second squared. How far will the car travel before it stops?

   (A) 5 m
   (B) 10 m
   (C) 15 m
   (D) 20 m
   (E) 25 m

Questions 3–5 relate to the velocity versus time graph below.

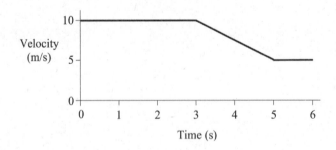

3. Determine the magnitude of acceleration from 2 to 3 seconds.

   (A) 0 m/s^2
   (B) 1.0 m/s^2
   (C) 5.0 m/s^2
   (D) 10 m/s^2
   (E) Cannot be determined.

4. Determine the displacement (change in position) from 3 to 5 seconds.

   (A) 1.0 m
   (B) 2.0 m
   (C) 5.0 m
   (D) 10 m
   (E) 15 m

5. Determine the magnitude of acceleration from 3 to 5 seconds.

   (A) 1.0 m/s^2
   (B) 2.5 m/s^2
   (C) 5.0 m/s^2
   (D) 5.5 m/s^2
   (E) Cannot be determined.

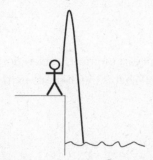

6. In the diagram above, a ball is thrown straight up. The ball slows on the way up, reaches an instantaneous stop at maximum height, increases speed on the way down, and finishes the motion at a lower final height. Which of the following graphs represents vertical displacement versus time?

(A)

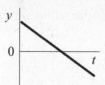

(B)

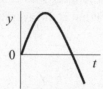

(C)

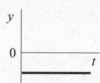

(D)

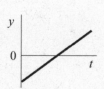

(E)

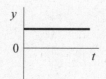

7. How high is the ceiling in a school gymnasium if a volleyball thrown upward from the floor at 10.0 meters per second barely touches the ceiling before returning to the ground?

(A) 4.0 m
(B) 5.0 m
(C) 6.0 m
(D) 7.0 m
(E) 8.0 m

8. A mass undergoes a constant acceleration from rest at 4.0 meters per second squared. What is the displacement of the mass at the end of 10 seconds?

(A) 100 m
(B) 150 m
(C) 200 m
(D) 400 m
(E) 450 m

9. A mass initially at rest experiences a uniform acceleration, a. During time interval t, the mass is displaced a distance, x. What is the displacement of the mass when the elapsed time doubles, and equals $2t$?

(A) x
(B) $\sqrt{2}(x)$
(C) $2x$
(D) $2\sqrt{2}(x)$
(E) $4x$

10. A runner is training on an oval-shaped track where each lap has a length of 400 meters. The runner maintains a constant speed and completes two laps, returning to the starting point in 100 seconds. What is the average velocity of the runner for this time interval?

(A) 0 m/s
(B) 2 m/s
(C) 4 m/s
(D) 8 m/s
(E) 12 m/s

ANSWERS EXPLAINED

		Key Words	Needed for Solution	Now Solve It
1.	**(C)**	Tossed upward; 24 meters above water; 3 seconds; the surface of the water; initial speed	Involves time, acceleration (gravity), and displacement: $x = v_i t + \frac{1}{2} at^2$ Motion is vertical: $y = v_i t + \frac{1}{2} gt^2$	Displacement and time are given. Initial velocity is requested. Displacement is negative as it is below the point of origin. Acceleration by gravity is also negative. $$y = v_i t + \frac{1}{2} gt^2$$ $$(-24 \text{ m}) = v_i(3 \text{ s}) + \frac{1}{2}(-10 \text{ m/s}^2)(3 \text{ s})^2$$ $$-24 \text{ m} = v_i(3 \text{ s}) + -45 \text{ m}$$ $$21 \text{ m} = v_i(3 \text{ s})$$ $$v_i = 7 \text{ m/s}$$
2.	**(D)**	Moving at 20 meters per second; deceleration of 10 meters per second squared; how far; before it stops	Involves kinematic variables, but time is not mentioned: $v_f^2 = v_i^2 + 2ax$ Hidden variable (stops): $v_f = 0$ m/s	Deceleration indicates acceleration in the opposite direction to motion. Initial velocity is positive, so acceleration is negative. $$v_f^2 = v_i^2 + 2ax$$ $$(0 \text{ m/s})^2 = (20 \text{ m/s})^2 + 2(-10 \text{ m/s}^2)x$$ $$-400 \text{ m}^2/\text{s}^2 = (-20 \text{ m/s}^2)x$$ $$x = 20 \text{ m}$$
3.	**(A)**	Velocity versus time graph; acceleration	Knowledge/definitions or Test if units of slope or area match the units of acceleration	The units of slope match the units needed in the answer. The slope of velocity versus time is acceleration. The slope for this time interval is zero.
4.	**(E)**	Velocity versus time graph; displacement	Knowledge/definitions or Test if units of slope or area match the units of displacement	The units of area match the units needed in the answer. The area under a velocity versus time graph is displacement. Determine the area between the graph and the x-axis from 3 to 5 seconds. Use the formula for the area of a trapezoid. $$\text{displacement} = \text{area of trapezoid}$$ $$= \frac{1}{2}(h_1 + h_2)(b)$$ $$\frac{1}{2}(10 \text{ m/s} + 5 \text{ m/s})(2 \text{ s}) = 15 \text{ m}$$
5.	**(B)**	Velocity versus time graph; acceleration	Knowledge/definitions or Test if units of slope or area match the units of acceleration	The units of slope match the units needed in the answer. The slope of velocity versus time is acceleration. The slope between $t = 3$ s and $t = 5$ s is: $$a = \text{slope} = \frac{\text{rise}}{\text{run}} = \frac{5 \text{ m/s} - 10 \text{ m/s}}{5 \text{ s} - 3 \text{ s}} = -2.5 \text{ m/s}^2$$ Magnitude is the absolute value, 2.5 m/s^2.

	Key Words	Needed for Solution	Now Solve It
6. **(B)**	Vertical displacement versus time	Knowledge/definitions and graphing	The phrase "vertical displacement versus time" indicates displacement is on the y-axis and time on the x-axis. The ball starts at the origin and initially moves upward $(+y)$. After reaching its maximum height, it moves downward. The ball passes its starting point $(y = 0)$ and finishes below its starting point $(-y)$. The graph of displacement appears nearly identical to the diagram of the throw shown in the problem.
7. **(B)**	How high is the ceiling; thrown upward from the floor at 10.0 meters per second; barely touches the ceiling	Involves kinematic variables, but time is not mentioned: $$v_f^2 = v_i^2 + 2ax$$ Hidden variable (barely touches ceiling): $$v_f = 0 \text{ m/s}$$	$$v_f^2 = v_i^2 + 2ax$$ $$(0 \text{ m/s})^2 = (10 \text{ m/s})^2 + 2(-10 \text{ m/s}^2)x$$ $$-100 \text{ m}^2/\text{s}^2 = (-20 \text{ m/s}^2)x$$ $$x = 5 \text{ m}$$
8. **(C)**	Constant acceleration; from rest; 4.0 meters per second squared; what is the displacement; 10 seconds	Involves time, acceleration, and displacement: $$x = v_i t + \frac{1}{2} a t^2$$ Hidden variable (from rest): $$v_i = 0 \text{ m/s}$$	Initially at rest $(v_i = 0 \text{ m/s})$ simplifies the equation. $$x = \frac{1}{2} a t^2$$ $$x = \frac{1}{2}(4 \text{ m/s}^2)(10 \text{ s})^2$$ $$x = 200 \text{ m}$$
9. **(E)**	Uniform acceleration; time doubles	Involves time, acceleration, and displacement: $$x = v_i t + \frac{1}{2} a t^2$$	Initially at rest $(v_i = 0 \text{ m/s})$ simplifies the equation. $$x = \frac{1}{2} a t^2$$ Time is squared. Doubling time quadruples the right side of the equation. The left side of the equation must also quadruple in order to maintain the equality. $$(4)x = \frac{1}{2} a (2t)^2$$
10. **(A)**	Oval-shaped track; constant speed; returning to the starting point; velocity	Knowledge/definitions	Constant speed does not necessarily mean constant velocity. Velocity is a vector relying on displacement. Displacement is the distance between the initial and final positions of an object. When an object returns to its original position, displacement is zero. $$v = \frac{\Delta x}{t} = \frac{0 \text{ m}}{t \text{ s}} = 0 \text{ m/s}$$

Kinematics in Two Dimensions

4

━━━

→ **INDEPENDENCE OF MOTION**

→ **TRUE VELOCITY AND DISPLACEMENT**

→ **RELATIVE VELOCITY**

→ **PROJECTILE MOTION**

The previous chapter introduced three kinematic equations for objects moving along one-dimensional lines. In one-dimensional kinematics, objects typically move along either the x-axis or the y-axis. This chapter will consider objects that move along the x-axis and the y-axis simultaneously: two-dimensional motion. The kinematic equations must now be used separately. You must analyze the x-motion and y-motion independently. This chapter focuses on the following objectives:

- Define and discuss the independence of motion.
- Determine the true velocity and displacement of objects.
- Solve projectile-motion problems.

Table 4.1 lists the variables that will be used in this chapter and their units.

Table 4.1 Variables Used in Two-Dimensional Kinematics

New Variables	Units
$\vec{v}_i$ = True, three-dimensional initial velocity	m/s (meters per second)
$\vec{v}_f$ = True, three-dimensional final velocity	m/s (meters per second)
$\vec{s}$ = True, three-dimensional displacement	m (meters)
v_{ix} = Initial velocity in the x-direction	m/s (meters per second)
v_{iy} = Initial velocity in the y-direction	m/s (meters per second)
v_{fx} = Final velocity in the x-direction	m/s (meters per second)
v_{fy} = Final velocity in the y-direction	m/s (meters per second)

INDEPENDENCE OF MOTION

Chapter 3 demonstrated how kinematic equations are used to determine the position, velocity, and acceleration of an object moving along a one-dimensional line. Consider an example of an astronaut in space throwing a ball horizontally with an initial velocity of $\vec{v}_i$. If the astronaut is very far from Earth, and gravity is negligibly small, the ball will continue to move in a straight line. The diagram below indicates the instantaneous velocity vectors on the ball at four different locations during its motion.

Figure 4.1 Velocity vectors

IF YOU SEE
two-dimensional motion

Solve the kinematic equations for x and y independently.

The motion in the x- and y-directions may be completely different.

Notice that all the velocity vectors have equal magnitudes and directions. If no external forces are acting on the ball after its release, the ball will continue moving with its initial velocity, $\vec{v}_i$, indefinitely.

Consider if someone throws the same ball horizontally near the surface of Earth. If the ball is given the same initial velocity, $\vec{v}_i$, its path will resemble a parabola.

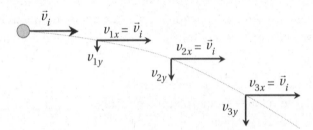

Figure 4.2 Projectile velocity vectors

The ball will continue to move in the x-direction at velocity $\vec{v}_i$. However, it will also experience a vertical acceleration due to gravity. As a result, the vertical velocity, v_y, will increase uniformly. The resulting path of the ball is a parabola.

Kinematic equations can be used for motion only along a straight line. Therefore, separate kinematic equations must be employed for x-variables and for y-variables. The resulting motion is described by two kinematic equations in combination. Adding x- and y-subscripts to the kinematic variables allows you to distinguish between similar variables acting in different directions. Table 4.2 compares the kinematic equations in one and in two dimensions.

Table 4.2 Kinematic Equations in One and Two Dimensions

One-dimensional Kinematics Equations	Two-dimensional Kinematic Equations	
	x-direction	y-direction
$x = v_i t + \frac{1}{2} at^2$	$x = v_{ix} t + \frac{1}{2} a_x t^2$	$y = v_{iy} t + \frac{1}{2} a_y t^2$
$v_f^2 = v_i^2 + 2ax$	$v_{fx}^2 = v_{ix}^2 + 2a_x x$	$v_{fy}^2 = v_{iy}^2 + 2a_y y$
$v_f = v_i + at$	$v_{fx} = v_{ix} + a_x t$	$v_{fy} = v_{iy} + a_y t$

Although the kinematic equations can never contain a mixture of x- and y-variables, the individual x- and y-equations do share one very important variable—time, t. The mathematically independent x- and y-motions take place simultaneously and share the same time, t.

IF YOU SEE
two-dimensional motion

Time is key.

Time is the common variable shared by the x- and y-equations.

TRUE VELOCITY AND DISPLACEMENT

The kinematic equations solve for x- and y-direction velocities and displacements. However, in two-dimensional motion problems, the path followed by the object does not lie solely along either the x- or y-axis. The ball in Figure 4.3 follows a parabolic path.

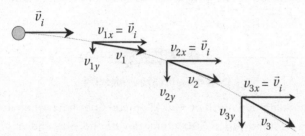

Figure 4.3 Projectile motion velocity vectors and their components

At any instant during this motion, there is the **true velocity**, $\vec{v}$, tangent to the motion of the object. The true velocity is found using vector mathematics, as described in Chapter 3. The x- and y-velocities, calculated by the separate one-dimensional kinematic equations, are the components of the true velocity. You can find the true velocity of the object by using the Pythagorean theorem.

$$|\vec{v}| = \sqrt{v_x^2 + v_y^2}$$

The **true displacement** of an object can be found in a similar manner. The kinematic equations solve separately for the x- and y-displacements. These displacements are also the vector components of the true displacement of an object. Use the Pythagorean theorem to find the true displacement.

$$|\vec{s}| = \sqrt{x^2 + y^2}$$

RELATIVE VELOCITY

The motion of an object, as described by two observers, may differ depending on the location of the observers. For example, a car reported as moving at 30 meters per second by a stationary observer will appear to be moving at 10 meters per second as observed by a driver in a car traveling alongside at 20 meters per second. Problems involving multiple velocities are known as **relative-velocity** problems. Common two-dimensional relative-velocity problems involve a boat moving across a river or an airplane flying through the air. In these problems, the velocities in both the x- and y-directions are constant. Therefore, the acceleration in the x-direction, a_x, and the acceleration in the y-direction, a_y, are both equal to zero, as shown in Table 4.3. This greatly simplifies the kinematic equations.

Table 4.3 Relative-Velocity Equations

Kinematic Equation Used for Relative Velocity	Modified for *x*-direction When $a_x = 0$	Modified for *y*-direction When $a_y = 0$
$x = v_i t + \frac{1}{2} at^2$	$x = v_{ix}t$	$y = v_{iy}t$

The velocities of river currents and the wind change the true velocity of boats and airplanes. Relative-velocity problems require you to understand both vector addition and independence of motion.

EXAMPLE 4.1

Determining True Velocity

An airplane is heading due north at 400 kilometers per hour when it encounters a wind from the west moving at 300 kilometers per hour, as shown in the following diagram.

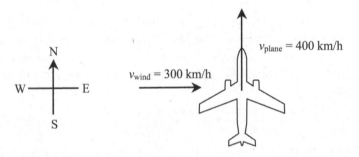

(A) Determine the magnitude of the true velocity of the plane with respect to an observer on the ground.

(WHAT'S THE TRICK?)

An observer on the ground will see the true velocity of the airplane. Mathematically, this is the resultant vector created by adding the airplane and wind velocity vectors tip to tail, as shown below, and using the Pythagorean theorem.

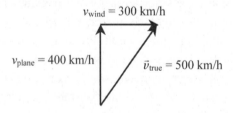

These vectors form a 3-4-5 right triangle. The airplane has a speed of 500 kilometers per hour relative to the ground.

(B) Which vector, of the choices given below, describes the direction the pilot must aim the plane in order for the plane to have a true velocity that points directly north?

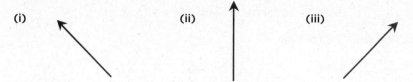

(i)   (ii)   (iii)

WHAT'S THE TRICK?

The velocity vectors for the plane and the wind must add together to create a true velocity that points north. Since the wind is blowing out of the west, the plane must have a component that moves toward the west to cancel the effect of the wind.

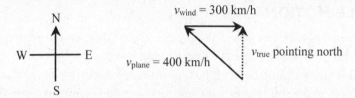

Answer (i) is the correct heading for the plane. When the plane is aimed northwest, it will actually move with a true velocity directly north.

EXAMPLE 4.2

Determining True Displacement

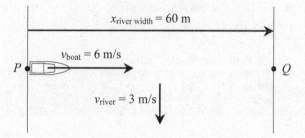

A boat capable of moving at 6 meters per second attempts to cross a 60-meter-wide river as shown in the diagram above. The river flows downstream at 3 meters per second. The boat begins at point P and aims for point Q, a point directly across the river. How far downstream from point Q will the boat drift?

WHAT'S THE TRICK?

The boat and the river both move at constant velocity. The motion of the boat and the river are mathematically independent but take place simultaneously. You should include subscripts to distinguish between the two velocities. The velocity of the boat, v_b, produces a cross-stream displacement, x. The velocity of the river, v_r, produces a downstream displacement, y.

$$x = v_b t \quad \text{and} \quad y = v_r t$$

IF YOU SEE
relative-
velocity
problems

x = vt

Solve this
equation
in both
the *x*- and
y-directions.

The variable shared by both equations is time. Simply substitute the known values into both equations. You can then solve one equation, followed by the next equation.

$$(60 \text{ m}) = (6 \text{ m/s})t \qquad \text{and} \qquad y = (3 \text{ m/s})t$$

Solve the first equation for time.

$$t = 10 \text{ s}$$

Now solve the second equation.

$$y = (3 \text{ m/s})(10 \text{ s}) = 30 \text{ m}$$

The boat will drift 30 m downstream from point Q.

PROJECTILE MOTION

Projectile motion describes an object that is thrown, or shot, in the presence of a gravity field. An object is considered a projectile only when it is no longer in contact with the person or device that has thrown it and before it has come into contact with any surfaces. As a result, the downward acceleration of gravity is the only acceleration acting on a projectile during its flight.

One of the most important aspects of projectile motion is the type of motion experienced in each direction. If the vertical acceleration of gravity is the only acceleration acting on a projectile, the horizontal speed of a projectile cannot change. Therefore, the horizontal component of velocity must always remain constant. Both the vertical velocity and the vertical displacement will be affected by the acceleration of gravity. When calculating the vertical portion of projectile motion, you must use the complete kinematic equations, as shown in Table 4.4. You should include additional subscripts to distinguish the x- and y-velocities from each other and from the true velocity, $\vec{v}$.

Table 4.4 Kinematic Equations with Gravity

x-direction Constant Velocity: $a_x = 0$	*y*-direction Gravity: $a_y = g$	True Velocity at Any Instant
$x = v_{ix}t$ $v_{fy} = v_{iy}$	$y = v_{iy}t + \dfrac{1}{2}gt^2$ $v_{fy}^2 = v_{iy}^2 + 2gy$ $v_{fy} = v_{iy} + gt$	$|\vec{v}| = \sqrt{v_x^2 + v_y^2}$

The first step in projectile-motion problems involves determining the initial velocity in the x-direction, v_{ix}, and the initial velocity in the y-direction, v_{iy}. These velocities are the components of the initial launch velocity, $\vec{v}_i$. Table 4.5 shows the velocities of two typical launches.

Table 4.5 Velocities of Two Typical Launches

Horizontally Thrown Projectile	Upward Launch of Projectile
The initial velocity, v_i, does not form the hypotenuse of the right triangle. $$v_{ix} = \vec{v}_i$$ $$v_{iy} = 0$$	The x- and y-velocities are vector components of the initial velocity, v_i. $$v_{ix} = \vec{v}_i \cos \theta$$ $$v_{iy} = \vec{v}_i \sin \theta$$

IF YOU SEE
projectile
motion

**Horizontal
velocity is
constant.**

This is true for
all projectiles
regardless of
launch angle.

The next step in projectile motion usually involves determining time, t. As with all two-dimensional motion problems, time is the only variable common to both the x- and y-directions. You should note that the time of flight, t, depends on y-direction variables, not on x-direction variables. Therefore, when the time of flight is not given, most problems begin by solving one of the three y-direction kinematic equations.

Horizontally Launched Projectiles

Horizontally launched projectiles are the most common projectile-motion problems encountered on introductory physics exams. As with any kinematics problem, identifying variables (especially hidden variables) is extremely important. In horizontal launches, the initial velocity in the y-direction is zero, as shown in Table 4.6. This simplifies the y-direction equations.

IF YOU SEE
projectile
motion

**Acceleration
of gravity
acts in the
y-direction.**

The y-velocity
decreases by
10 m/s^2 every
second.

Table 4.6 Kinematic Equations for Horizontal Launches

	x-equation	Simplified y-equations
	$x = v_{ix} t$	$y = \dfrac{1}{2} g t^2$ $v_{fy} = \sqrt{2gy}$ $v_{fy} = gt$

IF YOU SEE
projectile
motion

**Time is
controlled by
y-variables.**

Time is usually
solved using
y-component
equations.

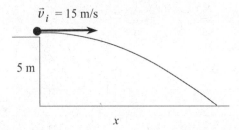

EXAMPLE 4.3

Horiztontally Launched Projectiles

$\vec{v}_i = 15$ m/s

5 m

x

A ball is thrown horizontally at 15 meters per second from the top of a 5-meter-tall platform, as shown in the diagram above. Determine the horizontal distance traveled by the ball.

WHAT'S THE TRICK?

First determine the x- and y-components of the initial velocity. This is easy for horizontal launches. All of the initial velocity is directed horizontally, and none is directed vertically.

$$v_{ix} = \vec{v}_i = 15 \text{ m/s} \qquad \text{and} \qquad v_{iy} = 0$$

The velocity in the y-direction is a hidden zero, which simplifies the y-equations.

If time is unknown, solve for time using y-direction equations. The vertical displacement of 5 meters is given. Use an equation containing both displacement and time.

$$y = v_{iy}t + \frac{1}{2}gt^2 \qquad \text{simplifies to} \qquad y = \frac{1}{2}gt^2$$

$$(-5 \text{ m}) = \frac{1}{2}(-10 \text{ m/s}^2)\, t^2$$

$$t = 1 \text{ s}$$

Finally, time is the one variable common to motion in any direction. Now use time in the horizontal equation to determine the horizontal distance.

$$x = v_{ix}t$$

$$x = (15 \text{ m/s})(1 \text{ s}) = 15 \text{ m}$$

Projectiles Launched at an Angle

Projectiles launched at angles are more difficult to solve mathematically. As a result, complex calculations for these projectiles may not appear on the SAT Subject Test in Physics. However, projectiles launched at upward angles do have several unique characteristics that will be tested conceptually.

Figure 4.4 depicts the flight path of a projectile launched with an initial speed of $\vec{v}_i = 50$ m/s at an upward launch angle. The projectile lands at the same height from which it was launched, $y = 0$, and has a final speed $v_f = 50$ m/s. The projectile is shown every second during its flight. In each of these positions, the instantaneous horizontal component of velocity, v_x, and the instantaneous vertical component of velocity, v_y, are shown.

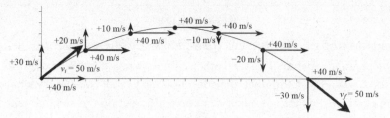

Figure 4.4 Projectile-motion vectors

Examination of the diagram reveals four key facts about projectiles launched at angles.

1. The horizontal component of velocity, v_x, remains constant.
2. When the projectile is moving upward, its vertical speed decreases by 10 meters per second every second until the projectile reaches an instantaneous vertical speed of zero at maximum height. The decreasing velocity then results in a changing downward speed that increases by 10 meters per second every second.
3. The projectile passes through each height twice, once on the way up and once on the way down (except the single point at maximum height). At points with equal height, the magnitude of the vertical velocity on the way up equals the magnitude of the vertical velocity on the way down.
4. The time the projectile rises equals the time the projectile falls, as long as the final height equals the initial height.

Retaining a mental image of the above diagram in your memory will help you answer conceptual problems for upwardly launched projectiles. The most common questions tend to focus on two key locations during the flight: the very top of the flight path (maximum height) and the landing point, as shown in Figure 4.5.

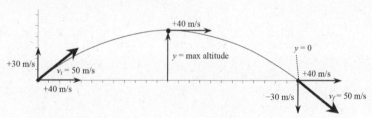

Figure 4.5 Two key instants during a projectile flight

Remember the following facts:

- The vertical component of velocity at maximum height is zero. Therefore, the true velocity at maximum height equals the horizontal component of the launch velocity: $\vec{v}_{\text{max height}} = v_{ix}$.
- Although the vertical component of velocity at maximum height becomes zero, the acceleration in the vertical direction remains a constant −10 meters per second squared.
- The maximum height, y, can be solved using $v_{fy}^2 = v_{iy}^2 + 2gy$. At maximum height, $v_{fy} = 0$.
- When a projectile lands at the same height as its launch point, the vertical displacement is zero: $y = 0$. As a result, the final speed of the projectile equals the launch speed: $v_f = v_i$.
- For a projectile landing at its initial launch height, the time to maximum height is half the total time of flight.
- To reach maximum range, x_{max}, a projectile must have a launch angle of 45 degrees.
- Any two launch angles totaling 90 degrees will have the same range. For example, if a projectile is launched at 30 degrees, it will reach the same impact point if it is launched at 60 degrees.

SUMMARY

1. **INDEPENDENCE OF MOTION.** The horizontal and vertical forces and motions on an object are independent of each other. Although one dimension does not affect the other, when taken together, they describe the overall two-dimensional motion of the object. Both motions occur simultaneously during the same time, t.

2. **TRUE VELOCITY, $\vec{v}$.** This is also referred to as the net velocity or resultant velocity. The true velocity is the velocity of the object as seen by a stationary observer. It is found by adding all the velocity vectors acting on an object by using vector addition.

3. **PROJECTILE MOTION.** In the horizontal direction, a projectile will always move at constant velocity. In the vertical direction, a projectile will accelerate at –10 meters per second squared due to gravity. The resulting motion of a projectile is a parabola. Kinematic equations can be used to describe the motion of the projectile as long as x- and y-variables are *not* used together in the same kinematic equation. One set of equations is used to calculate x-variables, and another separate set is used to calculate y-variables. However, time, t, is the one variable common to both x- and y-equations.

4. **RESOLVING VELOCITY VECTORS.** The initial velocity of a projectile thrown at an angle above the horizontal is the hypotenuse of a right triangle. The x- and y-component vectors are the two legs of the right triangle. The x-component velocity vector remains constant throughout the flight of the projectile. The y-component continually changes due to the acceleration of gravity.

If You See	Try	Keep in Mind
Any type of two-dimensional motion	Solve the x- and y-kinematic equations separately.	Time is the one variable shared by x- and y-variables.
Relative velocity	Use the constant-velocity formula in both the x- and y-directions. $$x = v_x t \qquad y = v_y t$$	Solve one equation first, followed by the other.
Projectile motion involving a horizontal launch	Solve for time using a simplified y-direction equation. $$y = \frac{1}{2} g t^2$$ Then use time in the x-direction equation to solve for range, x. $$x = v_{ix} t$$	$v_{iy} = 0$ is the hidden zero in horizontal launches. The y-variables control time. The motion in the x-direction is constant velocity.
A projectile launched at an angle greater than horizontal	Resolve the launch velocity into its components, v_{ix} and v_{iy}. There is no vertical velocity at the top of the flight path, $v_{fy} = 0$. Therefore, the true velocity at the top of the flight path is equal to the initial x-velocity. $$\vec{v}_{\text{max altitude}} = v_{ix}$$ When solving for maximum altitude, try $$v_{fy}{}^2 = v_{iy}{}^2 + 2gy$$ where $v_{fy} = 0$. If a projectile returns to its original launch height, its speed, v_f, equals the launch speed, v_i. $$v_f = v_i$$	The velocity in the x-direction, v_x, remains constant. The velocity in the y-direction, v_y, decreases by 10 m/s every second on the way up, is zero at the top of the flight, and increases by 10 m/s every second on the way down.

PRACTICE EXERCISES

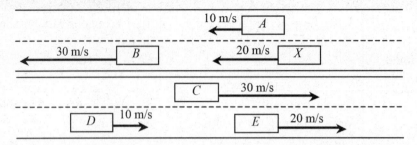

1. In the diagram above, six cars are shown driving on a four-lane highway. Two observers are riding in car *X*. When compared with the observers, which car appears to have a velocity of 10 meters per second in the opposite direction?

   (A) *A*
   (B) *B*
   (C) *C*
   (D) *D*
   (E) *E*

Questions 2–4

A boat with a speed of 4 meters per second will cross a 40-meter-wide river that has a current of 3 meters per second. The boat aims directly across the river, as shown in the diagram below.

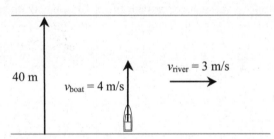

2. How long will the boat take to cross the river?

   (A) 10 s
   (B) 20 s
   (C) 30 s
   (D) 40 s
   (E) 50 s

3. How far downstream will the boat drift as a result of the river current?

   (A) 0 m
   (B) 15 m
   (C) 20 m
   (D) 30 m
   (E) 45 m

4. You want to cross the 40-meter-wide river, as shown, in the least amount of time. The location on the other side of the river is not important; only minimizing the time to cross the river matters. You should swim

   (A) completely upstream
   (B) slightly upstream
   (C) directly across the stream
   (D) slightly downstream
   (E) completely downstream

   ───────────────────────────

5. Which variable(s) remain(s) constant during the entire flight of a projectile?

   (A) Horizontal component of velocity only
   (B) Vertical component of velocity only
   (C) Acceleration only
   (D) Both A and C
   (E) Both B and C

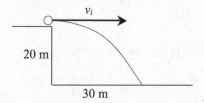

6. A ball thrown horizontally from the edge of a 20-meter-tall structure lands 30 meters from the base of the structure as shown above. Determine the speed at which the ball was thrown.

   (A) 5 m/s
   (B) 10 m/s
   (C) 15 m/s
   (D) 20 m/s
   (E) 25 m/s

7. A convertible car with an open top is driving at constant velocity when a ball is thrown straight upward by one of the passengers. If there is no air resistance, where will the ball land?

   (A) In front of the car
   (B) In the car
   (C) Behind the car
   (D) The answer depends on the speed of the projectile.
   (E) The answer depends on the speed of the car.

Questions 8–10

The diagram below depicts a projectile launched from point *A* with a speed *v* at an angle of *θ* above the horizontal. The projectile reaches its maximum height, *h*, at point *B*. The projectile impacts the ground at point *C*, achieving a final range of *x*. Points *A* and *C* are at the same height. The total time of flight from point *A* to point *C* is *t* seconds.

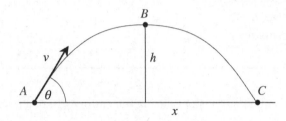

8. Determine the speed of the projectile at point *B*.

   (A) zero
   (B) v
   (C) $\frac{1}{2}v$
   (D) $v \cos \theta$
   (E) $v \sin \theta$

9. Determine the speed of the projectile at point *C*.

   (A) zero
   (B) v
   (C) $\frac{1}{2}v$
   (D) $v \cos \theta$
   (E) $v \sin \theta$

10. Which of the statements below is true regarding the acceleration acting on the projectile during its flight from point *A* to point *C*?

    (A) The acceleration decreases from *A* to *B* and increases from *B* to *C*.
    (B) The acceleration increases from *A* to *B* and decreases from *B* to *C*.
    (C) At *B*, the magnitude of the acceleration equals the *x*-component of the magnitude at *A*.
    (D) At *C*, the magnitude of the acceleration is the same as at *A*, but the direction is opposite.
    (E) The magnitude and direction of the acceleration remain constant from *A* to *C*.

ANSWERS EXPLAINED

		Key Words	Needed for Solution	Now Solve It
1.	**(A)**	Compared with the observers	Understanding of relative velocity	The observers in car *X* measure the motion of objects relative to themselves. When doing this, they see themselves as stationary. To have a speed of 0 m/s, the actual 20 m/s must be subtracted. Subtracting 20 m/s from every car will show their speeds relative to car *X*. When this is done, car *A* has a velocity of –10 m/s.
2.	**(A)**	Crossing a 40-meter-wide river	Kinematic equations solve independently in the *x*- and *y*-directions. $$y = v_{yi}t$$	The *y*-displacement is given, and the *x*- and *y*-variables must be used in separate equations. Use the *y*-velocity with the *y*-displacement to solve for the crossing time. $$y = v_{yi}t$$ $$t = \frac{y}{v_{yi}} = \frac{40 \text{ m}}{4 \text{ m/s}} = 10 \text{ s}$$
3.	**(D)**	How far downstream; as a result of the river current	Although the *x*- and *y*-directions solve independently, time is the key variable they share. $$x = v_{xi}t$$	Motion is simultaneous in the *x*- and *y*-directions, and time is the key variable shared by both equations. The time was determined in question 2. The current and downstream distance are both *x*-direction variables. $$x = v_{xi}t = (3 \text{ m/s})(10 \text{ s}) = 30 \text{ m}$$
4.	**(C)**	Cross a flowing river; minimizing the time	Crossing the river involves motion in the *y*-direction. $$y = v_{yi}t$$	Rearrange the equation to solve for cross-river time. $$t = \frac{y}{v_{yi}}$$ The cross-river distance, *y* = 40 m, is fixed. To minimize time, the velocity must be maximized. Therefore, try to swim directly across the river. If you try to swim either upriver or downriver, the cross-river component of velocity will be less than 4 m/s, causing you to take longer to cross the river.
5.	**(D)**	Variable(s) remain(s) constant; projectile	Projectile-motion definition	Projectiles are thrown or shot in the presence of gravity alone. Gravity is the only acceleration affecting a projectile; therefore, the acceleration of gravity remains constant during a projectile's flight. Answer C is correct. However, the *x*- and *y*-directions are independent, and gravity acts in only the *y*-direction. This implies that there is no acceleration in the *x*-direction, so the horizontal component of velocity must also remain constant. Answer A is also correct. Therefore, answer D is the most complete response.

	Key Words	Needed for Solution	Now Solve It
6. **(C)**	Thrown horizontally; 20-meter-tall structure; lands 30 meters from the base of the structure	Horizontally launched projectile $$y = \frac{1}{2}gt^2$$ $$x = v_{xi}t$$	Time is usually the key variable, and time is dependent on y-direction variables. $$y = \frac{1}{2}gt^2$$ $$t = \sqrt{\frac{2y}{g}} = \sqrt{\frac{2(20 \text{ m})}{\left(10 \text{ m/s}^2\right)}} = 2 \text{ s}$$ Once you have determined the time, solve the remaining equation. $$x = v_{xi}t$$ $$v_{xi} = \frac{x}{t} = \frac{(30 \text{ m})}{(2 \text{ s})} = 15 \text{ m/s}$$ For horizontally launched projectiles, $\vec{v}_i = v_{ix}$.
7. **(B)**	Car . . . constant velocity; ball is thrown straight upward; no air resistance	Projectile-motion knowledge, specifically the independence and types of x- and y-direction motions	Before the ball is thrown, it is moving at constant velocity in the x-direction. The throw gives the ball an additional velocity in the y-direction where gravity acts. However, gravity in the y-direction has no effect on the constant velocity in the x-direction. Therefore, the ball continues to move forward in the x-direction at the same speed as the car. The ball will land in the car as long as there is no air resistance.
8. **(D)**	Speed; projectile; maximum height	Projectile-motion knowledge or components of initial launch velocity. $$v_x = v \cos \theta$$ $$v_y = v \sin \theta$$	The velocity vector at any point during the flight of a projectile will be tangent to the parabolic path at that point. At maximum height, the vertical component of velocity will be zero, $v_y = 0$, and the true velocity will be completely horizontal, $\vec{v} = v_x$. The horizontal component of velocity is constant during the flight of a projectile, and the magnitude of velocity is speed, v. $$v = v_x = v \cos \theta$$
9. **(B)**	Speed; projectile; at the same height	Projectile-motion knowledge.	At any two points having the same height, a projectile will have the same speed. Point C has the same height as point A. $$v_C = v_A = v$$ If the problem had requested the velocity at C, the answer would have the same magnitude as the velocity at A. However, the direction would be negative θ since the projectile would strike the ground at a fourth-quadrant angle.

	Key Words	Needed for Solution	Now Solve It
10. **(E)**	Acceleration, projectile	Definition of a projectile's motion	The only acceleration in projectile-motion problems is gravity. Gravity has a constant magnitude, 10 m/s^2, and a constant downward direction at all times. Examining the other answers reveals that they seem to address the changing velocity of a projectile. If you read the question too quickly, you may be distracted into thinking about velocity instead of acceleration.

Dynamics

5

→ **INERTIA**

→ **FORCE**

→ **COMMON FORCES**

→ **FORCE DIAGRAMS**

→ **NEWTON'S LAWS OF MOTION**

→ **SOLVING FORCE PROBLEMS**

Dynamics and kinematics are explicitly linked. Together they are at the heart of Newtonian mechanics. **Dynamics** involves the actual causes of the motions represented by the kinematic equations.

- Objects with mass possess inertia, and their natural state of motion is constant velocity.
- A force is a vector quantity that can be thought of as a push or a pull acting to change the motion of an object.
- Sir Isaac Newton postulated three laws of motion summarizing the interactions among objects, forces, and motion.
- Forces and their influence on kinematics will be encountered in all areas of physics. Developing universal problem-solving strategies for all dynamics problems is an essential skill.

Table 5.1 lists the variables used in dynamics.

Table 5.1 Variables and Units for Forces in One Dimension

New Variables	Units
$\vec{F}$ = Applied force $\vec{F}_g$ = Force of gravity or weight $\vec{N}$ = Normal force $\vec{f}_s$ = Static friction $\vec{f}_k$ = Kinetic friction $\vec{T}$ = Tension $\vec{F}_s$ = Force of elastic devices, springs	N (newtons)
k = Spring constant	N/m (newtons per meter)
$\Sigma\vec{F}$ = Net force or sum of forces ΣF_x = Net force along the x-axis ΣF_y = Net force along the y-axis	N (newtons)

INERTIA

Inertia is the tendency of an object to resist changes in its natural motion. Galileo suggested that the natural state of motion of an object is constant velocity. Even an object at rest has a constant velocity of zero. Inertia is the tendency of a stationary object to remain at rest and for a moving object to continue moving at constant velocity. Inertia is a property of mass. The greater the mass of an object, the greater the inertia possessed by the object. Simply put, the more mass the object has, the harder it is to push around the object. Physicists regard mass as a quantity of inertia and a resistance to change in motion.

FORCE

An object cannot simply accelerate and change velocity on its own. In order to accelerate, an object must be pushed or pulled by some external source known as an **agent**. The push or pull of an agent that accelerates an object is known as a **force**. Forces are vector quantities measured in newtons.

Since forces are vector quantities, they can be added together to find a resultant sum. The sum of all forces, $\Sigma \vec{F}$, acting on an object is known as the **net force**. The direction of the net force is the same as the direction of the acceleration (change in velocity) of the object. The relationship between the direction of force and the direction of initial velocity of an object dictates the resulting change in motion, as shown in Table 5.2.

Table 5.2 Relationship Between Direction of Force and Direction of Initial Velocity

Direction of Force Matches Initial Motion	Direction of Force Opposes Initial Motion	Direction of Force Perpendicular to Initial Motion
Forces acting in the direction of motion attempt to increase the speed of objects. These forces can be considered positive.	Forces acting opposite the direction of motion attempt to decrease the speed of objects. These forces can be considered negative.	The force will cause the object to change direction.

Force problems usually focus on finding the magnitude of the net force and acceleration. To make them easier to solve, set the direction of the initial velocity as positive. Under these conditions, forces acting in the same direction as velocity create positive acceleration and attempt to increase the speed of objects. Forces acting in the opposite direction of velocity result in negative acceleration and attempt to slow down objects.

Setting the direction of initial motion as a positive value creates a dilemma for objects that are moving downward due to gravity. Although down is technically the negative direction, speed is increasing. When solving for the magnitude (absolute value) of force or acceleration for a falling object, it is easier to set down as the positive direction. Which direction is set as positive does not really matter. What matters is that the signs on all vector quantities match the convention chosen and remain consistent throughout the entire problem.

As a matter of convention, both one-dimensional force vectors and components of force vectors will be indicated in *italics*.

COMMON FORCES

Numerous agents, each with unique characteristics, are capable of generating forces. As a result, several important forces have been given unique variables.

Applied Force

The general variable letter, *F*, is used to represent any force that does not have its own variable designation. As an example, there is no specific variable for the force of a person pushing a box.

Force of Gravity (Weight)

Objects having mass are considered to be surrounded by a mathematical vector field known as the gravitational field, *g*. The gravity field is essentially the value of the acceleration of gravity at every point in space surrounding a mass. All masses are surrounded by a gravity field. However, when two masses interact, one of the masses (usually the larger mass) is thought of as the agent surrounded by a gravity field. The second mass (usually the smaller mass) is thought of as the object. When the second mass (object) is placed in the gravity field of the first mass (agent), the magnitude of the force of attraction on the object is the product of the object's mass, *m*, and the agent's gravity field, *g*.

$$F_g = mg$$

This scenario can also be reversed. The force of gravity acting on the agent due to the gravity field surrounding the object has the same magnitude, but acts in the opposite direction according to Newton's third law.

Most problems take place on Earth. Unless stated otherwise, Earth is the agent creating the gravity field, and $g = 10 \text{ m/s}^2$. As shown in Figure 5.1, the force of gravity pulls the object with mass *m* toward the agent, Earth. The gravity field and the force of gravity for objects that are not located on the surface of Earth will be address in greater detail in Chapter 9.

Figure 5.1 Force of gravity (weight)

The force of gravity is also the **weight** of an object. Some instructors, texts, and exams may use the variable *w* to represent weight (force of gravity).

$$w = F_g = mg$$

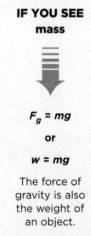

IF YOU SEE
mass

$F_g = mg$

or

$w = mg$

The force of gravity is also the weight of an object.

Inclines

online.barronsbooks.com

Normal Force

The **normal force**, N, is present whenever an object pushes on a surface. There is no specific formula to solve for the normal force. Rather, the normal force is a response force to an applied force. As will be shown shortly, the normal force is a result of Newton's third law. As shown in Figure 5.2, the normal force always acts perpendicular to a surface.

Figure 5.2 Normal force

IF YOU SEE

the words
rough or
friction

$f = F_{forward}$
or
$f = \mu N$

For objects at rest or moving at constant velocity, try the first equation. If this fails or if the object is accelerating, use the second equation.

Friction

The **force of friction**, f or F_f, is present when two conditions are met. First, an object must be pressed against a rough surface. Second, either a force or a component of force must be acting parallel to the surface. Figure 5.3 shows the applied force, F, the force due to friction, f, the force due to gravity, F_g, and the normal force, N, acting on an object with mass m.

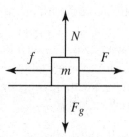

Figure 5.3 Free-body diagram

In Figure 5.3, the force of gravity, F_g, pulls the object onto the surface. This causes the surface to press back with the normal force N. When any force, in this case force F, is applied parallel to a surface, that force will attempt to accelerate the object. If the surface is rough, then a frictional force, f, will act opposite the applied force.

Friction is described by the following equation and is dependent on two quantities:

$$f \leq \mu N$$

1. The normal force, N, is the responding support force of a surface that results when an object presses against the surface. Note that surface area is not present in the above formula. Friction does not depend on surface area. For an object with constant mass, surface area is inversely proportional to the pressure acting per unit of surface area. These factors completely offset each other. As a result, rotating an object to a different surface with a different surface area will not change the magnitude of the normal force nor the magnitude of friction.

2. The coefficient of friction, μ (no units), is a ratio of the force of friction to the normal force. There are two categories of friction. When friction acts and an object remains stationary, it is subject to **static friction**, f_s, which involves the **coefficient of static friction**, μ_s. The friction acting on moving objects is known as **kinetic friction** and involves the **coefficient of kinetic friction**, μ_k. The coefficients of static and kinetic friction will have two different values for the same object, and the coefficient of static friction is greater than the coefficient of kinetic friction.

$$\mu_s > \mu_k$$

When an object is initially at rest on a rough surface and a force, or component of force, is applied parallel to the surface, an opposing friction force responds simultaneously. If the object remains at rest, then it is subject to static friction. There are two ways to solve for the magnitude of static friction. The first method is the most advantageous as it works consistently. When an object remains at rest it is in static equilibrium, and the forces acting on it must be equal and opposite. Therefore, the magnitude of static friction will be equal to the magnitude of the opposite applied force acting parallel to the surface.

$$f_s = F_{\text{applied}}$$

The second method involves the friction equation ($f \leq \mu N$). This is problematic due to the "less than or equal to" mathematical operator. When the equation is set as an equality, then it will only solve for the **maximum value of static friction**, $f_{s\,\text{max}}$.

$$f_{s\,\text{max}} = \mu_s N$$

However, static friction can be less than the value of maximum static friction. In static problems the friction equation can only be used when a problem indicates that static friction has reached its limit, and that even a slight increase in the forward applied force would cause the object to slip. To avoid mistakes in static problems, start by attempting the first method, which works consistently. If this does not arrive at an answer, then use the friction equation. When objects experience maximum static friction, solutions may require the use of both static friction equations.

Kinetic friction can act on objects that accelerate or that move at constant velocity. Fortunately, the friction equation always solves as an equality for moving objects.

$$f_k = \mu_k N$$

Kinetic-friction problems involving acceleration can only use the above equation. However, when an object is moving at constant velocity, the forces acting parallel to motion are equal and opposite. Therefore, kinetic-friction problems involving constant velocity can also be solved by setting kinetic friction equal to the applied forward force.

$$f_k = F_{\text{applied}}$$

When objects move at constant velocity, solutions may require the use of both kinetic friction equations.

Tension

Tension, T, is a force exerted by ropes and strings. There is no specific formula for tension. As with the normal force, tension is a response force and obeys Newton's third law. Tension acts along ropes or strings as shown in Figure 5.4.

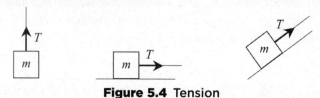

Figure 5.4 Tension

Tension has the same magnitude in every part of a particular rope or string.

Springs

When springs are stretched or compressed by an outside agent, a force is created in the spring. The force of a spring is known as a **restoring force**, and it acts to restore the spring to its original rest length. As a result, the force of a spring, F_s, always opposes the action of the agent. If an agent stretches a spring, the restoring force acts to compress the spring. If an agent compresses a spring, the restoring force acts to stretch the spring. See Figure 5.5.

Figure 5.5 The force of a spring

The magnitude of the force of a spring is described in **Hooke's law**:

$$F_s = |kx|$$

The variable k is the **spring constant** (units: N/m), and the variable x is the distance the spring is either stretched or compressed (units: m). Every spring has its own unique spring constant. Finding the spring constant, if it is not given, is often the first critical step in solving spring problems.

FORCE DIAGRAMS

The first step in solving a dynamics problem is to identify all forces, including the direction of each force, acting on an object. This can be accomplished by constructing a force diagram, known as a **free-body diagram**. A free-body diagram includes only the object and the forces acting on the object. It is free of clutter and does not include surfaces, strings, springs, or any other agents acting on the object. A free-body diagram serves two key purposes: First, it allows you to visualize how the forces will influence the motion. Second, it provides a frame of reference in which to work with the force vectors.

EXAMPLE 5.1

Free-Body Diagrams

A 10-kilogram mass is pulled at constant velocity to the right along a rough horizontal surface by a string. Construct a free-body diagram depicting all the forces acting on the mass.

WHAT'S THE TRICK?

Identify all the forces acting on the mass.

- Gravity is not mentioned in the problem. Unless specified otherwise, problems take place on Earth, so the force of gravity, F_g, is present.
- You may be able to deduce some forces from the diagram. Both a string and a surface are in contact with the object. Therefore, tension, T, and the normal force, N, are present.
- Identifying other forces requires you to read the text of the problem. This problem mentions a "rough" surface, implying that friction, f, is present.

The resulting free-body diagram is shown below left. Free-body diagrams are nearly identical to plotting vectors on coordinate axes, as shown below right.

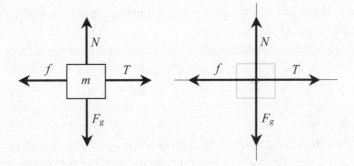

Including all force vectors and ensuring that they point in the correct direction are the most important aspects when drawing a free-body diagram.

IF YOU SEE constant velocity including a constant zero velocity

$\Sigma F = 0$
and
$a = 0$.

The object is in equilibrium. All forces cancel each other out.

NEWTON'S LAWS OF MOTION

Sir Isaac Newton deduced three laws of motion to describe the relationships among force, mass, and changes in velocity.

Newton's First Law

In his first law, Newton simply generalizes Galileo's principle of inertia. Essentially, the first law states that objects at rest will remain at rest, and objects in motion will remain at constant velocity unless acted upon by a net unbalanced force. This law is often referred to as the **law of inertia**.

When the word *equilibrium* is used in physics, the forces acting on an object are balanced (equal and opposite). Equilibrium is a stable condition where the net force is zero, $\Sigma \vec{F} = 0$, and

**IF YOU SEE
unbalanced
forces acting**

$$\Sigma \vec{F} = ma$$
and
$$a \neq 0$$

The object
will accelerate,
changing
either the
magnitude or
the direction
of velocity.

an object does not accelerate, $a = 0$. If the object is not accelerating, it must be experiencing one of only two types of equilibrium.

1. **Static equilibrium.** The object has a constant velocity equal to zero.
2. **Dynamic equilibrium.** The object has a constant velocity not equal to zero.

Newton's first law of motion governs objects in equilibrium.

Newton's Second Law

Newton's second law addresses changes to the inertia of an object. It states that when an object of mass m is acted upon by a net force of $\vec{F}_{net}$ or $\Sigma \vec{F}$ (the sum of all forces acting together), the object will accelerate. In addition,

- the acceleration of the object will be in the direction of the net force;
- the acceleration will be directly proportional to the net force; and
- the acceleration will be inversely proportional to the mass of the object.

These details are very concisely summarized in one the most famous equations in physics.

$$\Sigma \vec{F} = m \vec{a}$$

Newton's second law results in several important consquences.

1. When all the one-dimensional vector forces sum to a nonzero value, the forces are said to be unbalanced. The result is acceleration, which alters either the magnitude or the direction of velocity. When an object is accelerating, there must be a net, unbalanced force acting on the object.
2. The direction of acceleration is the same as the direction of the net force.
3. The direction of velocity is not necessarily in the same direction as the acceleration. An example would be the change in velocity of a car as it slows to a stop. The initial velocity is in one direction, but the acceleration is in the opposite direction, causing the car to slow down.
4. When vector forces add up to zero, the forces are balanced. There can be no acceleration. An object will continue at constant velocity based on the first law of motion. When an object is either at rest or moving at constant velocity, the one-dimensional forces acting on the object must cancel each other. The net force must equal zero.

**IF YOU SEE
two objects
interacting**

**There is an
equal and
opposite force
between them.**

The resulting
motion of the
objects may
not be equal.

Newton's Third Law

Newton's third law states that when objects interact, equal and opposite forces act simultaneously on both objects. Most problems focus only on the forces that act to determine the motion of a specific mass referred to as the "object." However, objects cannot exert a net force on themselves, which means they cannot accelerate themselves and cannot change their own velocities. Objects must be pushed or pulled by another entity known as the "agent." The forces that act on the object causing it to move are created by the agent. According to Newton's third law, when an agent pushes an object with an **action force**, the object simultaneously pushes back with an equal and opposite **reaction force**. This can also be viewed from a reversed frame of reference. When an object pushes on the agent with an **action force**, the agent simultaneously pushes back on the object with an equal and opposite **reaction force**. These simultaneous forces are referred to as an action-reaction pair. As an example, a person walking does not directly use their leg muscles to push themselves forward. They actually use their leg muscles to push their feet backward along the surface of the Earth (action), and simultaneously the surface of Earth pushes back with an equal and opposite friction force (reaction) that propels the person forward. While the problem may only focus on the force acting to move the person forward, this force cannot occur by itself. If you do not push your feet backward, the earth cannot push you forward, and if the surface of the earth were frictionless, no movement of the feet or legs would result in forward motion. Newton's third law is useful in determining a variety of forces such as the normal force, tension, and friction. It is also very evident in problems involving collisions and recoil, as seen in explosions. While a problem may only seem to focus on the forces acting on the object, it may also be important to consider the effect and consequences due to the equal and opposite force acting on the agent.

EXAMPLE 5.2

Newton's Third Law

An 800-newton person is involved in a tug of war with a 600-newton person. The 800-newton person pulls on the rope between them with 400 newtons of force. If the rope is not slipping, how much force does the 600-newton person pull with at the other end of the rope?

WHAT'S THE TRICK?

Newton's third law states that whenever two objects interact, there is an equal and opposite force between them. The problem is asking for the force in the rope. The first person pulls on the rope with 400 newtons. Therefore, the second person must also pull on the rope with 400 newtons. The weights of each person, 800 and 600 newtons, are distractors.

SOLVING FORCE PROBLEMS

The following four-step method is a suggested attack plan that breaks difficult force problems into a series of steps you can easily solve.

1. **Orient the problem.** Identify force vectors and the relevant directions. Then sketch a force diagram.
2. **Determine the type of motion.** Determine if the problem involves
   - equilibrium, constant velocity, balanced forces, $\Sigma \vec{F} = 0$ or
   - dynamics, acceleration, unbalanced forces, $\Sigma \vec{F} = m\vec{a}$.
3. **Sum the force vectors in the relevant direction.** Only forces, or components of forces, parallel to the direction of motion can change the speed of an object. Sum only the one-dimensional force vectors parallel to the object's motion. Any force vectors pointing in the direction of motion will increase the speed of the object and should be set as positive. Force vectors opposing the motion of the object should be set as negative.
4. **Substitute and solve.** In this last step, substitute known equations for specific forces and then substitute numerical values to find the solution.

The following sections include common example problems solved using this four-step strategy. With practice and experience, you will begin to discover shortcuts and will eventually solve problems without realizing you are actually doing each step.

Solve *x*-Forces and *y*-Forces Independently

The direction in which forces act is a key consideration when solving force problems. When adding vectors to determine the net force, solve one-dimensional *x*- and *y*-force vectors separately and independently. You must resolve any two-dimensional force vectors into separate one-dimensional *x*- and *y*-component vectors. For additional information on resolving vectors into other components, see Chapter 2. The following examples will illustrate the importance of force-vector direction and demonstrate the problem-solving strategy.

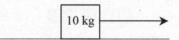

Solve *x*- and *y*- directions Independently

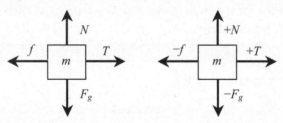

A 10-kilogram mass is pulled to the right along a rough horizontal surface by a string. The string has a tension of 20 newtons. Determine the acceleration of the mass if the coefficient of kinetic friction between the mass and the surface is 0.10.

(WHAT'S THE TRICK?)

Orient the problem: Sketching a free-body diagram, as shown in the diagram on the left below, clearly reveals the force relationships. The object is pulled to the right by a string. Set right as positive in the *x*-direction. The object is not moving vertically. Set up as positive in the *y*-direction. Directional plus and minus signs are shown in the figure on the right below.

Determine the type of motion: The *x*- and *y*-motions are different.

- *y*-direction: stationary, balanced forces, $\Sigma F_y = 0$
- *x*-direction: acceleration, unbalanced forces, $\Sigma F_x = ma$

Sum the force vectors in the relevant direction:

- *y*-direction: $\Sigma F_y = N - F_g$
- *x*-direction: $\Sigma F_x = T - f$

Substitute and solve: Substitute *ma* for ΣF_x and zero for ΣF_y into the above equations. Then substitute values and solve. The *y*-direction solves first and determines the normal force. This value is needed in the friction equation.

- *y*-direction: $0 = N - mg$

  $0 = N - (10 \text{ kg})(10 \text{ m/s}^2)$

  $N = 100 \text{ N}$

- *x*-direction: $ma = T - \mu N$

  $(10 \text{ kg})a = (20 \text{ N}) - (0.10)(100 \text{ N})$

  $a = 1.0 \text{ m/s}^2$

EXAMPLE 5.4

Components of Force

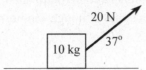

A 10-kilogram mass is pulled along a smooth horizontal surface by a 20-newton force acting 37° above the horizontal. Determine the acceleration of the mass.

WHAT'S THE TRICK?

Although the force is inclined at an angle, the mass is accelerating parallel to the surface. You must split the 20-newton force into its x- and y-components.

Orient the problem: The force diagram below left shows all the forces acting on the mass, including the components of the diagonal 20-newton force. The 37° angle indicates a 3-4-5 triangle. So, you can easily find the components.

$$F_y = F \sin \theta = (20)\left(\frac{3}{5}\right) = 12 \text{ N} \quad \text{and} \quad F_x = F \cos \theta = (20)\left(\frac{4}{5}\right) = 16 \text{ N}$$

Note: The components of the 20 N force are *not* included in a formal free-body diagram.

The object is accelerating in the x-direction. Only the x-force vectors matter, as shown in the diagram below right.

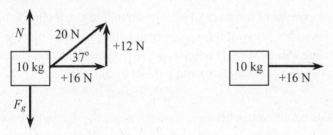

Determine the type of motion: The problem asks for acceleration. If the mass is accelerating then the mass obeys Newton's second law, $\Sigma \vec{F} = m\vec{a}$.

Sum the force vectors in the relevant direction: Since there is only one x-direction force vector, this problem is extremely simple.

$$\Sigma F_x = 16 \text{ N}$$

Substitute and solve: Substitute ma for ΣF_x in the above equation. Then substitute the numerical value for the mass and solve for acceleration.

$$ma = 16 \text{ N}$$
$$(10 \text{ kg})a = 16 \text{ N}$$
$$a = 1.6 \text{ m/s}^2$$

Inclines

Incline problems are a special case where the x- and y-axes are not useful. Instead, it is easier to work with tilted axes that are parallel and perpendicular to the incline, as shown in Figure 5.6(a). The left diagram also reveals that although the normal force lies on one of axes, the force of gravity does not.

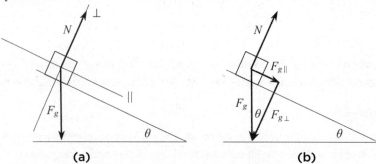

Figure 5.6 Solving incline problems

As a result, the force of gravity must be split into components, as shown in Figure 5.6(b). Solving the vector components of the force of gravity results in the following equations:

- Force of gravity parallel to the incline: $F_{g\parallel} = mg \sin \theta$
- Force of gravity perpendicular to the incline: $F_{g\perp} = mg \cos \theta$

The force of gravity perpendicular to the incline pushes the object into the incline. The incline pushes back with an equal and opposite force, creating the normal force, N.

$$N_{\text{incline}} = F_{g\perp} = mg \cos \theta$$

The component of the force of gravity parallel to the incline is unbalanced. If no other forces are present, this component will cause the object to accelerate down the incline. Whenever an incline is present, there is always a component of force acting to pull an object down and parallel to the incline.

$$F_{g\parallel} = mg \sin \theta$$

The relationship and difference between the force of gravity, F_g, and the force of gravity parallel to the incline, $F_{g\parallel}$, confuses many students. The force of gravity is the actual force acting on the object and is always included in free-body diagrams. However, it is not used to solve an incline problem mathematically. Calculating the mathematical solution for motion parallel to an incline requires force vectors or components of force vectors that are parallel to the incline. The force of gravity parallel to the incline, $F_{g\parallel}$, is a component of force that is always present in an incline problem. Since it is a component of a force vector, $F_{g\parallel}$ is never included in the free-body diagram. However, it is always included in calculations of the net force acting parallel to an incline. Figure 5.7(a) shows the free-body diagram on an incline. Figure 5.7(b) shows the vector component needed when solving equations for the same situation.

Figure 5.7 Force on an incline

EXAMPLE 5.5

Inclines

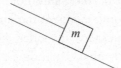

A 5.0-kilogram mass is positioned on a 30° frictionless incline. It is kept stationary by a string pulling parallel to the incline. Determine the tension in the string.

WHAT'S THE TRICK?

Work with forces and components that are parallel and perpendicular to the slope.

Orient the problem: The force diagram below left shows all forces and the components of the force of gravity parallel and perpendicular to the incline. However, to find the tension T, only forces parallel to the slope are needed. The diagram below right shows only the forces relevant to this problem.

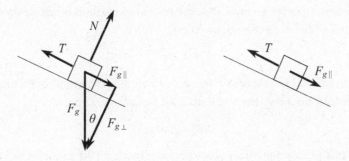

Note: The components of F_g are *not* included in a formal free-body diagram.

Determine the type of motion: The problem asks for the tension in the string that will keep the mass at rest. The mass obeys Newton's first law, $\Sigma F_{||} = 0$.

Sum the force vectors in the relevant direction: There are two vectors parallel to the incline. Normally, the direction of motion is set as positive. However, the masses are not moving. So, you can set either direction as positive.

$$\Sigma F_{||} = F_{g\,||} - T$$

Substitute and solve: Substitute zero for $\Sigma F_{||}$, substitute $mg \sin \theta$ for $F_{g\,||}$, substitute numerical values, and solve.

$$0 = mg \sin \theta - T$$
$$T = mg \sin \theta$$
$$T = (5.0 \text{ kg})(10 \text{ m/s}^2) \sin 30°$$

Remember: $\sin 30° = 0.5$

$$T = 25 \text{ N}$$

**IF YOU SEE
a net force**

**The object will
accelerate in
the direction
of the net
force.**

The resulting
displacement
and/or final
velocity is
determined
with
kinematics.

Dynamics and Kinematics

Dynamics and kinematics are explicitly linked. Students are commonly given a set of forces and asked to determine their effect on the motion of an object. Just as time was the key variable shared by both x- and y-components of motion in Chapter 4, acceleration is the key variable shared by both force equations and kinematics equations. Typically, the sum of forces helps students find the acceleration. Then the acceleration is used in the kinematic equations to determine displacement and final velocity.

EXAMPLE 5.6

Combining Dynamics and Kinematics

Mass m is initially at rest and then is pushed by force F through a distance, x, reaching a final velocity of v. What will be the new final velocity for mass m if the experiment is repeated with a force of $2F$?

WHAT'S THE TRICK?

To find the effect on velocity of a changing force, first determine how acceleration is altered. The force portion of this problem involves a single force whose direction is not important. In cases such as this, the four-step method is not needed. The lone force is responsible for all the acceleration.

$$F = ma$$

Mass remains constant. As a result, force and acceleration are directly proportional. Therefore, doubling the force doubles the acceleration.

$$(2F) = m(2a)$$

The rest of this problem involves analyzing kinematics. This problem has two important aspects. First, there is no mention of time. Second, the object starts at rest.

$$v_f^2 = v_i^2 + 2ax \quad \text{and} \quad v_i = 0$$
$$v_f = \sqrt{2ax}$$

From above, doubling the force doubles the acceleration.

$$\left(\sqrt{2}\right)v_f = \sqrt{2(2a)x}$$

Acceleration is under the square root. Doubling it is the same as multiplying the velocity by the square root of 2. The new final velocity is $\sqrt{2}\,v_f$.

Compound-Body Problems

A compound body consists of two or more separate masses experiencing the same motion. The masses may be pressing against one another or be tied together with strings so that they move as a single system. Figures 5.8(a) and 5.8(b) illustrate both types of problems.

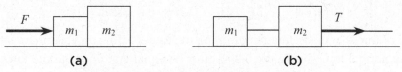

(a)   (b)

Figure 5.8 Compound-body problems

Both masses experience the same acceleration. The net force pushes or pulls on the total mass of the system. When solving for acceleration, you can sum the masses into a single system (sys).

$$\Sigma F_{sys} = (m_1 + m_2)a$$

In compound-body problems there are two categories of forces to consider. **Internal forces** are the forces acting between the masses that make up the system. In Figure 5.8(a) the masses push on each other, with their surfaces creating equal and opposite internal normal forces. In Figure 5.8(b) the masses are connected with a string, and each block is pulled by an equal and opposite internal tension force. Internal forces are always equal and opposite. When the system is viewed as a single entity the internal forces always cancel each other, and they have been left out of Figure 5.8 to simplify the diagrams. **External forces** are the forces that act on the entire system and contribute to the net force acting on all objects. In Figure 5.8(a) force F pushes both masses simultaneously, and in Figure 5.8(b) tension T pulls both masses simultaneously. A variety of variables, such as acceleration of the masses, can be solved very easily by summing the forces acting on the entire system. However, when solving for the magnitudes of the internal forces a different strategy is needed. If asked to determine the internal forces acting between two masses, work with only one of the masses individually. Draw a free-body diagram for one mass only, and then sum the forces for that mass. The internal forces will no longer cancel. While the internal forces act in opposite directions on each mass, they always act with equal magnitudes. As a result, it does not matter which mass you choose to work with when solving for internal forces.

Compound bodies can also be oriented vertically, as shown in Figures 5.9(a) and 5.9(b). In these problems, gravity is the external force pulling the masses downward.

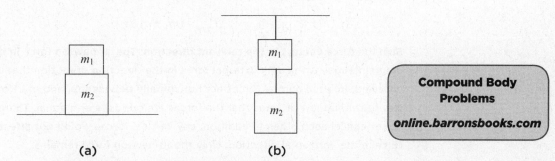

(a)   (b)

Figure 5.9 Vertical compound bodies

Compound Body
Problems

online.barronsbooks.com

If any horizontal external forces act on either mass in Figure 5.9(a), then friction may be present between the two masses. However, in beginning physics courses, stacked masses most often remain stationary. When the systems shown in Figure 5.9 remain stationary, the acceleration of the system, the net force acting on the system, and the net force acting on each mass are all zero.

$$a = 0 \qquad \Sigma F_{sys} = 0 \qquad \Sigma F_1 = 0 \qquad \Sigma F_2 = 0$$

If the sum of forces and acceleration are zero, the forces acting on each mass must be balanced. The normal forces in Figure 5.9(a) must be equal and opposite the force of gravity pulling down against each surface. The magnitude of the normal force will be equal to but opposite the force of gravity of all the masses stacked above a surface. In Figure 5.9(b) the tension vectors are determined in a similar manner. The tension in a string is equal to but opposite the force of gravity for all the masses hanging from a string.

EXAMPLE 5.7

Compound Bodies

A force of 18 newtons pushes two masses (m_1 = 2 kilograms and m_2 = 4 kilograms) horizontally to the right.

(A) Determine the acceleration of mass m_2.

WHAT'S THE TRICK?

Both masses are moving together as a system. The acceleration of mass m_2 equals the acceleration of the system.

Orient the problem: Visualize the masses as a single system.

Determine the type of motion: The problem asks for acceleration.

$$\Sigma F_{sys} = (m_1 + m_2)a$$

Sum the force vectors in the relevant direction: The 18-newton force in the diagram above is the only external force in the direction of motion that is not canceled. Internal normal forces act horizontally between m_1 and m_2. However, Newton's third law dictates that the forces are opposite and equal. Therefore, the forces cancel each other. In addition, any vertical forces would not affect the net force in the horizontal direction. Only the 18-newton force remains.

$$\Sigma F_{sys} = 18 \text{ N}$$

Substitute and solve: Combine the above equations and substitute values.

$$18 = (m_1 + m_2)a$$
$$18 = (2 + 4)a$$
$$a = 3 \text{ m/s}^2$$

(B) Determine the magnitude of the force between the masses.

WHAT'S THE TRICK?

According to Newton's third law, the two interacting masses push on each other with opposite and equal force. Therefore, solve for the force on either mass. The forces acting between the blocks are normal forces since the surfaces of the masses push against each other.

Orient the problem: Again, the motion is horizontal, and only horizontal force vectors contribute forces for this motion.

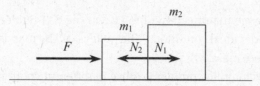

Determine the type of motion: Acceleration obeys Newton's second law.

$$\Sigma F_1 = m_1 a \qquad \text{and} \qquad \Sigma F_2 = m_2 a$$

Work with the masses individually and not as part of a system.

Sum the force vectors in the relevant direction: Force F and the normal force of m_2, which is N_2, push on m_1. Only the normal force of m_1, which is N_1, pushes on m_2. With only one force acting on it, m_2 will be easier to solve. However, the solution for both masses is shown to prove that either mass can be used.

m_1 m_2
$\Sigma F_1 = F - N_2$ $\Sigma F_2 = N_1$

Substitute and solve: Substitute $m_1 a$ for ΣF_1 and $m_2 a$ for ΣF_2. Then substitute values and solve. The acceleration, $a = 3 \text{ m/s}^2$, was determined in part (A).

$m_1 a = F - N_2$ $m_2 a = N_1$
$(2 \text{ kg})(3 \text{ m/s}^2) = (18 \text{ N}) - N_2$ $(4 \text{ kg})(3 \text{ m/s}^2) = N_1$
$N_2 = 12 \text{ N}$ $N_1 = 12 \text{ N}$

The normal forces pressing between the masses are opposite and equal.

Pulley Problems

Pulley problems are compound-body problems where more than one mass is connected together by a string draped over a pulley. Two common pulley problems are shown in the following diagram. Figure 5.10(a) depicts an **Atwood machine** designed by George Atwood to test constant acceleration and Newton's laws of motion. Figure 5.10(b) shows a modified Atwood machine.

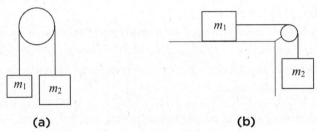

(a)   (b)

Figure 5.10 An Atwood machine

Pulleys are used to change the direction of force. On the SAT Subject Test in Physics, these devices will operate under ideal conditions. The pulleys will be massless and frictionless. The strings will also have zero mass.

Although the masses in pulley problems move in different coordinate directions, they do share one common motion. The masses in pulley problems always move in the same direction as the string. Any force aligned with the motion of the string can influence acceleration. When summing the forces in the relevant direction, set all the forces pointing in the direction of the string's motion as positive and all the forces opposing the string's motion as negative.

EXAMPLE 5.8

Atwood Machine

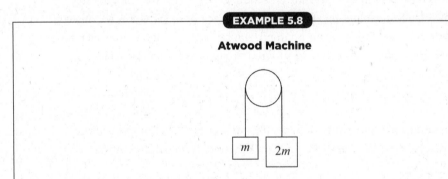

Masses m and $2m$ are connected by a string, which is draped over a pulley. The masses are released from rest. Determine the magnitude of acceleration of mass m.

WHAT'S THE TRICK?

When solving for pulley problems, determine the direction the string is moving. Make this the positive direction and determine all forces parallel to the string. If solving for acceleration, treat both masses as a single system.

Orient the problem: Although the force of gravity acts on both masses, gravity pulls down the $2m$ mass. This causes the smaller mass m to move upward. The diagram below is not a formal free-body diagram. However, viewing the problem in this manner shows a consistent direction of motion that makes it easier to sum the forces for the system.

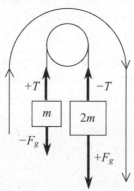

Pulley Problems

online.barronsbooks.com

Forces in the direction of motion increase the speed of an object. They have a positive influence on acceleration and are set as positive. Vectors opposing motion slow down an object and are set as negative.

Determine the type of motion: The forces are unbalanced, resulting in the acceleration of the entire system. Applying Newton's second law of motion to the entire system results in the following:

$$\Sigma F_{sys} = (m_1 + m_2)a$$

Sum the force vectors in the relevant direction: Include all the forces parallel to the motion of the string. The sum of the vector forces will be:

$$\Sigma F_{sys} = F_{g_m} - T + T - F_{g_{2m}}$$

For the SAT Subject Test in Physics, the pulleys themselves will always be massless. Under these ideal conditions, the tension in a string will be the same everywhere. The equal and opposite tensions will cancel. As a result, tensions can be ignored when solving for the acceleration of the entire system. (Important: tensions cancel only when summing the forces for an entire system. Do not cancel tensions when summing the forces acting on a single independent mass.)

$$\Sigma F_{sys} = F_{g_{2m}} - F_{g_m}$$

Substitute and solve: Substitute $(m_1 + m_2)a$ for ΣF_{sys}. Substitute known equations and values. Then solve.

$$\Sigma F_{sys} = F_{g_{2m}} - F_{g_m}$$
$$(m + 2m)a = (2m)g - (m)g$$

In this problem, the numerical values associated with each mass are coefficients rather than subscripts. Mass $2m$ is not a mass with a value of 2 units. Instead, it is a mass that is twice as large as mass m. This problem will not have a numerical answer. Instead, the solution will be a simplified equation using variables.

$$3ma = 2mg - mg$$
$$a = g/3$$

EXAMPLE 5.9

Modified Atwood Machine

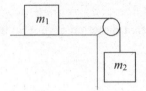

Two masses, m_1 = 2 kg and m_2 = 3 kg, are connected by a string, which is draped over a pulley as shown above. Mass 1 is positioned on a horizontal surface, while mass 2 hangs freely. The masses are released from rest.

(A) Determine the acceleration of mass 2.

WHAT'S THE TRICK?

To solve for acceleration, treat the masses as a single system and sum the forces acting on both masses simultaneously.

Orient the problem: Mass m_2 will move downward, dragging mass m_1 along the surface to the right. Friction is never mentioned or alluded to in the problem, so consider its effect to be negligible.

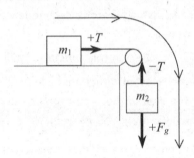

Modified Atwood's Machine

online.barronsbooks.com

Determine the type of motion: The forces are unbalanced, resulting in the acceleration of the entire system. Newton's second law of motion says:

$$\Sigma F_{sys} = (m_1 + m_2)\, a$$

Sum the force vectors in the relevant direction: The tensions cancel, leaving only one force acting to accelerate the entire system.

$$\Sigma F_{sys} = F_{g_{m_2}}$$

Substitute and solve: Substitute $(m_1 + m_2)\, a$ for ΣF_{sys}. Substitute known equations and values. Solve.

$$(m_1 + m_2)a = (m_2)g$$
$$(2 \text{ kg} + 3 \text{ kg})a = (3 \text{ kg})(10 \text{ m/s}^2)$$
$$a = 6 \text{ m/s}^2$$

(B) Determine the tension in the string.

WHAT'S THE TRICK?

To solve for the tension in the string, sum the forces for only one of the two masses. You *cannot* sum the forces for the system since this causes the tension to cancel. We *can* sum the forces for either mass as both will result in the same answer. Solutions using both masses are shown below. Only one is needed.

Orient the problem: Separate the masses into separate diagrams.

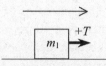

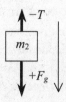

Determine the type of motion: Both masses will accelerate. Newton's second law of motion applied to each mass results in:

$$\Sigma F_1 = m_1 a \qquad \text{and} \qquad \Sigma F_2 = m_2 a$$

Sum the force vectors in the relevant direction:

$$\Sigma F_1 = T \qquad\qquad \Sigma F_2 = F_{g_{m_2}} - T$$

Substitute and solve:

$m_1 a = T$ $m_2 a = m_2 g - T$

$(2 \text{ kg})(6 \text{ m/s}^2) = T$ $(3 \text{ kg})(6 \text{ m/s}^2) = (3 \text{ kg})(10 \text{ m/s}^2) - T$

$T = 12 \text{ N}$ $T = 12 \text{ N}$

Using either mass arrives at the same answer. The easiest solution usually involves selecting the mass with the least amount of forces acting on it.

SUMMARY

1. **WHEN THE FORCES ACTING ON AN OBJECT CANCEL OUT EACH OTHER**, the object is in a state of equilibrium. In order for forces to cancel out, they must be equal and opposite. The forces are balanced, and the net force (sum of forces) is equal to zero. There is no acceleration, and objects move at constant velocity, which includes a velocity of zero when at rest. When an object remains at rest, it is in static equilibrium. When an object moves at constant velocity, it is in dynamic equilibrium. The motion of these objects is described by Newton's first law of motion, known as the law of inertia.

2. **WHEN THE FORCES ACTING ON AN OBJECT *DO NOT* CANCEL**, the forces are unbalanced. The net force (sum of forces) will have a nonzero value. According to Newton's second law, the net force acts to accelerate the object. The direction of the net force is the same as the direction of the resulting acceleration.

3. **THE ACCELERATION DETERMINED USING FORCES CAN BE USED IN KINEMATIC EQUATIONS** to determine the distance moved by the object and its final velocity.

4. **A FORCE ACTING ON AN OBJECT IS CREATED BY AN OUTSIDE AGENT.** Many types of agents generate different forces, each with unique characteristics. Newton's third law states that when an object and an agent interact, there is always an equal and opposite force exerted between them. The object and agent form an action-reaction pair and experience opposite reactions. The actual motion of the object and agent may not be equal. It depends on the mass of each object composing the action-reaction pair.

5. **FORCE PROBLEMS OFTEN INVOLVE MORE THAN ONE FORCE ACTING SIMULTANEOUSLY.** The forces will have different characteristics, and they will act in different directions. Adopting a problem-solving strategy that breaks complex problems into a series of solvable steps is extremely advantageous.

   ■ Orient the problem. Draw a free-body diagram, and determine the direction of motion. Set all force vectors that match the object's motion as positive and those opposing its motion as negative.

   ■ Determine the type of motion. Is the object accelerating, $\Sigma F = ma$, or does it have a constant velocity (including $v = 0$ m/s), $\Sigma F = 0$?

   ■ Sum the vector forces in the relevant direction.

   ■ Substitute known force equations, substitute given values, and solve.

If You See	Try	Keep in Mind
Mass	$F_g = mg = w$	The force of gravity is also an object's weight.
Rough or friction	At rest or at constant velocity: $$f = F_{forward}$$ If this fails or if accelerating: $$f = \mu N$$	Static friction is stronger than kinetic friction.
Stretched or compressed spring	Hooke's law: $$F_s = kx$$	Spring force acts opposite the stretch or compression.
Constant velocity, including rest or the word *equilibrium*	$\Sigma F = 0$ $a = 0$	Forces are balanced: equal and opposite.
Acceleration	$\Sigma F = ma$	Forces are unbalanced.
Two objects interacting	There is an equal and opposite force acting on each object.	The size of the object has no effect on the force, only on the resulting motion.
Incline	$F_{g\parallel} = mg \sin \theta$ $N = mg \cos \theta$	$F_{g\parallel}$ is not drawn in free-body diagrams.
Compound bodies	The total force acts on all the masses simultaneously, and all the masses have the same acceleration. All the masses can be added together and treated as one large mass.	
Pulleys	Pulley problems are compound-body problems where the pulley changes the direction of a force without affecting the magnitude of the force. Sum the forces acting at the opposite ends of the string. Set forces in the same direction as the motion as positive and those opposite the motion as negative.	

PRACTICE EXERCISES

1. A car accelerates from rest to a velocity of 20 meters per second. When it reaches this velocity, the driver removes his foot from the accelerator. Assume air resistance is negligible. The car will

    (A) speed up slightly due to its inertia
    (B) slow due to the lack of forward force
    (C) come to a stop
    (D) continue at 20 m/s briefly and then slow
    (E) continue at 20 m/s until an unbalanced force acts on the car

2. A small object is attracted to Earth by a 50-newton gravitational force. The object pulls back on Earth with

(A) a force of 0 N
(B) a negligible, small, nonzero force
(C) a force of 5 N
(D) a force of 50 N
(E) a force of 500 N

3. When an object is thrown upward in the absence of air resistance and reaches the top of its trajectory, it has an instantaneous velocity of zero. At this point, the net force acting on the object is

(A) zero
(B) equal to the weight of the object
(C) equal to the mass of the object
(D) *g*
(E) changing direction

Tension Problems

online.barronsbooks.com

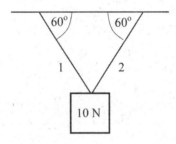

4. Two strings support a 10-newton object, as shown above. The tension in string 1 is

(A) zero
(B) between 0 and 5 N
(C) 5 N
(D) between 5 and 10 N
(E) 10 N

5. A 70-kilogram person rides in an elevator that is accelerating upward at 2.0 meters per second squared. What is the normal force acting on the person?

(A)   560 N
(B)   700 N
(C)   840 N
(D)   960 N
(E)   1,000 N

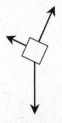

6. A mass remains at rest on an incline, as shown above. Which free-body diagram is correct?

(A)

(B)

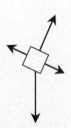

(C)

(D)

(E)

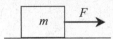

7. A mass, m, is pulled along a rough horizontal surface, as shown above. If the mass is accelerating, then the friction force, f, is equal to

(A) F only
(B) N only
(C) μmg only
(D) both A and C
(E) both B and C

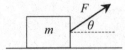

8. A mass, m, is pulled along a horizontal surface by force F, as shown above. If the resulting motion is constant velocity, then the frictional force is equal to

(A) zero
(B) μmg
(C) $mg \cos \theta$
(D) $F \cos \theta$
(E) $F \sin \theta$

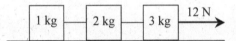

9. Three masses are connected with strings, as shown above. Determine the acceleration of the 2-kilogram mass.

(A) 0 m/s^2
(B) 1 m/s^2
(C) 2 m/s^2
(D) 4 m/s^2
(E) 6 m/s^2

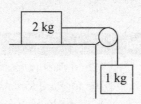

10. As shown in the diagram above, a 2-kilogram mass lies on a horizontal rough surface, while a 1-kilogram mass hangs vertically. The string between them is massless, and the pulley is both massless and frictionless. What minimum coefficient of friction will keep the blocks at rest?

(A) 0.1
(B) 0.2
(C) 0.3
(D) 0.4
(E) 0.5

ANSWERS EXPLAINED

	Key Words	Needed for Solution	Now Solve It
1. **(E)**	Removes his foot from the accelerator; air resistance is negligible	Newton's first law	When the driver stops pressing the accelerator, the forward force and acceleration of the engine become zero. If air resistance is negligible, then it is so small that it is treated as zero. With no horizontal forces acting on the car, the automobile will follow Newton's first law. The car will continue moving forward at its current speed.
2. **(D)**	A small object is attracted to Earth; the object pulls back on Earth with a force of B	Newton's third law	Whenever two objects interact, there is an equal and opposite force between them. Two objects of different size are given in an attempt to trick students into matching the size of the force with the size of the object.
3. **(B)**	Thrown upward in the absence of air resistance; net force	Force diagram:	Drawing a free-body diagram reveals only one force: the force of gravity acting in the downward direction. The force of gravity is also the weight of an object. By stating that the object stops instantaneously at the top of its flight, this problem distracts the reader with the velocity of the object. Read all questions carefully to ensure your answers always match the questions.
4. **(D)**	Two strings support; the tension in the string is 1	Draw a force diagram, and split the diagonal tension vectors into components:	**Orient the problem:** y-direction. Gravity pulls the object downward, and the strings are pulling upward. **Determine the type of motion:** Stationary, $\Sigma F = 0$. **Sum the forces in the relevant direction:** Two tensions are pulling upward, and the diagonal tension vectors must be split into component vectors. $$\Sigma F = 2T_y - F_g$$

	Key Words	Needed for Solution	Now Solve It
			Substitute and solve: The weight of the object is 10 newtons, $F_g = 10$ N. $$0 = 2T_y - 10$$ $$T_y = 5\,\text{N}$$ The actual tension is directed diagonally, and this vector is longer than the vertical 5-newton component. Therefore, the force of tension in each rope must be greater than 5 newtons. However, it cannot be greater than the 10-newton total weight of the object. As a result, the tension in each string is between 5 and 10 newtons.
5. **(C)**	70-kilogram person; accelerating upward at 2 meters per second squared; normal force	Draw a force diagram, and then sum the forces acting in the vertical direction:  	**Orient the problem:** y-direction. The floor of the elevator pushes the person upward with a normal force, and gravity acts downward. **Determine the type of motion:** Acceleration, $\Sigma F = ma$. **Sum the forces in the relevant direction:** $$\Sigma F = N - F_g$$ **Substitute and solve:** $$ma = N - mg$$ $$N = ma + mg$$ $$N = (70)(2) + (70)(10)$$ $$N = 840\,\text{N}$$
6. **(A)**	Remains at rest on an incline; free-body diagram	Create a force diagram:  	The force of gravity acts downward, while the normal force acts perpendicular to the incline (Figure 1 below). These two forces are out of balance. They add together to create a net force acting parallel to the incline (Figure 2). This net force is not drawn in free-body diagrams since it is the sum of two acting forces. Essentially, the force of gravity and the normal force acting together pull objects down the incline. In this problem, the object remains at rest. This means a third force has to act upward along the incline to cancel the effect of gravity and the normal force (Figure 3). 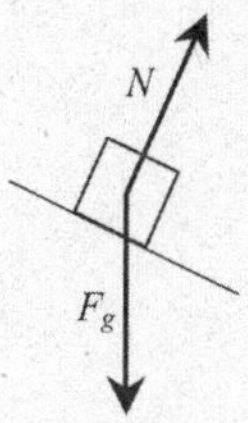 Figure 1 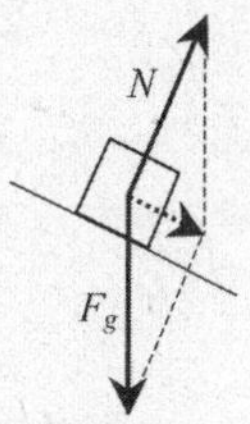 Figure 2 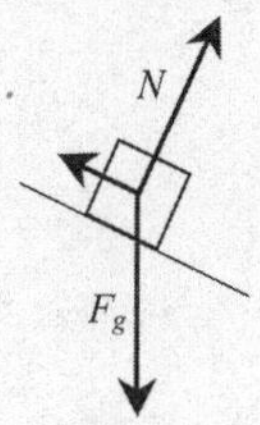 Figure 3

	Key Words	Needed for Solution	Now Solve It
7. **(C)**	Pulled along a rough horizontal surface; accelerating; friction force	Use the force problem-solving method, including a force diagram.	**Orient the problem:** Friction, f, acts in the x-direction and opposes the forward applied force, F. However, friction also depends on the normal force, N, which acts in the y-direction.

Determine the type of motion: Stationary in the y-direction, $\Sigma F_y = 0$, while accelerating in the x-direction, $\Sigma F_x = ma$.

Sum the force vectors in the relevant direction: Forces must be summed in the y-direction to find the normal force. However, it is not necessary to sum the forces in the x-direction. Simply use the friction formula.

$$\Sigma F_y = N - F_g \quad \text{and} \quad f = \mu N$$

Substitute and solve:

$$0 = N - mg$$
$$N = mg$$
$$f = \mu N$$
$$f = \mu(mg)$$

(Force diagram for item 7: block m with $+N$ up, $-F_g$ down, $+F$ right, $-f$ left.)

	Key Words	Needed for Solution	Now Solve It
8. **(D)**	Pulled along a horizontal surface; constant velocity; the frictional force	Use the force problem-solving method, including a force diagram.	**Orient the problem:** The friction force, f, acts in the x-direction. The applied force, F, acts at an angle and must be resolved into its components. The x-component of the applied force, F_x, pulls the object in the positive direction. Friction slows the object and is negative.

Determine the type of motion: Constant velocity, $\Sigma F_x = 0$.

Sum the force vectors in the relevant direction:

$$\Sigma F_x = F_x - f$$
$$0 = F \cos \theta - f$$
$$f = F \cos \theta$$

(Force diagram for item 8: block m with N up, F_g down, f left, and F at angle θ with components F_x and F_y.)

	Key Words	Needed for Solution	Now Solve It
9. **(C)**	Three masses are connected with a string; acceleration of the 2-kilogram mass	This is a compound-body problem. To solve for acceleration, treat all the masses as one large mass (one system).	**Orient the problem:** Visualize the masses as a single system. **Determine the type of motion:** acceleration $$\Sigma F_{sys} = (m_1 + m_2 + m_3)a$$ **Sum the force vectors in the relevant direction:** Only the 12-newton force creates acceleration. $$\Sigma F_{sys} = 12 \text{ N}$$ $$(m_1 + m_2 + m_3)a = 12$$ $$(1 + 2 + 3)a = 12$$ $$a = 2 \text{ m/s}^2$$
10. **(E)**	2-kilogram mass lies on a horizontal rough surface; 1-kilogram mass hangs vertically; minimum coefficient of friction; rest	This is a compound-body problem involving a pulley. Treat the two masses as a single system. Sum the forces acting parallel to the string.	**Orient the problem:** The force of gravity acting on the 1-kg mass pulls it downward while simultaneously dragging the 2-kg mass along the rough horizontal surface. Friction acts on the 2-kg mass, opposing the effect of gravity. **Determine the type of motion:** Friction is strong enough to keep the system at rest, $\Sigma F_{sys} = 0$. **Sum the force vectors in the relevant direction:** The tension vectors cancel. The force of gravity acts on the 1-kg mass, while friction acts on the 2-kg mass. $$F_{sys} = F_{g_{m_1}} - f_{m_2}$$ **Substitute and solve:** $$\Sigma F_{sys} = m_1 g - \mu m_2 g$$ $$0 = (1)(10) - \mu(2)(10)$$ $$\mu = 0.5$$

Circular Motion

6

→ **UNIFORM CIRCULAR MOTION**

→ **ANGULAR DISPLACEMENT**

→ **DYNAMICS IN CIRCULAR MOTION**

→ **ANGULAR VELOCITY**

→ **THE FALSE CENTRIFUGAL FORCE**

Forces acting perpendicular to the velocity of an object will change the direction of the object. In problems involving the currents in rivers, the force of the wind, and projectiles in motion, the perpendicular force always acts in a constant direction. As a result, motion can be split into independent x- and y-components, which can be analyzed as two separate linear motions taking place simultaneously. However, circular motion presents unique challenges. In circular motion, the direction of force is continually changing and always acts perpendicular to the motion of an object. Therefore, circular motion requires a unique set of equations and its own problem-solving strategies in order to do the following:

- Describe the characteristics of uniform circular motion.
- Distinguish between the period and frequency of a circling object.
- Determine the magnitude and direction of the tangential velocity and centripetal acceleration for objects in circular motion.
- Solve dynamics problems involving uniform circular motion.

Table 6.1 lists the variables and units that will be discussed in this chapter.

Table 6.1 Variables Used in Circular Motion

New Variables	Units
T = Period	s (seconds)
f = Frequency	Hz (hertz) = 1/seconds
v_T = Tangential velocity	m/s (meters per second)
a_c = Centripetal acceleration	m/s^2 (meters per second squared)
F_c = Centripetal force	N (newtons)
ω = Angular velocity	rad/s (radians per second)

UNIFORM CIRCULAR MOTION

Uniform circular motion involves objects moving at a constant speed but with changing velocity. When an object moves at constant speed, the magnitude of velocity is also constant. How then

**Constant
speed and
uniform
acceleration**

Acceleration
has constant
magnitude, but
its direction
is changing
such that the
acceleration
always points
to the center
of its motion.

magnitude and direction. Objects moving in circular motion are continually changing direction. As a result, the velocity of the object is changing even though it moves at a constant speed.

Like velocity, the acceleration of an object in uniform circular motion has constant magnitude but has changing direction. Acceleration is the rate of change in velocity (the change in velocity during a time interval). If velocity has a constant magnitude, the acceleration will also have a constant value. However, the direction of acceleration constantly changes as an object moves in circular motion. In uniform circular motion, the direction of the acceleration vector is toward the center of the circle. This acceleration is said to be uniform: having a constant magnitude and applied in the same manner (toward the center) at all times.

Period and Frequency

All linear-motion quantities are based on the linear meter. However, the linear meter is not very useful when describing motion that does not follow a straight path. All circles have one thing in common. Objects moving in circles return to the same location every time they complete one **cycle** (one circle, one revolution, one rotation, and so on). A cycle consists of one circumference, and this is the basis for all circular-motion quantities.

The time to complete one cycle is known as the **period**, T. The period of a circling object can be calculated by dividing the time of the motion, t, by the number of cycles completed during time t.

$$T = \frac{t}{\text{number of cycles}}$$

The number of cycles does not have any units, and the units of a period are seconds.

The **frequency** of an object is the number of cycles an object completes during one second. Think of the frequency as how frequently the object is cycling. Mathematically, frequency is the inverse of the period. The units of frequency are inverse seconds ($1/s$ or s^{-1}). These units are also known as Hertz (Hz). Any of these units may be used on exams, and the formula for frequency is simply the inverse of the formula for the period.

$$f = \frac{\text{number of cycles}}{t}$$

The relationship between the period and frequency is expressed in the following equation:

$$T = \frac{1}{f}$$

EXAMPLE 6.1

Period and Frequency

An object completes 20 revolutions in 10 seconds. Determine the period and frequency of this motion.

(WHAT'S THE TRICK?)

A revolution is another way to indicate a cycle, and a cycle is simply an event. The number of cycles is just the count of an event, and therefore it has no units. Period is the time for one complete cycle. Time has the units of seconds, and therefore time must be in the numerator.

$$T = \frac{t}{\text{number of cycles}} = \frac{10 \text{ s}}{20} = 0.5 \text{ s}$$

Frequency is the inverse formula.

$$f = \frac{\text{number of cycles}}{t} = \frac{20}{10 \text{ s}} = 2 \text{ Hz} \qquad \text{or} \qquad f = \frac{1}{T} = \frac{1}{(0.5)} = 2 \text{ Hz}$$

Tangential Velocity and Centripetal Acceleration

Velocity is not constant in uniform circular motion. However, when any moving object is paused (frozen for an instant of time), it will have an instantaneous velocity vector with a specific magnitude and direction. When objects follow a curved path, the instantaneous velocity is tangent to the motion of the object. Thus, the instantaneous velocity is referred to as the **tangential velocity**. Several instantaneous velocity vectors are shown for the object circling in Figure 6.1.

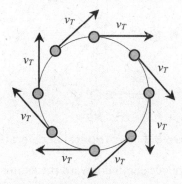

Figure 6.1 Instantaneous velocity vectors

In Figure 6.1, it is apparent that although the direction of the tangential velocity is continually changing, its magnitude remains constant. The magnitude of the tangential velocity is also equal to the speed of the circling object. They are both determined using a modified version of the constant-speed formula:

$$v = \frac{d}{t} \quad \text{becomes} \quad v = \frac{2\pi r}{T}$$

As previously stated, circular motion is based on one complete cycle. In one complete cycle, an object travels one circumference, $d = 2\pi r$, in one period, $t = T$.

One important aspect of tangential velocity involves objects leaving the circular path. Forces are responsible for creating circular motion. If the forces causing circular motion stop acting, the object will leave the circular path. When this happens, the tangential velocity becomes the initial velocity for the object's subsequent motion. If no forces act on the object, it will move in a straight line matching the tangential velocity at the time of release, as shown in Figure 6.2(a). However, if another force acts on the object, the object will become subject to the new force. In Figure 6.2(b), the object leaves the circle and is acted upon by gravity, causing projectile motion.

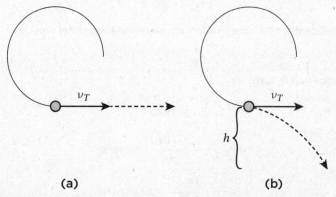

(a)   (b)

Figure 6.2 Tangential velocity. In (a) the object is initially circling horizontally on a frictionless surface. In (b) the object is initially circling vertically at a distance _h_ above the surface.

IF YOU SEE
a circling
object

Instantaneous
velocity is
tangent to the
motion

and

acceleration
is directed
toward the
center of
the circular
motion.

Since circling objects have a changing velocity vector, they are continuously accelerating. This type of acceleration is known as **centripetal acceleration**, a_c. *Centripetal* means "center seeking." Centripetal acceleration is directed toward the center of the circle, as shown in Figure 6.3.

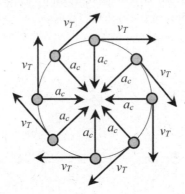

Figure 6.3. Centripetal acceleration

Even though circling objects are accelerated toward the center of the circle, their tangential velocities prevent them from ever reaching the center. Since the acceleration vectors lie along the radii of the circle, the centripetal acceleration may also be referred to as the **radial acceleration**. The centripetal acceleration can be determined with the following formula:

$$a_c = \frac{v^2}{r}$$

EXAMPLE 6.2

Tangential Velocity and Centripetal Acceleration

Determine the acceleration of an object experiencing uniform circular motion. It is moving in a circle with a radius of 10 meters and a frequency of 0.25 Hertz.

WHAT'S THE TRICK?

The period is the inverse of the frequency.

$$T = \frac{1}{f}$$

Solving for the tangential velocity requires the period.

$$v = \frac{2\pi r}{T}$$

Solving for the centripetal acceleration requires the tangential velocity.

$$a_c = \frac{v^2}{r}$$

Solve each equation in turn.

$$T = \frac{1}{f} = \frac{1}{0.25\,\text{Hz}} = 4\ \text{s}$$

$$v = \frac{2\pi r}{T} = \frac{2\pi(10)}{4} = 5\pi\ \text{m/s}$$

$$a_c = \frac{v^2}{r} = \frac{(5\pi)^2}{10} = 2.5\pi^2\ \text{m/s}^2$$

Answers may be expressed in terms of π in order to avoid multiplying by 3.14.

ANGULAR DISPLACEMENT

Angular displacement, $\Delta\theta$, is the angle, measured in radians, subtended by a circling object. The list of published topics covered in the SAT Subject Test in Physics does not include rotation. As a result, angular displacement is most likely to be reported as the number of revolutions (rev) the object completes during the specified motion. However, in most courses angular displacement is reported in radians. Switching between these units involves the following conversion factor.

$$1 \text{ rev} = 2\pi \text{ rad}$$

DYNAMICS IN CIRCULAR MOTION

In order for objects to experience acceleration, a net force (sum of forces) must act in the direction of the acceleration. Objects in circular motion are being accelerated toward the center of a circular path. This implies that the net force is also directed toward the center of the circle. In uniform circular motion, the net force is known as the **centripetal force**, F_c. In linear-motion problems, the net force is represented by ΣF and Newton's second law dictates the relationship between the net force and acceleration: $\Sigma F = ma$. In circular-motion problems, F_c replaces ΣF and Newton's second law applied to circular motion becomes

$$F_c = ma_c$$

Frequently, circular-motion force problems involve the speed and/or tangential velocity of the circling object. Since $a_c = \dfrac{v^2}{r}$, the formula for centripetal force is frequently written as:

$$F_c = m\frac{v^2}{r} \qquad \text{or} \qquad \frac{mv^2}{r}$$

Dynamics problems in circular motion are solved in a similar manner as all other force problems.

1. **Orient the problem.** Draw a force diagram. If the motion is circular, any force vectors pointing toward the center of the circle are considered positive. Those pointing away from the center are negative.
2. **Determine the type of motion.** If an object is moving along a circular path, recognize that the circular net force, F_c, is used in place of ΣF.
3. **Sum the force vectors.** Add all forces pointing toward the center of the circle, and then subtract all forces pointing away from the center.

$$F_c = + F_{\text{toward center}} - F_{\text{away from center}}$$

4. **Substitute and solve.** In this last step, substitute known equations for specific forces and substitute numerical values to find in the solution.

Keep in mind that circular-motion problems can incorporate elements from problems learned in other chapters.

ANGULAR VELOCITY

Angular velocity, ω, is the rate of angular displacement.

$$\omega = \frac{\Delta\theta}{t}$$

The units of angular velocity may appear as revolutions per second (rev/s), revolutions per minute (rpm), or radians per second (rad/s). Working in radians per second has many advantages, and switching to these units involves unit conversion.

$$1\frac{\text{rev}}{\text{s}}\left(\frac{2\pi\ \text{rad}}{1\ \text{rev}}\right) = 2\pi\frac{\text{rad}}{\text{s}} \qquad 1\frac{\text{rev}}{\text{min}}\left(\frac{2\pi\ \text{rad}}{1\ \text{rev}}\right)\left(\frac{1\ \text{min}}{60\ \text{s}}\right) = \frac{\pi}{30}\frac{\text{rad}}{\text{s}}$$

Angular velocity is also related to the tangential velocity, v, of a point located a radius, r, from the axis about which the object circles.

$$v = \omega r$$

However, this equation requires the angular velocity to be in radians per second.

EXAMPLE 6.3

Angular Displacement and Angular Velocity

An object in circular motion with a radius of 2.0 meters completes 30 revolutions in 10 seconds.

(A) Determine the object's angular velocity.

(WHAT'S THE TRICK?)

The object's angular displacement is given as 30 revolutions. Solve the angular velocity.

$$\omega = \frac{\Delta\theta}{t} = \frac{30}{10} = 3\ \text{rev/s} = 3\ \text{s}^{-1}$$

Note: Revolution (rev) is technically a count of a repetitive event and is not actually a unit. It may appear in the units of some problems and be omitted in others. Both are shown above.

(B) Determine the object's tangential velocity.

(WHAT'S THE TRICK?)

To switch to tangential velocity, the angular velocity must be in radians per second.

$$3.0\frac{\text{rev}}{\text{s}}\left(\frac{2\pi\ \text{rad}}{1\ \text{rev}}\right) = 6.0\pi\ \text{rad/s}$$

$$v = \omega r = (6.0\pi\ \text{rad/s})(2.0\ \text{m}) = 37.7\ \text{m/s}$$

THE FALSE CENTRIFUGAL FORCE

When a compound body or an object that can stretch experiences circular motion, there is a tendency for a portion of the object to move away from the center of the turn, as though a force acting away from the center is present. Examples include watching a package in the back of a pickup truck slide left, relative to the truck, as the truck turns right, and a mass at the end of an elastic cord moving through larger and larger circles as tangential speed is increased. This outward motion, which appears to be caused by a force directed away from the center of a turn, has been called the **centrifugal force**. However, there is no such force, and it is referred to as a false force. The actual cause of this apparent outward motion is inertia, the tendency of an object to continue moving with constant velocity when no net forces act upon it.

Examine the example of a package in the back of a pickup truck sliding to the left side of the truck when the truck turns rightward. Prior to the turn both the truck and the package are moving at a constant velocity in a straight line. As the truck negotiates the turn, friction between the road and tires acts centripetally to turn the truck rightward. The package's mass (inertia) attempts to keep the package moving forward in a straight line. As a result, the package appears to move leftward relative to the truck. However, if the package is sliding toward the left side of the truck, then kinetic friction would be directed rightward. Eventually the package would reach the left side of the truck and collide with the left wall of the truck. The wall of the truck would then press the package with a normal force directed rightward, toward the center of the turn. Both the rightward kinetic friction and eventual rightward normal force act on the package centripetally in the direction of the rightward turn.

In the example with the elastic cord, the velocity of the mass at the end of the cord is directed tangent to its circular motion. The mass is attempting to leave the circular path due to its inertia. As a result, the elastic cord is stretched.

The restoring force of an elastic device acts in a direction opposite to its displacement. Since the elastic cord is stretched outward, the restoring force is directed inward toward the center of the circular motion. Conceptual problems will attempt to make students incorrectly state that force is directed outward in circular motion. This is false. Always choose the answer that correctly identifies force as centripetal, acting toward the center of a turn.

EXAMPLE 6.4

Horizontal Circular Motion

A car moving at 10 meters/second completes a turn with a radius of 20 meters. Determine the minimum coefficient of friction between the tires and the road that will allow the car to complete the turn without skidding.

WHAT'S THE TRICK?

Orient the problem: Picture a car making a turn.

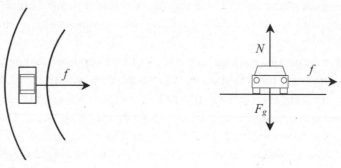

| Viewed from above | Viewed head on |

Gravity is pulling down, and the normal force is acting upward. Neither of these forces is in the direction of motion. The force acting to keep the car in the turn is friction. Without friction acting toward the center of the turn, the car would slide in a straight path following the tangential velocity.

Determine the type of motion: A turning car is in circular motion.

Sum the force vectors in the relevant direction: In circular motion F_c is used instead of ΣF and friction is pointing toward the center of the circle.

$$F_c = f$$

Substitute and solve:

$$m\frac{v^2}{r} = \mu N$$

The normal force is acting vertically in the y-direction. It is equal and opposite the force of gravity: $N = F_g = mg$.

$$m\frac{v^2}{r} = \mu mg$$

$$\mu = \frac{v^2}{rg}$$

$$\mu = \frac{(10 \text{ m/s})^2}{(20 \text{ m})(10 \text{ m/s}^2)}$$

$$\mu = 0.5$$

EXAMPLE 6.5

Vertical Circular Motion

A 2.0-kilogram mass is attached to the end of a 1.0-meter-long string. When the apparatus is swung in a vertical circle, the mass reaches a speed of 10 meters per second at the bottom of the swing.

(A) Determine the tension in the string at the bottom of the swing.

WHAT'S THE TRICK?

Orient the problem: Sketch the scenario, including force vectors.

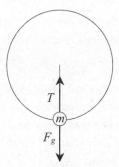

Determine the type of motion: This is circular motion.

Sum the force vectors in the relevant direction: In circular motion, F_c is used instead of ΣF. Tension is pointing to the center of the circle and is positive. The force of gravity is pointing away from the center of the circle and is negative.

$$F_c = T - F_g$$

$$T = F_c + F_g$$

Substitute and solve:

$$T = m\frac{v^2}{r} + mg$$

$$T = (2.0 \text{ kg})\frac{(10 \text{ m/s})^2}{(1.0 \text{ m})} + (2.0 \text{ kg})(10 \text{ m/s}^2)$$

$$T = 220 \text{ N}$$

(B) Determine the minimum speed at the top of the loop that allows the mass to make one complete cycle.

WHAT'S THE TRICK?

The minimum speed at the top of a loop indicates a special case where the forces contributing to the overall centripetal force must be at an absolute minimum. Begin by solving the problem the same as any other force problem.

Orient the problem: Sketch the scenario, including force vectors.

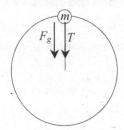

Determine the type of motion: This is circular motion.

Sum the force vectors in the relevant direction: Tension and gravity are both pointing toward the center of the circle. Both are set as positive.

$$F_c = T + F_g$$

Now you must consider how to make the centripetal force, F_c, as small as possible. The force of gravity cannot be altered. However, tension can be reduced by swinging the mass more slowly. When the mass is slowed to the point where it is just barely completing a full circle, the tension will be zero, $T = 0$, at the exact instant that the mass is at the very top of the circle.

Substitute and solve.

$$m\frac{v^2}{r} = (0) + mg$$

$$(2.0 \text{ kg})\frac{v^2}{(1.0 \text{ m})} = (2.0 \text{ kg})(10 \text{ m/s}^2)$$

$$v = \sqrt{10} \text{ m/s}$$

SUMMARY

1. **UNIFORM CIRCULAR MOTION INVOLVES MOTION AT A CONSTANT SPEED IN A CIRCULAR PATH.** Because direction is constantly changing, velocity is continuously changing, and thus circling objects are accelerating. Since the magnitudes of velocity and acceleration remain constant despite the changing direction, this motion is said to be uniform.

2. **ALL QUANTITIES IN CIRCULAR MOTION ARE BASED ON ONE COMPLETE CYCLE.** The time, in seconds, to complete one cycle is known as the period, T. The frequency, f, is the number of cycles completed each second and is the inverse of the period.

3. **SPEED AND TANGENTIAL VELOCITY HAVE THE SAME VALUE.** The tangential velocity is an instantaneous value and is directed tangent to the circular path. Centripetal force and centripetal acceleration, which cause the continuous change in direction, are both directed toward the center of the circle.

4. **IN DYNAMICS PROBLEMS, THE CENTRIPETAL FORCE, F_c, IS THE SUM** of the all the perpendicular forces acting to cause the circular motion. In circular-motion problems, F_c is used instead of the linear sum of forces, ΣF.

If You See	Try	Keep in Mind
Circular motion, even a fraction of a circle	Be aware of the relationships among speed, velocity, and acceleration.	Speed and the magnitude of velocity are constant while direction changes continually. This results in a uniform acceleration.
Period and frequency	Period:$$T = \frac{t}{\text{number of cycles}}$$Frequency:$$f = \frac{\text{number of cycles}}{t}$$Relationship:$$T = \frac{1}{f}$$	All variables and equations use one complete cycle as the standard for measurements.
Speed and/ or tangential velocity	$$v = \frac{2\pi r}{T}$$	Tangential velocity is tangent to the circular path.
Centripetal or radial acceleration	$$a_c = \frac{v^2}{r}$$	Centripetal acceleration points toward the center.
Circular motion and forces are mentioned	Draw a force diagram. Substitute previously learned force formulas and solve.$$F_c = F_{\text{toward center}} - F_{\text{away from center}}$$	Set forces toward the center as positive and those pointing away from the center as negative.
An object in a vertical circle and minimum speed at the top is requested	Draw a force diagram.$$F_c = F_{\text{toward center}} - F_{\text{away from center}}$$	Centripetal force must be as small as possible. Gravity cannot change, so the other forces acting on the object must be zero.

PRACTICE EXERCISES

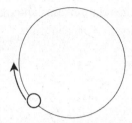

1. The object in the diagram above is in uniform circular motion. Which vectors show the tangential velocity and centripetal acceleration for the object at the instant diagrammed?

(A)

(B)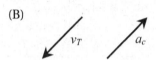

(C)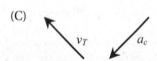

(D)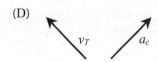

(E)

Questions 2–3

A 5.0-kilogram mass is moving in uniform circular motion with a radius of 1.0 meter and a frequency of 3.0 hertz.

2. Determine the tangential velocity of the mass.

   (A) $\frac{1}{6}\pi$ m/s

   (B) $\frac{2}{3}\pi$ m/s

   (C) $\frac{3}{2}\pi$ m/s

   (D) 3π m/s

   (E) 6π m/s

3. Determine the centripetal acceleration of the mass.

   (A) $3\pi^2$ m/s^2

   (B) $6\pi^2$ m/s^2

   (C) $12\pi^2$ m/s^2

   (D) $36\pi^2$ m/s^2

   (E) $72\pi^2$ m/s^2

---

4. A 30-kilogram child is sitting 2.0 meters from the center of a merry-go-round. The coefficients of static and kinetic friction between the child and the surface of the merry-go-round are 0.8 and 0.6, respectively. Determine the maximum speed of the merry-go-round before the child begins to slip.

   (A) $\sqrt{12}$ m/s

   (B)   4 m/s

   (C)   8 m/s

   (D) 12 m/s

   (E) 16 m/s

5. A roller coaster needs to complete a vertical loop that has a radius of 10 meters. What must its minimum speed be at the top of the loop?

   (A)   5 m/s

   (B)   7 m/s

   (C) 10 m/s

   (D) 14 m/s

   (E) 100 m/s

ANSWERS EXPLAINED

	Key Words	Needed for Solution	Now Solve It
1. **(D)**	Tangential velocity; centripetal acceleration	Knowledge/definitions	Tangential velocity is in the direction of motion and is tangent to the curving motion. Centripetal acceleration is directed toward the center of a circular path.
2. **(E)**	Tangential velocity	Use either kinematics: $$v = \frac{2\pi r}{T}$$ or force: $$F_c = m\frac{v^2}{r}$$	Given the available variables, using kinematics is the only possible option. However, this requires you to find the period, but the problem has given the frequency: $$T = \frac{1}{f}$$ $$v = \frac{2\pi r}{T} = \frac{2\pi r}{1/f} = 2\pi rf$$ $$v = 2\pi(1.0 \text{ m})(3.0 \text{ s}^{-1}) = 6\pi \text{ m/s}$$
3. **(D)**	Centripetal acceleration	Use either kinematics: $$a_c = \frac{v^2}{r}$$ or force: $$F_c = ma_c$$	Given the available variables, using kinematics is the only possible option. This also requires you have the answer to the previous problem. $$a_c = \frac{v^2}{r} = \frac{(6\pi \text{ m/s})^2}{1.0 \text{ m}} = 36\pi^2 \text{ m/s}^2$$
4. **(B)**	Coefficients of static and kinetic friction; maximum speed; before the child begins to slip.	Sum the forces for circular motion: $$F_c = F_{\text{to center}} - F_{\text{away from center}}$$	**Orient the problem:** The only force acting in the plane of the circle is friction. The force of gravity and the normal force are equal, oppose each other, and act perpendicular to the circling child: $$N = F_g = mg$$ **Determine the type of motion:** This is circular motion. **Sum the force vectors in the relevant direction:** $$F_c = f$$ **Substitute and solve:** $$m\frac{v^2}{r} = \mu N$$ $$m\frac{v^2}{r} = \mu mg$$ Use the coefficient of static friction. While the merry-go-round is moving, the child is not moving relative to its surface: $$v = \sqrt{\mu gr} = \sqrt{(0.8)(10)(2.0)} = 4 \text{ m/s}$$

	Key Words	Needed for Solution	Now Solve It
5. **(C)**	Vertical loop; minimum speed; top of the loop	Sum the forces for circular motion: $F_c = F_{\text{to center}}$ $\quad - F_{\text{away from center}}$	**Orient the problem:** Both the force of gravity and the normal force (acting from the track on the roller coaster) point down toward the center of the circle. **Determine the type of motion:** This is circular motion. **Sum the force vectors in the relevant direction:** $$F_c = N + F_g$$ To solve for the minimum speed, F_c must be as small as possible. The force of gravity cannot change. However, the normal force can be reduced to zero: $N = 0$. **Substitute and solve:** $$m\frac{v^2}{r} = (0) + mg$$ $$v = \sqrt{rg} = \sqrt{(10)(10)} = 10 \text{ m/s}$$

Energy, Work, and Power

7

→ **MECHANICAL ENERGY**

→ **WORK**

→ **POWER**

→ **CONSERVATION OF ENERGY**

Energy is difficult to define. It is easier to understand the definition of energy by exploring the characteristics of energy. Several fundamental concepts govern energy. This chapter will review the skills needed to do the following:

- Identify and distinguish among the various forms of energy.
- Understand how constant and variable forces change the energy of an object.
- Calculate and interpret the rate of energy change.
- Analyze energy transformations from one type to another and energy transfers from one object to another.

Table 7.1 lists the variables that will be discussed in this chapter.

Table 7.1 Variables That Describe Energy, Work, and Power

New Variables	Units
K = Kinetic energy	J (joules) or N • m (newton meter)
U = Potential energy	J (joules) or N • m (newton meter)
E = Total mechanical energy	J (joules) or N • m (newton meter)
W = Work	J (joules) or N • m (newton meter)
P = Power	W (watts) or J/s (joules per second)

MECHANICAL ENERGY

Several forms of energy are addressed in a first-year physics course. These include the different types of mechanical energy, electrical energy, thermal energy, heat, light energy, and nuclear energy. At this point, only the different types of mechanical energy will be discussed. The other forms of energy will be explored in subsequent chapters. Mechanical energy is the sum of the kinetic and potential energy of a system. The system is simply the object, or mass, under investigation.

IF YOU SEE
speed, *v*

**Kinetic
energy**

Direction of
motion is not
important.

Kinetic Energy

Kinetic energy, K, is the energy possessed by moving objects. If an object with mass m is moving at a speed v, its kinetic energy is

$$K = \frac{1}{2}mv^2$$

Kinetic energy is directly proportional to mass, m, and to the square of velocity, v^2. Kinetic energy is a scalar. Therefore, the direction of an object's velocity is not important. Only the magnitude of velocity (speed) is significant. In addition, kinetic energy is always positive.

Potential Energy

Potential energy, U, is the energy possessed by an object based on its position. To possess useful potential energy, an object must be in a position where it will move when released from rest. When the object is released, it gains kinetic energy at the expense of potential energy. If an object has the potential to create kinetic energy, the object has potential energy. There are two mechanical forms of potential energy featured on the SAT Subject Test in Physics: gravitational potential energy and elastic potential energy.

IF YOU SEE
height, *h*

**Gravitational
potential
energy**

Only vertical
height matters.

Gravitational Potential Energy

Gravitational potential energy, U_g, is due to the position of an object in a gravity field. A mass, m, in Earth's gravity field, g, is pulled toward Earth by the force of gravity, F_g. When the mass is raised to a height of h, the mass is in a position where it will accelerate toward Earth if released. At a height of h, the mass has the potential to create kinetic energy. If the mass is moved to a greater height, the mass will have a greater potential to increase both the final speed and the final kinetic energy. Height is the key factor determining the magnitude of gravitational potential energy.

$$U_g = mgh$$

Gravitational potential energy is a scalar. Only the height is important. Technically, both height and gravitational potential energy can have negative values. This depends on the location defined as zero height. To avoid working with negative heights, set the zero point for height measurements at the lowest point that the object reaches during a problem.

IF YOU SEE
a spring
stretched or
compressed
a distance *x*

**Elastic
potential
energy and
Hooke's law**

Energy is
proportional to
the square of
x, while force
is proportional
to *x*.

Elastic Potential Energy

Elastic potential energy, U_s, is the potential energy associated with elastic devices, such as springs. If a spring is stretched or compressed a distance of x, a restoring force, F_s, is generated in the spring according to Hooke's law.

$$F_s = |kx|$$

When a spring is displaced a distance of x and is subsequently released, the restoring force accelerates the spring back to its original length. Thus, displacing a spring has the potential to create kinetic energy.

$$U_s = \frac{1}{2}kx^2$$

Elastic potential energy is directly proportional to the spring constant, k, and to the square of displacement, x^2. Elastic potential energy is a positive scalar. The direction the spring is displaced (stretched or compressed) does not matter.

Total Mechanical Energy

Total mechanical energy, ΣE, is the sum of the kinetic and potential energies in a system. Total mechanical energy is an instantaneous value. Freeze the action in a problem and assess if the object you are interested in has height (h), has speed (v), and/or is attached to a spring displaced a distance (x). Next, sum the energy equations corresponding to the key variables that appear in the problem. This is most often done at the start and/or end of a problem.

$$E = K + U$$

Regardless of its form, energy is expressed in units of joules (N • m). As will soon be demonstrated, mechanical energy can change forms through a process known as work. Energy is known as a state function. Work is a process through which the state of energy may be changed.

EXAMPLE 7.1

Recognizing and Calculating Mechanical Energy

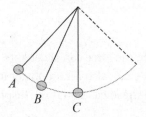

A pendulum initially at rest at point *A* is released and swings through points *B* and *C*. What types of mechanical energy are present at points *A*, *B*, and *C*? What formulas could be used to calculate the total mechanical energy at each point?

WHAT'S THE TRICK?

Look for height (h), speed (v), and spring displacement (x). If any of these are present, include the corresponding energy in the answer.

Point *A*: The pendulum has height above the lowest point in the swing but it is at rest. Only gravitational potential energy is present. The total mechanical energy at point *A* is

$$E = U_g = mgh$$

Point *B*: The pendulum has both speed and height. Both kinetic energy and gravitational potential energy are present. The total mechanical energy at point *B* is

$$E = K + U_g = \frac{1}{2}mv^2 + mgh$$

Point *C*: The pendulum has lost all its height. Only speed remains, so only kinetic energy is present. The total mechanical energy at point *C* is

$$E = K = \frac{1}{2}mv^2$$

EXAMPLE 7.2

Assessing the Effect of Changing Values on Energy

A ball has mass and speed and therefore has kinetic energy. What is the effect on the kinetic energy of the ball when its speed is doubled?

WHAT'S THE TRICK?

The key is assessing the relationships in the formula.

$$K = \frac{1}{2}mv^2$$

This problem never mentions changing the mass. Mass must remain constant and so is not a factor. Kinetic energy is proportional to the square of speed $K \propto v^2$.

$$(4K) = \frac{1}{2}m(2v)^2$$

Doubling the speed, which is squared, quadruples the right side of the equation. To maintain equality, the kinetic energy must also quadruple.

IF YOU SEE
any force
and
displacement
d

Work

F and *d* must
be parallel.

Work requires
motion (*d*).

WORK

When a force is applied to an object, the force can accelerate the object, changing the object's speed and its kinetic energy. A force can also lift an object to a height h, or it can displace a spring through a distance x. In all these cases, a force displaces an object and changes the object's total energy.

Work is the process of applying a force through a distance to change the energy of an object. Positive work is associated with an increase in the speed of an object. Negative work is associated with a decrease in speed. You can solve for work in two principal ways. The first method involves force and displacement. The second focuses on the change in energy of the object.

Work, W, is the product of the average force, F_{avg}, applied to an object and the component of displacement parallel to the average force, $d_{parallel}$.

$$W = F_{avg}\, d_{parallel}$$

- In most problems, the given force will be equal to the average force needed in the formula. The single exception in introductory physics courses is the restoring force of springs.
- Work requires a change in position known as displacement, which may be indicated in several ways (d, x, or h). It is important to recognize that work involves motion and requires objects to be displaced.
- Both force and displacement are vectors, and vectors at angles can be split into component vectors. The most important aspect of solving for work is remembering to **use parallel force and displacement vectors** or the parallel components of these vectors.
- Forces do work only when they are parallel to the displacement of the object. Forces perpendicular to displacement result in no work.
- Work is positive when the components of force and displacement both point in the same direction. Work is negative when the components of these vectors point in opposite directions.

EXAMPLE 7.3

Solving for Work Using Force and Displacement

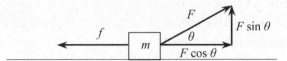

In the figure above, mass m is being pulled horizontally to the right by force F, inclined at angle θ. Force F and its components are shown in the diagram. The surface is rough. Friction, f, opposes the motion.

(A) Determine the work done by the applied force, W_F, as it moves the mass m through displacement d.

WHAT'S THE TRICK?

The applied force, F, is not parallel to the displacement. However, the x-component of the applied force, $F \cos \theta$, is parallel to the displacement and has the same direction as the displacement. $F \cos \theta$ is capable of doing positive work, increasing the speed and kinetic energy of the object.

$$W_F = (+F \cos \theta)d$$

(B) Determine the work of friction, W_f.

WHAT'S THE TRICK?

The friction vector, f, is opposite the displacement. It opposes motion, slows the object, decreases the object's kinetic energy, and performs negative work.

$$W_f = (-f)d$$

(C) Determine the net work, W_{net}.

WHAT'S THE TRICK?

The net work is the total work done on the object. One way to solve for the net work is to use the net force (sum of parallel forces, ΣF) acting on the object.

$$W_{net} = (\Sigma F)d$$
$$W_{net} = (F \cos \theta - f)d$$

A second method, using energy, will be discussed in the "Work–Kinetic Energy Theorem" section of this chapter.

All other forces (force of gravity, normal force, and $F \sin \theta$) are perpendicular to motion. Perpendicular forces and perpendicular components of force do no work.

Work by Gravity

Gravity near the surface of Earth is a good example of the **work done by a constant force**. Earth's gravity, g, is essentially constant for small changes in height. When a force is constant, the average force equals the value of the constant force, $F_{avg} = F_g = mg$. Displacement parallel to the force of gravity is equal to the change in height, $d = \Delta h$. Changes in height are also associated with changes in gravitational potential energy. The work done by gravity can be solved either as a force through a distance or as a change in gravitational potential energy.

$$W = F_{avg} d_{parallel} \qquad\qquad W = -\Delta U_g$$
$$W_g = -mg\,\Delta h \qquad\qquad W_g = -mg\,\Delta h$$

IF YOU SEE
a change in
height, Δh

Gravity is
doing work.

Ignore
horizontal
motion. Only
a change in
height matters.

The work done by gravity depends on only a change in height, Δh. As a result, the work done by gravity for horizontal motion is always zero. When objects follow any path consisting of a combination of vertical and horizontal motion, the work done by gravity depends on only the vertical change in height. Although there is a negative sign in the formula, the resulting sign on work also depends on the sign of the change in height, $\Delta h = h_f - h_i$. Ultimately, the work done by gravity is positive for downward motion, when force and vertical displacement have the same direction. It is negative for upward motion, when force and vertical displacement oppose each other.

Most questions regarding vertical motion involve the **work done by an external applied force that acts to lift an object**. When an object is lifted, the work done by an external applied force, F, is done against gravity. If the object starts at rest and finishes at rest or if the object moves at constant velocity, the work done by the external applied force will have the same magnitude as the work done by gravity. However, work will have the opposite sign.

$$W_F = mg\,\Delta h$$

The work done by an external force to lift an object is positive since force and displacement have the same direction.

EXAMPLE 7.4

The Work of Gravity

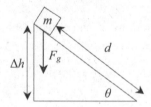

In the figure above, mass m is positioned at the top of a frictionless incline with an angle of θ. The mass slides down the incline a distance d. Determine the work done by gravity.

WHAT'S THE TRICK?

Two displacements are occurring, Δh and d. Use the change in height, Δh, since it is parallel to the force of gravity.

$$W = F_{avg}\, d_{parallel}$$
$$W = F_g \Delta h$$
$$W_g = mg\Delta h$$

The sign on work is positive in this case. Why? You can determine the sign on work using two methods.

- **Vector method:** If the force and the displacement vectors point in the same direction, work is positive. When they are opposite each other, work is negative. F_g and Δh have the same direction. So the work is positive.
- **Energy method:** If the kinetic energy increases during a displacement, the work is positive. Gravity increases the speed of the mass, so the work of gravity is positive.

Work by a Spring

The work by a spring, W_s, is done when a spring is stretched or compressed through a displacement x, thereby changing its length. The work by a spring is a good example of the **work done by a variable force**. Springs obey Hooke's law, $F_s = kx$. The restoring force, F_s, is the instantaneous force in a spring at a specific spring length, x. Changing spring length, Δx, changes the restoring force, as shown in Figure 7.1.

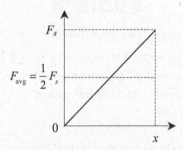

Figure 7.1 Restoring force versus displacement

The work formula uses the force acting when distance changes. What force should you use for a spring? Since the restoring force changes in a linear manner, you can use the average force, F_{avg}. The average force is simply half of the force needed to change the length of the spring.

$$F_{avg} = \frac{1}{2}F_s$$

$$F_{avg} = \frac{1}{2}k\,\Delta x$$

As with the work by gravity, there are two ways to determine the work done by a spring. The work done by a spring is the product of the average restoring force, $\frac{1}{2}F_s$, and the change in length of the spring, Δx. It is also equal to the change in elastic potential energy.

$$W = F_{avg}d_{parallel} \qquad\qquad W = -\Delta U$$

$$W_s = -\left(\frac{1}{2}k\Delta x\right)(\Delta x) \qquad\qquad W_s = -\Delta U_s$$

$$W_s = -\Delta\left(\frac{1}{2}kx^2\right) \qquad\qquad W_s = -\Delta\left(\frac{1}{2}kx^2\right)$$

IF YOU SEE
a change
in spring
length, Δx

Work by a spring

The average force is half of the restoring force.

The work done by a spring is positive when the spring is released and moves to restore to its original unstretched or uncompressed length. (In this case, force and displacement have the same direction.) The work is negative when the spring is being stretched or compressed. (In this case, force and displacement oppose each other.)

Questions may instead focus on the **work done on the spring** by an external applied force, F, which acts to stretch or compress the spring from its equilibrium position. The external applied force that stretches or compresses a spring must be equal in magnitude, but opposite in direction, to the resulting restoring force created as the length of the spring changes. As a result, the magnitude of the work done on a spring by an external applied force is equal to the work done by the spring when it is released and moves to restore to equilibrium.

$$W_F = \Delta\left(\frac{1}{2}kx^2\right)$$

The work done on a spring is positive when the spring is being stretched or compressed (force and motion have the same direction). The external applied force must be removed in order for the spring to restore to its original length. Thus, the work done on a spring by an external force is applicable only when the spring is being stretched or compressed.

EXAMPLE 7.5

Work by a Spring

A 4.0-kilogram mass is attached to a spring with a spring constant of 20 newtons per meter. The mass is lowered 0.50 meters to equilibrium, where it remains at rest. How much work was done stretching the spring?

WHAT'S THE TRICK?

This problem involves a spring undergoing a displacement (stretch or compression). Work always involves change. The work done to a spring depends on the change in length of the spring, $\Delta x = x_f - x_i$.

$$W_s = \Delta\left(\frac{1}{2}kx^2\right)$$

$$W_s = \left(\frac{1}{2}kx_f{}^2\right) - \left(\frac{1}{2}kx_i{}^2\right)$$

The spring was not stretched initially, $x_i = 0$.

$$W_s = \left(\frac{1}{2}(20 \text{ N/m})(0.50 \text{ m})^2\right) - (0) = 2.5 \text{ J}$$

The mass attached to the spring is not needed in the formula and is simply a distracter.

IF YOU SEE
Force versus
displacement
graph

Slope =
spring constant

Area =
work done by
a spring

Interpreting a Force versus Displacement Graph

Recall from Chapter 1 that slopes of lines and areas under curves have significance. The slope of a force versus displacement graph (Figure 7.1) has units of N/m. These are the same units as the spring constant (k). The area underneath Figure 7.1 has units of N • m, or joules. These are the same units as for work and energy.

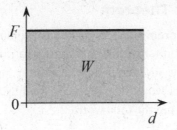

Figure 7.2 Force-displacement graph

In the graph in Figure 7.2, the height is equal to the force and the base is equal to the displacement.

$$W = F_{avg} \, d = \text{height} \times \text{base}$$

EXAMPLE 7.6

Work and Force-Displacement Graphs

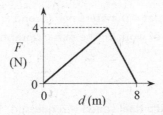

A variable force acts on a 5.0-kilogram mass, displacing the mass 8.0 meters. The force and displacement are graphed above. Determine the work done on the mass by the variable force.

WHAT'S THE TRICK?

Determine the area of the triangle formed by the graph.

$$W = \frac{1}{2}(\text{height} \times \text{base})$$

$$W = \frac{1}{2}(4)(8) = 16 \text{ J}$$

Solving this example required you to find the area of a triangle in a graph. The one-half in the formula actually solves for the average force.

IF YOU SEE
a change in
speed, Δ*v*

Work–kinetic
energy
theorem

Only a change
in speed
matters.

Work-Kinetic Energy Theorem

The **work–kinetic energy theorem** states that work is equal to the change in kinetic energy. When one or more unbalanced forces act on a mass, the mass will accelerate, changing in both speed and kinetic energy. Several forms of work may be present (W_F, W_g, W_s, and W_f) and can be calculated. The net work, W_{net}, done on the mass will be equal to the change in kinetic energy, ΔK, resulting from the total work.

$$W_{net} = \Delta K$$

Using the work–kinetic energy theorem is a quick way to find the net work done on a mass when the initial and final speeds are known. It can also be used to find the final speed of a mass if the net work and initial velocity are known. Since the net work is tied to changes in kinetic energy and changes in speed, a mass must accelerate in order for net work to be non-zero. Thus, when an object moves at constant velocity, the net work is always equal to zero.

EXAMPLE 7.7

Changes in Speed and Net Work (Total Work)

A 2.0-kilogram mass moving at 4.0 meters per second is acted upon by a force, which does 20 joules of work on the mass. Determine the final speed of the object.

WHAT'S THE TRICK?

An initial speed is given, and a final speed is requested. This implies a change in speed. There are two ways to solve for final speed. One involves force and kinematics. The other involves work and energy. The key in this problem is the 20 joules of work. No value for force is given. Try using the work–kinetic energy theorem:

$$W_{net} = \Delta K$$
$$W_{net} = K_f - K_i$$
$$W_{net} = \left(\frac{1}{2}mv_f^2\right) - \left(\frac{1}{2}mv_i^2\right)$$
$$(20 \text{ J}) = \left(\frac{1}{2}(2.0 \text{ kg})v_f^2\right) - \left(\frac{1}{2}(2.0 \text{ kg})(4.0 \text{ m/s})^2\right)$$
$$v_f = 6.0 \text{ m/s}$$

Work in Uniform Circular Motion

When a mass follows a curved path, such as a circle, the instantaneous-velocity and displacement vectors are directed tangent to the curved path. The net force is centripetal and is directed toward the center of the circle. As result, the net (centripetal) force and the displacement are always perpendicular to each other, as shown in Figure 7.3.

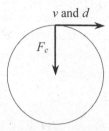

v and *d*

F_c

Figure 7.3 Vectors in uniform, circular motion

Perpendicular forces do no work. So, the work in uniform circular motion is zero.

Using the work–kinetic energy theorem, $W_{net} = \Delta K$, results in the same conclusion. In uniform circular motion, speed is constant. If speed is constant, then the change in kinetic energy and the net work are both zero.

POWER

Power, *P*, is the amount of work done over a period of time. It is the rate at which work is done. It is also the rate of energy use or energy generation. It can be calculated using the following equation:

$$P = \frac{W}{t} = \frac{\Delta E}{t}$$

Since $W = Fd$, the above formula can be written as

$$P = \frac{Fd}{t}$$

Distance divided by time should be very familiar to you. It is velocity. Therefore, power can also be expressed as

$$P = Fv$$

Power has the units of watts (W). When you look at the top of a lightbulb, it is labeled in watts. A watt is the rate at which the lightbulb uses energy. Another way to express a watt is a joule per second. A 100-watt lightbulb uses 100 joules of energy every second. Power is a scalar quantity. It has magnitude but no direction.

EXAMPLE 7.8

Determining the Power Required to Lift an Object

Determine the power required to lift a 10-newton crate up to a 400-centimeter-high shelf in 2.0 seconds.

WHAT'S THE TRICK?

Power is the rate at which work is done. Solve for work using the methods described in the previous section, and then divide by time. In this case, force and displacement are given. However, displacement is in units of centimeters. You must first convert centimeters into meters.

$$P = \frac{W}{t}$$

$$P = \frac{Fd}{t}$$

$$P = \frac{(10 \text{ N})(4.0 \text{ m})}{(2.0 \text{ s})} = 20 \text{ W}$$

EXAMPLE 7.9

Find the Power at a Constant Velocity

Determine the power required to lift a 10-newton crate vertically at a constant speed of 2.0 meters per second.

WHAT'S THE TRICK?

This problem is similar to Example 7.8. However, this time the speed is given instead of the distance and time.

$$P = Fv$$
$$P = (10 \text{ N})(2.0 \text{ m/s}) = 20 \text{ W}$$

IF YOU SEE
Δh, Δx,
and/or Δv

Conservation of energy

The total energy is constant. The total energy at any two points in a problem is the same.

CONSERVATION OF ENERGY

Energy is always conserved in an isolated system. The amount of energy present at the start of a problem must remain constant throughout the entire problem. You must consider some important properties of energy to understand conservation of energy fully.

1. Energy can transform from one type to another.
2. Energy can transfer from one object to another.
3. Energy can leave and enter a closed system as work.

Questions concerning the conservation of energy will ask you to account for all of the joules of energy being transferred and transformed. The unit for energy, joules, can be used for problems involving kinetic energy, gravitational potential energy, elastic potential energy, electrical energy, light energy, nuclear energy, heat, and work. Typically, questions will require you to account for where the energy goes in the given scenario.

The actual values of gravitational potential energy and kinetic energy depend on a comparison to a zero-reference height and a zero-reference velocity in order to quantify an object's height and speed. For objects moving on Earth, the earth itself is a commonly used zero-reference point. Conservation-of-energy problems involving an object, such as a roller coaster, should actually specify a roller coaster–Earth system as opposed to mentioning only the roller coaster by itself. However, in beginning physics problems Earth as a commonly used zero-reference point is often taken for granted. It is wise to understand that the roller coaster does not have quantifiable mechanical energy without regarding it as part of a larger system.

Conservative Forces

The force of gravity, F_g, and the force of springs, F_s, are examples of conservative forces. When work is done only "on" or "by" conservative forces, and all other forces do no work, then the total mechanical energy of a system remains constant. Take, for example, a roller coaster with respect to Earth. At the top of a hill, the roller coaster will have maximum gravitational potential energy, $U_g = mgh$. Should the roller coaster be moving as well, it will also possess kinetic energy, $K = \frac{1}{2}mv^2$. The sum of these energies, $K + U$, is the total mechanical energy of the roller coaster. As the roller coaster descends the hill, it will lose height and gravitational potential energy. However, the roller coaster will simultaneously gain speed and kinetic energy. During the descent, energy transforms from gravitational potential energy into kinetic energy. The force of gravity is a conservative force. When work is done solely by conservative forces, the total mechanical energy of a system (roller coaster with respect to Earth) remains constant.

$$K_i + U_i = K_f + U_f$$

Nonconservative Forces

When nonconservative forces act, energy transfers into or out of a system. As a result, the total mechanical energy of the system is not conserved. At first glance, this seems to violate the conservation of energy. When energy is gained by the system, the energy comes from the environment. When energy is lost by the system, the energy moves to the environment. Energy is conserved when the system and environment are examined together. However, most problems deal with only a specific system. When nonconservative forces act, the energy of a system is not conserved. The transfer of energy into and out of a system is known as the work of nonconservative forces. It is equal to the change in energy of the system.

$$W_{\text{nonconservative force}} = \Delta E_{\text{sys}}$$
$$W_{\text{nonconservative force}} = E_f - E_i$$

The most commonly encountered nonconservative force is kinetic friction, f_k. Kinetic friction acts on moving objects, such as the roller coaster described in the section "Conservative Forces." When kinetic friction acts, some of the total mechanical energy of the system (roller coaster) is lost (not conserved). Kinetic friction transfers a portion of the initial energy to the environment as heat, resulting in a lower final energy for the system. As a result, the final kinetic energy and speed are less than they would have been in a frictionless environment. The decrease in kinetic energy when kinetic friction slows a moving object is referred to as **kinetic energy lost**. As with other forms of work, the work done by kinetic friction is equal to the product of the force of kinetic friction and displacement. In addition, the work done by friction is equal to the kinetic energy lost by the system and the resulting heat transfer to the environment.

$$W_f = -f_k d$$

$$|W_f| = K_{\text{lost by the system}} = \text{Heat}_{\text{gained by environment}}$$

IF YOU SEE friction, f

Energy lost

Lost energy moves to the environment as heat and equals the work of friction.

EXAMPLE 7.10

Conservation of Energy When Conservative Forces Act

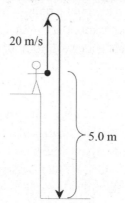

A boy standing on a ledge throws a ball with a mass of 0.50 kilograms straight up into the air with an initial velocity of 20 meters per second. The ball is initially 5.0 meters above the ground. It rises to a maximum height and then falls, missing the ledge on the way down. The ball impacts the ground 5.0 meters below its initial starting point.

(A) What is the total mechanical energy of the ball as measured from the ground?

(WHAT'S THE TRICK?)

All of the energy is present in the ball at the moment the ball is released. Sum the kinetic energy of the throw and the gravitational potential energy due to the height above the ground. The ground is the lowest point reached by the ball. Set the ground as zero height and zero gravitational potential energy.

$$E_i = K_i + U_i$$

$$E_i = \frac{1}{2}mv_i^2 + mgh_i$$

$$E_i = \frac{1}{2}(0.50 \text{ kg})(20 \text{ m/s})^2 + (0.50 \text{ kg})(10 \text{ m/s}^2)(5.0 \text{ m}) = 125 \text{ J}$$

(B) What is the total mechanical energy of the ball 2.0 meters above its release point?

(WHAT'S THE TRICK?)

The total mechanical energy is conserved and constant.

$$E_i = E_{\text{any point}} = 125 \text{ J}$$

(C) What is the maximum height reached by the ball as measured from the ground?

(WHAT'S THE TRICK?)

The kinetic energy of the boy's throw will be converted into potential energy at the top of the ball's trajectory. In addition, when an object reaches maximum height, its vertical speed is zero.

METHOD 1: Energy Solution

$$K_i + U_i = K_f + U_f$$

$$\frac{1}{2}mv_i^2 + mgh_i = \frac{1}{2}mv_f^2 + mgh_f$$

The left side of the equation was calculated in part (A), and the final velocity is zero.

$$125 = 0 + (0.50 \text{ kg})(10 \text{ m/s})h_f$$

$$h_f = 25 \text{ m}$$

The height of the ledge was included in the initial energy. Therefore, the calculated final height is the height above the ground.

METHOD 2: Kinematics Solution

Time is unknown. Use the kinematic equation that does not include time.

$$v_f^2 = v_i^2 + 2g\Delta y$$

$$v_f^2 = v_i^2 + 2g(y_f - y_i)$$

$$(0)^2 = (20 \text{ m/s})^2 + 2(-10 \text{ m/s}^2)(y_f - 5.0 \text{ m})$$

$$y_f = 25 \text{ m}$$

This is the height as measured from the ground. The height of the ledge was included as the initial height, $y_i = 5$ m.

(D) What is the impact velocity when the ball reaches the ground?

(WHAT'S THE TRICK?)

Again, the total mechanical energy remains constant. When the ball reaches the ground, the height and gravitational potential energy will be zero, $mgh_f = 0$.
All of the ball's initial energy will be completely converted into kinetic energy.

$$K_i + U_i = K_f + U_f$$

$$\frac{1}{2}mv_i^2 + mgh_i = \frac{1}{2}mv_f^2 + mgh_f$$

$$125 \text{ J} = \frac{1}{2}(0.50 \text{ kg})\, v_f^2 + 0$$

$$v_f = 22.4 \text{ m/s}$$

EXAMPLE 7.11

Conservation of Energy and Pendulums

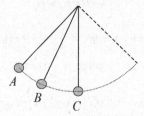

A 0.02-kilogram pendulum bob is released from position A. As the pendulum passes position C, located 1.25 meters below position A, what is its velocity?

WHAT'S THE TRICK?

In position A, the pendulum is at rest and is located at the maximum height. All of the energy is stored as gravitational potential energy. In position C, the height and gravitational potential energy are zero. Potential energy has been entirely transformed into kinetic energy. The total mechanical energy remains constant.

$$K_i + U_i = K_f + U_f$$

$$\frac{1}{2}mv_i^2 + mgh_i = \frac{1}{2}mv_f^2 + mgh_f$$

$$(0) + mgh_i = \frac{1}{2}mv_f^2 + (0)$$

Rearrange and simplify.

$$v_f = \sqrt{2gh_i}$$

$$v_f = \sqrt{2(10 \text{ m/s})(1.25 \text{ m})}$$

$$v_f = 5.0 \text{ m/s}$$

This equation is the same as the equation for an object dropped straight down from a height h. How is this possible? The pendulum bob does fall a height h, but it is attached to a string that changes the pendulum's direction. The string changes the direction of the pendulum bob without affecting the change in energy from potential to kinetic.

EXAMPLE 7.12

Conservation of Energy and Springs

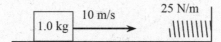

A 1.0-kilogram block moves with a speed of 10 meters per second and strikes a spring that has a spring constant of 25 newtons per meter. Determine the maximum compression of the spring.

WHAT'S THE TRICK?

Initially, the moving block has only kinetic energy and the spring is at rest. When the block strikes the spring, the spring compresses and the block slows. The kinetic energy of the moving block is transferred to the spring and stored as elastic potential energy. At maximum compression, all of the kinetic energy initially present has been entirely converted to elastic potential energy.

$$K_i + U_i = K_f + U_f$$

$$\frac{1}{2}mv_i^2 + \frac{1}{2}kx_i^2 = \frac{1}{2}mv_f^2 + \frac{1}{2}kx_f^2$$

$$\frac{1}{2}mv_i^2 + (0) = (0) + \frac{1}{2}kx_f^2$$

Substitute values and solve.

$$\frac{1}{2}(1.0 \text{ kg})(10 \text{ m/s})^2 + (0) = (0) + \frac{1}{2}(25 \text{ N/m})x_f^2$$

$$x_f = 2.0 \text{ m}$$

EXAMPLE 7.13

Energy Lost by Nonconservative Forces

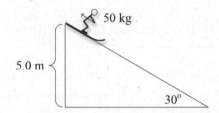

A 50-kilogram skier, starting at rest, skis down a 5.0-meter-tall hill inclined at 30°. The skier arrives at the bottom with 2,000 joules of kinetic energy. How much energy was lost due to friction?

WHAT'S THE TRICK?

If the hill were frictionless, the skier should arrive at the bottom with the kinetic energy equal to the initial potential energy. When friction acts, energy is lost from the system and transfers to the environment as heat. Energy lost due to friction is the difference between the expected and the actual kinetic energy.

$$K_i + U_i = K_f + U_f$$
$$0 + U_i = K_f + 0$$
$$K_f = U_i = mgh_i = (50 \text{ kg})(10 \text{ m/s}^2)(5.0 \text{ m}) = 2{,}500 \text{ J}$$

At the bottom of the hill, the skier has an actual kinetic energy of 2,000 J. The difference between the expected and the actual kinetic energies is the energy lost due to friction.

$$K_{lost} = (2{,}500 \text{ J}) - (2{,}000 \text{ J}) = 500 \text{ J}$$

SUMMARY

What is energy? Many texts define energy as the ability to do work. They then define work as a change in energy, creating a circular argument. The exact definition of energy may not be as important as understanding the characteristics of energy.

1. **THERE ARE MANY FORMS OF ENERGY.** The forms of energy studied so far are kinetic energy, gravitational potential energy, and elastic potential energy. We will learn about other forms of energy, such as thermal energy and electrical energy, in the chapters ahead. Think of energy as a quantity of motion (kinetic energy) or as the potential to create motion (potential energy). There are many types of motion. They include moving masses in mechanics, the motion of gases in thermodynamics, and the motion of charges in electricity.

2. **A FORCE THAT DISPLACES AN OBJECT CAN DO WORK ON THE OBJECT, CHANGING THE AMOUNT AND TYPE OF ENERGY POSSESSED BY THE OBJECT.** Only forces acting parallel to the displacement of an object are capable of doing work. Forces acting in the direction of displacement will increase the speed and kinetic energy of an object. They do positive work on the object. Forces opposing motion decrease speed, decrease kinetic energy, and do negative work on objects. Calculations of work require you to use the average force acting parallel to the motion of the object.

3. **THE RATE OF ENERGY CHANGE DETERMINES HOW QUICKLY ENERGY IS USED OR GENERATED.** At times, you need to know how quickly energy is being used or generated. Power is the rate of change in energy. It is the amount of work done every second.

4. **ENERGY CAN TRANSFORM FROM ONE TYPE TO ANOTHER AND BE TRANSFERRED FROM ONE OBJECT TO ANOTHER.** Energy is always conserved. The magnitude of energy (joules) at the beginning and at the end of a problem are equal. Most energy problems involve the change of one form of energy into another. For example, the gravitational potential energy of a rock at the top of a cliff is converted into kinetic energy as the rock falls, or a mass with kinetic energy compresses a spring and creates elastic potential energy. Work is the process through which mechanical energy can be transformed or transferred.

5. **NONCONSERVATIVE FORCES CHANGE THE TOTAL ENERGY OF SYSTEMS.** The change in the system's energy is equal to the work of the nonconservative forces acting on the system. Friction is the most commonly encountered nonconservative force. When kinetic friction acts on a moving object, the object will lose kinetic energy and slow down. The lost kinetic energy is transferred to the environment as heat. The work of friction is equal to the kinetic energy lost and to the heat moving into the environment. Although energy is not conserved in the system, it is conserved when the system and environment are analyzed together.

If You See	Try	Keep in Mind
Speed (v), height (h), or spring displacement (x)	$$K = \frac{1}{2}mv^2$$ $$U_g = mgh$$ $$U_s = \frac{1}{2}kx^2$$	Energy is scalar. The direction of speed does not matter. Only vertical height is important.
Work, work of gravity, or work of a spring	$$W = F_{avg}d_{parallel}$$ $$W_g = mg\,\Delta h$$ $$W_s = \left(\frac{1}{2}kx^2\right)$$	Choose the force parallel to the displacement or the displacement parallel to force.
Net work	$$W_{net} = \Sigma Fd$$ $$W_{net} = \Delta K = \Delta\left(\frac{1}{2}mv^2\right)$$	The net force is equal to the sum of the forces (total force). $W_{net} = 0$ for constant velocity and uniform circular motion.
Rate of work or rate of energy use/generation	$$P = \frac{W}{t} = \frac{\Delta E}{t} \text{ or } P = \frac{Fd}{t} \text{ or } P = Fv$$	Power is measured in watts or joules of work per second.
A change in height, Δh, or a change in spring displacement, Δx, causing or being caused by a change in velocity, Δv	Conservation of energy $$K_i + U_i = K_f + U_f$$	Energy is a scalar, and direction does not matter. The total mechanical energy at one point must equal the total mechanical energy at any other point.
Nonconservative forces such as kinetic friction	Work of nonconservative forces $$W = F_{avg}\,d_{parallel}$$ $$W_f = -f_k\,d$$ Effect on system energy $$|W_f| = K_{\text{lost by the system}}$$	Although energy is always conserved, it is not necessarily conserved in the system being investigated. When nonconservative forces act, energy transfers between the system and the environment. The work done by the nonconservative force equals the change of the system.

PRACTICE EXERCISES

1. In order for an object to have potential energy capable of doing work

   (A) the potential energy must have a positive value
   (B) the object must have energy associated with its position
   (C) the object must have kinetic energy
   (D) the object must be in a position where it can lose its potential energy when released
   (E) the object must be stationary

2. Stretching a spring a distance of x requires a force of F. In the process, potential energy, U, is stored in the spring. How much force is required to stretch the spring a distance of $2x$, and what potential energy is stored in the spring as a result?

   (A) F and $2U$
   (B) $2F$ and $2U$
   (C) $2F$ and $4U$
   (D) $4F$ and $4U$
   (E) $4F$ and $8U$

3. The total mechanical energy of a system

   (A) always remains constant
   (B) is equal to the net work done on a system
   (C) is the energy stored in the system
   (D) is the kinetic energy of the system
   (E) is the sum of the kinetic and potential energies in the system

4. A 1,000-kilogram satellite completes a uniform circular orbit of radius 8.0×10^6 meters as measured from the center of Earth. The mass of Earth is approximately 6.0×10^{24} kilograms, and the universal gravity constant is approximately 7.0×10^{-11} N • m^2/kg^2. Determine the work done by gravity as the satellite completes one full orbit around Earth.

   (A) zero
   (B) 5.3×10^{10} J
   (C) 8.0×10^{10} J
   (D) 3.3×10^{11} J
   (E) 5.0×10^{11} J

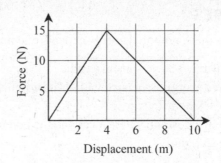

5. The force-displacement graph above depicts the force applied to a 2.0-kilogram mass as it is displaced 10 meters. The initial speed of the mass is 5.0 meters per second. The final speed of the mass is most nearly

(A) 7 m/s
(B) 10 m/s
(C) 12 m/s
(D) 16 m/s
(E) 20 m/s

Questions 6–7

The diagram below depicts a 3.0-newton block sliding down a frictionless 30° incline. The mass is initially at rest at the top of the incline.

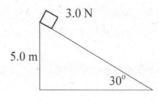

6. The work done by gravity is

(A)   15 J
(B)   30 J
(C)   60 J
(D) 150 J
(E) 300 J

7. The final speed of the mass when it reaches the bottom of the incline is most nearly

(A)   1.0 m/s
(B)   1.4 m/s
(C)   5.0 m/s
(D)   10 m/s
(E) 100 m/s

8. You want to lift an object, changing its height by 1 meter. Pushing the object up a 30° incline to achieve a 1 meter change in height takes twice as long as lifting it straight upward. When you use the incline, compared to lifting it straight up, you do

(A) half the work and use half the power
(B) half the work and use the same power
(C) the same work and use half the power
(D) the same work and use the same power
(E) twice the work and use half the power

9. A rock of mass 5.0 kilograms is dropped from a height of 10 meters. What is the kinetic energy of the rock when it reaches the halfway point of its descent?

(A)   15 J
(B)   25 J
(C)   50 J
(D) 250 J
(E) 500 J

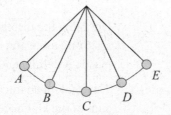

10. As shown in the diagram above, a pendulum swings from point *A* to point *E*. Identify the statement below that is NOT correct.

(A) The energy at both points *A* and *E* is entirely potential energy.
(B) The total energy is decreasing from *A* to *C* and is increasing from *C* to *E*.
(C) The kinetic energy is greatest at point *C*.
(D) The total energy at point *A* equals the total energy at point *C*.
(E) Potential energy is decreasing and kinetic energy is increasing from point *A* to *C*.

ANSWERS EXPLAINED

	Key Words	Needed for Solution	Now Solve It
1. **(D)**	Potential energy capable of doing work	Knowledge/definitions	**Potential energy** is related to an object's location or position. To be capable of doing work, an object must be able to move through a distance and be able to generate kinetic energy by losing potential energy. Choice B is almost correct, but the object is not necessarily in a position that is capable of transforming potential energy to kinetic energy.
2. **(C)**	Spring; distance of x; force of F; potential energy, U	Hooke's law $$F_s = kx$$ Elastic potential energy $$U_s = \frac{1}{2} kx^2$$	**Spring force** follows Hooke's law, where force is directly proportional to spring displacement. $$(2F_s) = k(2x)$$ **Spring potential energy** is directly proportional to the square of displacement. $$(4U_s) = \frac{1}{2} k(2x^2)$$ Doubling the displacement doubles the restoring force and quadruples the spring's elastic potential energy.
3. **(E)**	Total mechanical energy	Knowledge/definitions Total energy $$E = K + U$$	The total mechanical energy of a system is the sum of the kinetic and potential energies at a specific instant in time. Why are the other answers wrong? (A) The total energy does not always remain constant. Nonconservative forces, such as friction, can change the energy of the system. (B) The net work equals the change in kinetic energy only. This answer overlooks potential energy. (C) Stored energy is potential energy. This answer is missing kinetic energy. (D) This answer does not include potential energy.
4. **(A)**	Uniform circular orbit; work done by gravity	Parallel force and distance $$W = F_{avg}d_{parallel}$$ Changing height $$W_g = mg\,\Delta h$$	You can arrive at the answer in two ways: ■ Force/displacement method: In uniform circular motion, such as circular orbits, force and displacement are perpendicular. Perpendicular forces do no work. ■ Energy/work method: The work of gravity requires a change in height, Δh. In a uniform circular orbit, the height above Earth's surface is constant.

	Key Words	Needed for Solution	Now Solve It
5. (B)	Force-displacement graph; initial speed; final speed	Graphs often involve slope or area. The key is $$W = \text{area}$$ Work–kinetic energy theorem $$W_{net} = \Delta K$$	The work done by force F must be responsible for the change in speed of the mass. Set the two work equations in the middle column equal to each other. $$\Delta K = \text{area}$$ $$\frac{1}{2}mv_f^2 - \frac{1}{2}mv_i^2 = \frac{1}{2}(\text{height})(\text{base})$$ $$\frac{1}{2}(2.0 \text{ kg})v_f^2 - \frac{1}{2}(2.0 \text{ kg})(5.0 \text{ m/s})^2$$ $$= \frac{1}{2}(15 \text{ N})(10 \text{ m})$$ $$v_f = 10 \text{ m/s}$$
6. (A)	Sliding down a frictionless 30° incline; work done by gravity	Work of gravity $$W_g = mg\,\Delta h$$ Weight/force of gravity $$F_g = mg$$	The work done by gravity depends on only the object's change in height, Δh. Be careful. In this problem, the mass is not given. Instead, you are given the weight of the object in newtons. $$W_g = F_g \Delta h$$ $$W_g = (3.0 \text{ N})(5.0 \text{ m}) = 15 \text{ J}$$
7. (D)	Sliding down a frictionless 30° incline; initially at rest; final speed	Conservation of energy $$K_i + U_i = K_f + U_f$$	Initially, the mass starts at rest and possesses only gravitational potential energy due to its initial height. At the end of the motion, all of the potential energy is gone, and the block now possesses only kinetic energy. $$0 + mgh_i = \frac{1}{2}mv_f^2 + 0$$ $$v_f = \sqrt{2gh}$$ $$v_f = \sqrt{2(10 \text{ m/s}^2)(5.0 \text{ m})} = 10 \text{ m/s}$$
8. (C)	Lift an object 1 meter; incline takes twice as long as lifting it straight up; compared to	Work $$W_g = mg\,\Delta h$$ Power $$P = \frac{W}{t}$$	The work against gravity is associated with only the change in vertical height, $W_g = mg\,\Delta h$. Therefore, the work done is the same whether the object is lifted straight up or follows a diagonal, inclined, path. Power is inversely proportional to time, $P = \frac{W}{t}$. Pushing the object up the incline uses twice the time and requires only half the power.

	Key Words	Needed for Solution	Now Solve It
9. **(D)**	10-meter-tall ledge; kinetic energy; halfway point	Conservation of energy $$K_i + U_i = K_f + U_f$$	Initially, the rock is at rest and possesses only potential energy. At the halfway point it has half its initial height and now has kinetic energy. $$0 + mgh_i = K_f + mgh_f$$ $$(5.0 \text{ kg})(10 \text{ m/s}^2)(10 \text{ m})$$ $$= K_f + (5.0 \text{ kg})(10 \text{ m/s}^2)(5 \text{ m})$$ $$K_f = 250 \text{ J}$$ An alternate solution comes from realizing that the rock has lost half of its potential energy when it is halfway to the ground. The lost potential energy transforms into kinetic energy. $$K_f = (5.0 \text{ kg})(10 \text{ m/s}^2)(5 \text{ m}) = 250 \text{ J}$$
10. **(B)**	Pendulum; NOT correct	Total energy $$\Sigma E = K + U$$ Conservation of energy $$K_i + U_i = K_f + U_f$$	A pendulum illustrates the transformation of energy from potential to kinetic, and back again, in a repeating process. During the swing, energy is changing forms between potential and kinetic. However, total mechanical energy is conserved and remains constant. Answer B is incorrect because it states that the total energy is changing from A to C and C to E. This does not occur.

Momentum and Impulse

8

→ **MOMENTUM**

→ **IMPULSE**

→ **CONSERVATION OF MOMENTUM**

→ **ENERGY IN COLLISIONS**

Objects having both mass and velocity are said to possess momentum. Changing the momentum of an object involves delivering an impulse, a force during a time interval, to the object. When objects interact with one another, momentum can be transferred from one object to another. Thus, momentum, like energy, can be conserved. In this chapter we will do the following:

- Define momentum.
- Define impulse and examine impulse-momentum theory.
- Solve conservation-of-momentum problems for collisions.
- Determine energy changes during collisions.

Table 8.1 lists the variables that will be discussed in this chapter.

Table 8.1 Variables for Momentum and Impulse

New Variables	Units
$\vec{p}$ = Momentum	kg • m/s (kilograms • meters per second) or N • s (newton • seconds)
$\vec{J}$ = Impulse	kg • m/s (kilograms • meters per second), or N • s (newton • seconds)

MOMENTUM

There are two types of momentum—linear momentum and angular momentum. A 2017 College Board publication does not list rotation and rotational quantities, such as angular momentum and its conservation, as topics covered on the SAT Subject Test in Physics. In addition, released College Board exams as of 2017 have not included questions regarding these topics. The exam seems to focus exclusively on the conservation of linear momentum.

Linear Momentum

Linear momentum is the product of the mass and velocity of an object.

$$\vec{p} = m\vec{v}$$

Momentum has units of kilograms • meters per second. Momentum is related to the inertia of a moving object. The greater the momentum of a moving object, the more difficult the object is to stop. Momentum is a vector quantity. The direction of the momentum vector is the same as the direction of the velocity of the object.

Total Momentum

Many problems involve a system (more than one object) whose objects move simultaneously. The total momentum of a system can be found by adding the individual momentums of all the objects making up the system.

$$\Sigma \vec{p} = \vec{p}_1 + \vec{p}_2 + \cdots$$
$$\Sigma \vec{p} = m\vec{v}_1 + m\vec{v}_2 + \cdots$$

Note that momentum and velocity are both vector quantities. Vector direction for one-dimensional motion can be annotated with a positive or a negative sign as shown in the formula below.

$$\Sigma p = m_1(\pm v_1) + m_2(\pm v_2) + \cdots$$

By adding plus and minus signs, the vector velocity, $\vec{v}$, becomes a scalarlike quantity, v, allowing simple addition. You must note the direction of velocity and add the correct sign when solving a problem for momentum.

IMPULSE

Impulse is a force that is applied to an object over a period of time. A kick or a shove would be considered an impulse.

$$\vec{J} = \vec{F}\Delta t$$

From this equation, the units of impulse will be newtons • seconds (N • s). This is also equivalent to the units of momentum, kg • m/s. Both momentum and impulse can be expressed in either of these units. Impulse is a vector quantity, and the impulse vector points in the direction of the force acting on the object.

When analyzing the equation for impulse, it is apparent that if the duration of a collision can be lengthened, the force of the impact can be lessened. For example, air bags in an automobile are designed to increase the amount of time needed for a passenger to come to a complete stop, thereby decreasing the force exerted on the passenger during a collision.

Impulse–Momentum Theorem

When an impulse acts on an object, the momentum of the object will change. Note that impulse does not equal momentum even though their units are the same. Impulse causes and is equal to the change in the momentum of an object. Impulse is similar to work (discussed in the previous chapter) in that they both create a change. Work changes the energy of a mass, and impulse changes the momentum of a mass. Impulse and work are both processes of change, while momentum and energy are both state functions.

$$\vec{J} = \Delta \vec{p} = \vec{p}_f - \vec{p}_i = m\vec{v}_f - m\vec{v}_i = \vec{F}\Delta t$$

IF YOU SEE
a need to
reduce the
force during
a collision

Increase the
elapsed time
during the
collision.

The applied force and the elapsed time are inversely proportional.

IF YOU SEE
a change
in speed or
momentum

Impulse is
equal to the
change in
momentum.

An impulse causes the change in momentum.

From the impulse-momentum theorem, the equation describing Newton's second law of motion can be derived.

$$m\Delta v = F\Delta t$$
$$F = m\frac{\Delta v}{\Delta t} = ma$$

The impulse-momentum theorem and Newton's second law of motion are strongly linked and provide two ways to examine force and motion problems.

EXAMPLE 8.1

Change in Momentum

(A) Determine the change in momentum (impulse) for a 0.5-kilogram lump of clay striking a wall at 15 meters per second.

WHAT'S THE TRICK?

The lump of clay will stick to the wall and come to a stop.

$$\Delta \vec{p}_{\text{clay}} = m\vec{v}_f - m\vec{v}_i$$
$$\Delta \vec{p}_{\text{clay}} = (0.5 \text{ kg})(0) - (0.5 \text{ kg})(15 \text{ m/s})$$
$$\Delta \vec{p}_{\text{clay}} = -7.5 \text{ kg} \cdot \text{m/s}$$

The negative sign in the answer indicates that the change in momentum is opposite the initial velocity. The minus sign may not appear in the available answer choices since problems of this type are often concerned with just the value.

(B) Determine the change in momentum (impulse) for a 0.5-kilogram rubber ball striking a wall at 15 meters per second and bouncing off the wall in the opposite direction.

WHAT'S THE TRICK?

The rubber ball will bounce off of the wall. Unless told otherwise, assume that the final speed of the ball leaving the wall is the same as the initial speed of the ball.

$$\Delta \vec{p}_{\text{ball}} = m\vec{v}_f - m\vec{v}_i$$
$$\Delta \vec{p}_{\text{ball}} = (0.5 \text{ kg})(-15 \text{ m/s}) - (0.5 \text{ kg})(15 \text{ m/s})$$

Note the minus sign on the final velocity of the ball. Set the initial velocity as positive. Since the ball reversed direction during the bounce, the final velocity is in the opposite direction and must have the opposite sign.

$$\Delta \vec{p}_{\text{ball}} = -15 \text{ kg} \cdot \text{m/s}$$

The change in momentum (impulse) for an object that bounces with no loss in speed is twice as large as the change in momentum for an object coming to a stop.

Force-Time Graph

Impulse is equal to the area under a force versus time function. The graph in Figure 8.1(a) is an example of the impulse delivered to a soccer ball when the ball is kicked. However, this graph will probably appear in the simplified form of Figure 8.1(b) to allow you to calculate the area under the function with ease.

In the graphs in Figure 8.1, the force is continually changing. Therefore, the formula for impulse requires you to find the average force, F_{avg}, delivered during the time interval, Δt. The magnitude of impulse in the graph in Figure 8.1(b) is:

$$J = \frac{1}{2}(\text{height} \times \text{base})$$

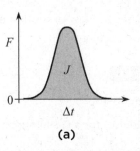

(a)

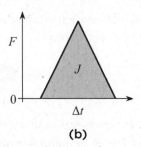

(b)

Figure 8.1 Force-time graphs

You can solve for impulse in a variety of ways. Impulse is equal to the following:

- The applied force multiplied by the elapsed time, $\vec{F}\Delta t$.
- A change in momentum, $m\vec{v}_f - m\vec{v}_i$.
- The area under a force-time graph.

$$\vec{J} = \vec{F}\Delta t = m\vec{v}_f - m\vec{v}_i = \text{area}_{F\text{-}t\,\text{graph}}$$

The most challenging problems may require you to set one formula for impulse equal to another.

EXAMPLE 8.2

Impulse

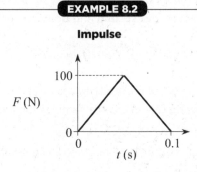

The above graph illustrates the force acting on a 0.20-kilogram soccer ball as it is kicked. Determine the final speed of the ball.

(WHAT'S THE TRICK?)

A force-time graph indicates that this problem involves impulse. You can solve an impulse problem in many possible ways.

$$\vec{J} = \vec{F}\Delta t = m\vec{v}_f - m\vec{v}_i = \text{area}_{F\text{-}t\,\text{graph}}$$

You need a method that lets you use the graph to solve for speed. The above formula can be simplified to include speed and the area of a triangle.

$$m\vec{v}_f - m\vec{v}_i = \frac{1}{2}(\text{height} \times \text{base})$$

$$(0.2\ \text{kg})v_f - (0.2\ \text{kg})(0) = \frac{1}{2}(100\ \text{N})(0.1\ \text{s})$$

$$v_f = 25\ \text{m/s}$$

CONSERVATION OF MOMENTUM

When objects interact in a closed system the total momentum of the objects is conserved. This means that the total momentum at the beginning of a problem must equal the total momentum at the end of a problem. This is known as the conservation of momentum. It is most often used to solve problems involving collisions and explosions. Mathematically, conservation of momentum can be expressed as follows:

$$\Sigma \vec{p}_i = \Sigma \vec{p}_f$$
$$\vec{p}_{1i} + \vec{p}_{2i} + \cdots = \vec{p}_{1f} + \vec{p}_{2f} + \cdots$$
$$m_1 \vec{v}_{1i} + m_2 \vec{v}_{2i} + \cdots = m_1 \vec{v}_{1f} + m_2 \vec{v}_{2f} + \cdots$$

ENERGY IN COLLISIONS

Kinetic energy shares the same variables as momentum. Changes in momentum can cause changes in kinetic energy. When it comes to conservation of energy, only total energy is conserved. Kinetic energy is only one of many energies composing total energy, and kinetic energy can change during a problem. Therefore, kinetic energy is not always conserved.

Kinetic Energy Lost

During a collision where no external force acts, the kinetic energy after the collision will either be equal to or less than the kinetic energy before the collision. When kinetic energy decreases during a collision, kinetic energy is said to be lost. Conservation of energy dictates that energy cannot be lost, which means that the lost kinetic energy must have gone somewhere. During a collision, objects contact each other, causing the molecules in the objects to vibrate. The microscopic, and invisible, random motion of molecules is known as thermal energy. If you touch a nail after striking it repeatedly with a hammer, you will feel the increase in temperature. The thermal energy generated during the collision then radiates into the environment as heat. Kinetic energy lost in collisions becomes equal to the heat generated. The equation for kinetic energy during a collision is very similar to the equation for conservation of energy. The main difference is the subtraction of kinetic energy lost, K_{lost}, from the left side of the equation.

$$K_{1i} + K_{2i} - K_{lost} = K_{1f} + K_{2f}$$

$$\frac{1}{2} m_1 v_{1i}^2 + \frac{1}{2} m_2 v_{2i}^2 - K_{lost} = \frac{1}{2} m_1 v_{1f}^2 + \frac{1}{2} m_2 v_{2f}^2$$

Collisions are sorted into two main categories, elastic and inelastic, depending on whether kinetic energy is lost during the collision.

Kinetic energy can increase during a collision if previously stored potential energy is released during the collision. Imagine a compressed spring attached to a cart, and the spring is released during a collision with a second cart. As the spring restores to its equilibrium length it produces an external force that acts on the two-cart system doing work, which increases kinetic energy.

Elastic Collisions

In elastic collisions, objects bounce off of each other. In the process, kinetic energy is conserved. In order for this to occur, the collision must not create any vibrations in the colliding objects, which would imply that the objects never touch each other. Examples may include

IF YOU SEE
An elastic collision

$K_{lost} = 0$
Kinetic energy is conserved.

particles, such as two colliding protons, whose repulsion would prevent them from hitting one another. A larger example would be two objects with a spring mounted on one of them. During the collision, the spring would temporarily store and then release the kinetic energy that would have been lost during the collision. Any energy loss at the spring may be negligible and can be ignored. Very hard objects such as billiard balls or steel spheres will also be nearly elastic with minimal energy loss. For testing purposes, if a problem states that a collision is elastic, then kinetic energy is conserved and there is no kinetic energy lost. Linear momentum is always conserved in any type of collision where NO external forces, such as friction, act.

$$m_1\vec{v}_{1i} + m_2\vec{v}_{2i} = m_1\vec{v}_{1f} + m_2\vec{v}_{2f}$$

IF YOU SEE
An inelastic collision

Kinetic energy is not conserved.

K_{lost} becomes thermal energy.

Inelastic Collisions

If kinetic energy is lost during a collision, then the collision is inelastic. There are two types of inelastic collisions. In the first, the objects may bounce off of each other. In the second, the objects may stick together. When objects stick together, the collision is said to be perfectly (completely, totally) inelastic. The majority of collisions involve an energy loss, making them inelastic collisions.

INELASTIC (ORDINARY) COLLISIONS

In ordinary inelastic collisions, the objects bounce off of one another as they do in elastic collisions. Momentum is conserved as before, but now kinetic energy is lost.

$$m_1\vec{v}_{1i} + m_2\vec{v}_{2i} = m_1\vec{v}_{1f} + m_2\vec{v}_{2f}$$

$$\frac{1}{2}m_1v_{1i}^2 + \frac{1}{2}m_2v_{2i}^2 - K_{lost} = \frac{1}{2}m_1v_{1f}^2 + \frac{1}{2}m_2v_{2f}^2$$

Both elastic and ordinary inelastic collisions involve objects that bounce off of each other. How are these two collisions distinguished on an exam? Conservation of linear momentum is the same for both of these collisions. The only aspect that differs is kinetic energy. When the collision is elastic (kinetic energy conserved), exam questions will specifically state that the collision is elastic or will indicate that kinetic energy loss is negligible. When the collision is inelastic or perfectly inelastic, exam questions may focus on solving the amount of kinetic energy that is lost.

PERFECTLY INELASTIC COLLISIONS

In these collisions, the objects fuse to become one larger combined mass. To combine, they touch and vibrate, resulting in lost kinetic energy. The conservation of momentum and kinetic energy lost equations are simplified slightly to account for the combined final mass with a single velocity.

$$m_1\vec{v}_{1i} + m_2\vec{v}_{2i} = (m_1 + m_2)\vec{v}_f$$

$$\frac{1}{2}m_1v_{1i}^2 + \frac{1}{2}m_2v_{2i}^2 - K_{lost} = \frac{1}{2}(m_1 + m_2)v_f^2$$

EXAMPLE 8.3

Inelastic Collisions

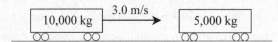

A 10,000-kilogram railroad freight car is moving at 3.0 meters per second when it strikes and couples with a 5,000-kilogram freight car that is initially stationary. What is the resulting speed of the railroad freight cars after the collision?

WHAT'S THE TRICK?

In collisions, momentum is conserved. Since the freight cars combine, this is an inelastic collision.

$$m_1 \vec{v}_{1i} + m_2 \vec{v}_{2i} = (m_1 + m_2)\vec{v}_f$$
$$(10{,}000 \text{ kg})(3.0 \text{ m/s}) + (1{,}000 \text{ kg})(0 \text{ m/s}) = (10{,}000 \text{ kg} + 5{,}000 \text{ kg})\vec{v}_f$$
$$\vec{v}_f = 2.0 \text{ m/s}$$

Explosions

In an explosion, a large mass separates into smaller masses. Aside from events caused by actual explosive devices, there are other examples of explosions: ice skaters pushing each other apart and a rocket launch where the rocket is pushed forward as exhaust gases are pushed backward. Essentially, explosions are the reverse of inelastic collisions.

$$(m_1 + m_2)\vec{v}_i = m_1 \vec{v}_{1f} + m_2 \vec{v}_{2f}$$

Since the initial mass is often stationary, $\vec{v}_i = 0$, this formula can often be simplified for most exam problems.

$$0 = m_1 \vec{v}_{1f} + m_2 \vec{v}_{2f}$$

In addition, the masses move in opposite directions. Therefore, one of the two velocities must be negative, resulting in a negative momentum. When this momentum is moved to the other side of the equation, it becomes positive.

$$0 = m_1(+\vec{v}_{1f}) + m_2(-\vec{v}_{2f})$$
$$m_1 \vec{v}_{1f} = m_2 \vec{v}_{2f}$$

Energy is not lost in an explosion. In order for an explosion to occur, energy stored in the system must be released. In mechanics, this is often accomplished by using a compressed spring between two masses or by having a person throw an object. Positioning a spring between masses and then releasing it converts the potential energy stored in the spring into the total kinetic energy of the now-moving masses.

$$\Delta U_s = K_1 + K_2$$

IF YOU SEE collisions or explosions

Conservation of momentum

Identify collisions or explosions as elastic or inelastic, and use the correct equation.

SUMMARY

1. **MOMENTUM AND IMPULSE ARE VECTOR QUANTITIES.** The direction of these vectors is the same as the direction of the velocity of the object. Even though they have the same units, momentum and impulse are not equal to each other. Rather, impulse equals a change in momentum. The impulse–momentum theorem describes the mathematical connection between these two quantities:

$$\vec{J} = \vec{F}\Delta t = \Delta \vec{p} = m\Delta \vec{v}.$$

2. **IMPULSE CAN BE SOLVED FOR IN A VARIETY OF WAYS.** The different equations solving for impulse can be set equal to each other to explore the relationships among impulse, force, changes in momentum, and their graphical representation.

$$\vec{J} = \vec{F}\Delta t = m\vec{v}_f - m\vec{v}_i = \text{area}_{F\text{-}t \text{ graph}}$$

3. **THE FORCE ACTING ON AN OBJECT AND THE ELAPSED TIME ARE INVERSELY PROPORTIONAL.** Allowing an object to slow down over a longer period of time will reduce the force on the object.

4. **MOMENTUM IS CONSERVED IN COLLISION PROBLEMS.** For elastic and inelastic collisions, the total momentum of the objects before the collision must equal the total momentum after the collision.

5. **KINETIC ENERGY IS CONSERVED DURING ELASTIC COLLISIONS.** For elastic collisions, the kinetic energy after the collision is equal to the kinetic energy before the collision. The kinetic energy lost is equal to zero.

6. **KINETIC ENERGY IS LOST DURING INELASTIC COLLISIONS.** For inelastic collisions, some of the kinetic energy is transformed into thermal energy. As a result, the total energy is conserved. However, kinetic energy is lost during the collision. Perfectly inelastic collisions are those where objects stick together as a result of the collision.

If You See	Try	Keep in Mind
A change in speed	Impulse–momentum theorem $$\vec{J} = \Delta \vec{p} = m\vec{v}_f - m\vec{v}_i$$	Impulse is not equal to momentum. It is equal to a change in momentum, which requires a change in velocity and/or mass.
A need to alter the force acting on an object during a collision	Change the elapsed time of the collision	During a change in momentum, the force acting on an object and the elapsed time are inversely proportional.
Force-time graph	The area under the graphed function $$\vec{J} = \Delta \vec{p} = \text{area}_{F\text{-}t}$$ $$\vec{J} = m\vec{v}_f - m\vec{v}_i = \text{area}_{F\text{-}t}$$	The area solves for impulse and can also be used to solve for final velocity.

If You See	Try	Keep in Mind
A collision or an explosion	Conservation of momentum Elastic and inelastic: $m_1\vec{v}_{1i} + m_2\vec{v}_{2i} = m_1\vec{v}_{1f} + m_2\vec{v}_{2f}$ Perfectly inelastic: $m_1\vec{v}_{1i} + m_2\vec{v}_{2i} = (m_1 + m_2)\vec{v}_f$ Explosion: $(m_1+m_2)\vec{v}_i = m_1\vec{v}_{1f} + m_2\vec{v}_{2f}$	Add plus and minus signs, indicating vector direction, to the velocities. This allows you to solve the equations as scalar quantities.
An elastic collision	$K_{lost} = 0$ $K_i + K_f$	Kinetic energy is conserved.
An inelastic collision or a perfectly inelastic collision	Kinetic energy is lost $K_{lost} = K_i - K_f$	When kinetic energy is lost, it becomes thermal energy.

PRACTICE EXERCISES

1. Which of the following quantities is conserved in perfectly inelastic collisions?

    (A) The velocities of the individual colliding objects
    (B) The linear momentum of each individual object
    (C) Total linear momentum of the system
    (D) The kinetic energy of each individual object
    (E) The total kinetic energy of the system

2. During an inelastic collision, which of the following quantities decreases?

    (A) Linear momentum
    (B) Total energy
    (C) Mass
    (D) Kinetic energy
    (E) Both B and D

3. The magnitude of impulse is equal to:

    I. Fd
    II. $F\Delta t$
    III. mv
    IV. $m\Delta v$

    (A) I only
    (B) II only
    (C) II and III only
    (D) II and IV only
    (E) II, III, and IV

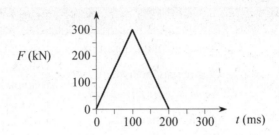

4. The force versus time graph above shows the force acting to stop a 1,500-kilogram car during an accident. Determine the initial speed of the car.

(A) 10 m/s
(B) 15 m/s
(C) 20 m/s
(D) 25 m/s
(E) 30 m/s

5. Two ice skaters, one with a mass of 75 kg and the other with a mass of 50 kg, are initially stationary and standing together on the ice. If they push off of each other, which statement below will be true?

(A) The 50-kg skater receives a larger impulse than the 75-kg skater.
(B) The 50-kg skater receives a smaller impulse than the 75-kg skater.
(C) The 50-kg skater experiences a greater change in momentum.
(D) The 50-kg skater experiences a greater change in velocity.
(E) A, C, and D are all true

6. An 80-kilogram person running at 2.0 meters per second jumps onto a 20-kilogram stationary cart. What is the resulting speed of the combined cart and person?

(A) 1.2 m/s
(B) 1.6 m/s
(C) 1.8 m/s
(D) 2.0 m/s
(E) 2.4 m/s

7. Five objects have the same initial speed of 10 meters per second. After striking a wall, some of them are stopped and some of them bounce back with varying speeds. Which object experiences the greatest change in momentum?

(A) A 1.0-kilogram mass stopping
(B) A 1.0-kilogram mass bouncing back at 10 meters per second
(C) A 2.0-kilogram mass stopping
(D) A 2.0-kilogram mass bouncing back at 5.0 meters per second
(E) A 2.0-kilogram mass bouncing back at 10 meters per second

ANSWERS EXPLAINED

		Key Words	Needed for Solution	Now Solve It
1.	**(C)**	Conserved; perfectly elastic collision	Knowledge/definitions	As long as no external forces act, total linear momentum is conserved in all collisions and explosions. Velocity is not a conserved quantity, eliminating answer A. The individual momentums of objects may change during collisions, which eliminates answer B. In any type of inelastic collision, kinetic energy is not conserved, eliminating answers D and E.
2.	**(D)**	Inelastic collision; quantities decrease	Knowledge/definitions	Linear momentum, total energy, and mass are always conserved in collisions. However, kinetic energy is not conserved in inelastic collisions. During these collisions, some kinetic energy is lost. This energy becomes thermal energy. This conversion to another form of energy keeps the total energy constant.
3.	**(D)**	Impulse is equal to	Knowledge/definitions	Answer I is the expression for work, not impulse. Answer III is not correct because impulse does not equal momentum. Impulse is equal to both $\vec{F}\Delta t$ and a change in momentum, $m\Delta\vec{v}$.
4.	**(C)**	Force versus time graph; initial speed	Graphing knowledge and the impulse–momentum theorem	$$\vec{J} = m\vec{v}_f - m\vec{v}_i = \vec{F}\Delta t = \text{area}_{F\text{-}t\text{ graph}}$$ Use the two expressions on the right. $$m\vec{v}_f - m\vec{v}_i = \frac{1}{2}(\text{height} \times \text{base})$$ $$(1{,}500\text{ kg})(0) - (1{,}500\text{ kg})\,v_i = \frac{1}{2}(300\text{ kN})(200\text{ ms})$$ $$v_i = 20\text{ m/s}$$ **Note:** The conversion factors for kilonewtons and milliseconds cancel each other. These units are frequently encountered in force versus time graphs.
5.	**(D)**	Ice skaters; initially stationary; push off of each other	Several possible solutions exist involving Newton's third law, the impulse–momentum theorem, and the conservation of momentum in an explosion	The easiest solution involves seeing the separating skaters as an explosion starting from rest. $$m_1 v_{1f} = m_2 v_{2f}$$ The skater with less mass must have a greater velocity in order to have equal and opposite momentum. Answers A, B, and C cannot be correct. Newton's third law dictates that the force on each skater will be identical. If the force is the same, then the impulse and the change in momentum of each skater is the same.

	Key Words	Needed for Solution	Now Solve It
6. **(B)**	Person running; stationary cart; resulting speed of combined cart and person	Conservation of momentum in an inelastic collision	$m_{\text{person}}\vec{v}_{\text{person}} + m_{\text{cart}}\vec{v}_{\text{cart}}$ $= (m_{\text{person}} + m_{\text{cart}})\vec{v}_{\text{person-cart}}$ $(80\text{ kg})(2.0\text{ m/s}) + (20\text{ kg})(0\text{ m/s})$ $= (80\text{ kg} + 20\text{ kg})\vec{v}_{\text{person-cart}}$ $\vec{v}_{\text{person-cart}} = 1.6\text{ m/s}$
7. **(E)**	Striking a wall; some of them are stopped; some of them bounce back; greatest change in momentum	Change in momentum $\Delta\vec{p} = m\Delta\vec{v}$ $\Delta\vec{p} = m(\vec{v}_f - \vec{v}_i)$	If the initial velocity vector is regarded as positive, then an object that reverses direction will have a final velocity vector that is negative. Whether the object is stopped or reverses direction the vector describing the change in momentum will calculate as a negative value. The negative sign indicates that the change in momentum is directed opposite the initial velocity, which is consistent with the force acting opposite the motion of the object. The magnitude of the change in momentum vector is the absolute value of the quantity calculated.

Choice	Change in Momentum
A	$\Delta p = 1.0\,[(0) - (10)] = \lvert -10\text{ kg} \cdot \text{m/s} \rvert$
B	$\Delta p = 1.0\,[(-10) - (10)] = \lvert -20\text{ kg} \cdot \text{m/s} \rvert$
C	$\Delta p = 2.0\,[(0) - (10)] = \lvert -20\text{ kg} \cdot \text{m/s} \rvert$
D	$\Delta p = 2.0\,[(-5) - (10)] = \lvert -10\text{ kg} \cdot \text{m/s} \rvert$
E	$\Delta p = 2.0\,[(-10) - (10)] = \lvert -40\text{ kg} \cdot \text{m/s} \rvert$

Objects that bounce have a greater change in momentum. The object with the largest combination of mass and change in velocity will have the greatest change in momentum.

Gravity

9

→ **UNIVERSAL GRAVITY**

→ **GRAVITATIONAL FIELD**

→ **CIRCULAR ORBITS**

→ **KEPLER'S LAWS**

On Earth, the gravity at the planet's surface has a constant value of 10 meters per second squared in the downward direction. It can be considered to be a constant field over the surface of Earth with minor fluctuations due to elevation.

The actual value of Earth's surface gravitational field is the result of the mass of Earth and the distance of the surface from the center of the planet. Although Earth has varying surface elevations, these are actually relatively minor. So, it is acceptable to use the constant value of 10 meters per second squared for all calculations.

Gravity is an important concept and links together ideas from linear motion as well as circular motion. This chapter will review the following concepts:

- Understand Newton's law of universal gravitation and its inverse-square-law relationship.
- Visualize the gravity field surrounding masses, such as planets, and calculate its value at specific points in space.
- Calculate circular orbits governed by Newton's law of universal gravitation.
- Understand how Kepler's laws describe orbital motion.

Table 9.1 lists the variables that will be studied in this chapter.

Table 9.1 Variables Used with Gravity

New Variables	Units
F_g = Force of gravity	N (newtons)
G = universal gravitational constant	m^3/kg • s^2 (meters cubed per kilogram seconds squared) or $\dfrac{\text{N} \cdot \text{m}^2}{\text{kg}^2}$ (newtons • meters squared per kilograms squared)

IF YOU SEE
two planets
separated by
a distance

Newton's law
of universal
gravitation

$$F_g = G\frac{mM}{r^2}$$

The force
acting on
both planets
is opposite
and equal
(Newton's
third law).

UNIVERSAL GRAVITY

Isaac Newton determined that the **force of gravitational attraction**, F_g, between two masses was directly proportional to the product of their masses and inversely proportional to the square of the distance between them. **Newton's universal law of gravitation** can be expressed as an equation.

$$F_g = G\frac{m_1 m_2}{r^2}$$

The letter G represents the universal gravitational constant, $G = 6.67 \times 10^{-11}$ m^3/kg • s^2.

Inverse-Square Law

As the distance between two masses increases, the force of gravity between them decreases by the square of that distance. This means that a doubling of distance would result in a quartering of the gravitational force between the masses.

EXAMPLE 9.1

Calculating the Change in the Force of Gravity

Calculate the resulting force of gravitational attraction between two masses if one of the masses was to double and the distance between them were to triple.

WHAT'S THE TRICK?

The original force of attraction can be calculated using Newton's law of gravity.

$$F_g = G\frac{m_1 m_2}{r^2}$$

Substituting $2m_1$ for m_1 and $(3r)^2$ for r^2 represents the doubling of the mass and the tripling of the distance, respectively.

$$\left(\frac{2}{3^2}\right)F_g = G\frac{(2m_1)m_2}{(3r)^2}$$

Note that the quantity $3r$ is squared.

$$\left(\frac{2}{9}\right)F_g = \left(\frac{2}{9}\right)G\frac{m_1 m_2}{r^2}$$

The force of gravity between the masses will be two-ninths of its original value.

GRAVITATIONAL FIELD

The influence, or alteration, of space surrounding a mass is known as a **gravitational field**. The gravitational field of Earth can be visually represented as several vector arrows pointing toward the surface of Earth, labeled g, as shown in Figure 9.1. The surface gravity field for Earth, g, is 10 meters per second squared. When an object of mass m is placed in Earth's gravity field, it experiences a force of gravity, F_g, in the same direction as the gravity field.

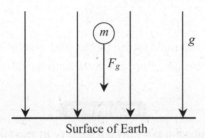

Surface of Earth

Figure 9.1 The gravitational field of Earth

The effect on mass m in the gravity field can be solved as a force problem.

$$\Sigma F = F_g$$
$$ma = mg$$
$$a = g$$

It now becomes apparent that the acceleration of a mass in a gravity field equals the magnitude of the gravity field. Although there is a conceptual difference between the gravity field and the acceleration of gravity, they both have the same magnitude and direction.

Finding Surface Gravity

The gravity field of any planet is a function of the mass of the planet and the distance of the planet's surface from its center. The exact relationship can be derived using two formulas for the force of gravity. When a mass m rests on the surface of Earth, the force of gravity can be determined using either the weight formula or Newton's law of gravitation.

$$F_g = mg \qquad \text{or} \qquad F_g = G\frac{mM_{\text{Earth}}}{r_{\text{Earth}}^2}$$

When these formulas are set equal to one another, mass m cancels. This results in a formula that describes both the strength of the gravity field and the acceleration due to gravity. Note that the mass of Earth is 5.98×10^{24} kg and that the radius is 6.37×10^6 m.

$$mg = G\frac{mM_{\text{Earth}}}{r_{\text{Earth}}^2}$$

$$g = G\frac{M_{\text{Earth}}}{r_{\text{Earth}}^2}$$

$$g = (6.67 \times 10^{-11} \text{ N} \cdot \text{m}^2/\text{kg}^2)\,(5.98 \times 10^{24} \text{ kg})/(6.37 \times 10^6 \text{ m})^2$$
$$g = 9.8 \text{ m/s}^2$$

IF YOU SEE
a question
regarding
the gravity
of a planet

$$g = G\frac{M}{r^2}$$

Remember
that r is the
distance from
the center of
the planet.
To find the
gravity at the
surface of a
planet, r is the
radius of
the planet.

The gravity field near the surface of Earth is considered uniform. It has the same value at all points close to the surface of Earth. The magnitude of mass m placed on the surface of Earth does not matter since it cancels. Elephants and feathers are both in the same gravitational field and experience the same acceleration of 9.8 m/s^2. For the SAT Subject Test in Physics, the value for g will be rounded to 10 m/s^2 in order to make calculations easier.

Although the formula for g was derived on the surface of Earth, it can be generalized to solve for gravity on any planet or at a point in space near any planet.

$$g = G\frac{M}{r^2}$$

In its generalized form, M is the mass of the planet and r is the distance measured from the center of the planet to the location where gravity, g, is to be calculated.

EXAMPLE 9.2

Calculating the Surface Gravity of the Moon

Calculate the surface gravity of the Moon. The Moon has a mass of 7.36×10^{22} kilograms and a radius of 1.74×10^6 meters.

WHAT'S THE TRICK?

Use the formula for finding the gravity of a planet and substitute in the values for the Moon.

$$g_{Moon} = G\frac{M_{Moon}}{r^2_{Moon}}$$

$$g_{Moon} = 1.62 \text{ m/s}^2$$

This value is roughly one-sixth of the value of the surface gravity of Earth. The SAT Subject Test in Physics does not allow you to use a calculator, and it is unlikely to present a problem requiring calculations that will be this involved. This example serves to demonstrate the universality of the equation for finding the gravitational field and the acceleration of gravity for any celestial object.

CIRCULAR ORBITS

When a ball is thrown horizontally on Earth, it will follow a parabolic path toward the ground. During its flight, the ball simultaneously experiences a constant downward acceleration and a constant forward velocity. To an observer, it appears that the ball is moving in a parabola relative to a flat Earth.

Newton hypothesized that if a ball could be thrown with sufficient forward velocity, it would travel so quickly that the acceleration pulling it downward would not bring it to Earth. This is because the spherical Earth would curve out of the ball's way as it fell. Today, satellites in orbit are able to do this with speeds exceeding 7,900 meters per second (17,500 miles per hour).

As discussed in Chapter 6, "Circular Motion," objects experiencing an acceleration perpendicular to their motion will move in a circle. The magnitude of their acceleration remains constant. However, their direction is constantly changing so that it always points toward the center of rotation.

Calculating Tangential Orbital Velocities

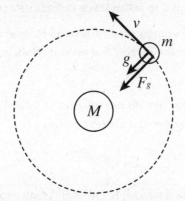

Figure 9.2 Satellite of mass *m* orbiting central body *M*

Equations for gravity and circular motion can be combined to determine the velocity of a satellite in a circular orbit. In Chapter 7, the following circular-motion equations for centripetal acceleration and centripetal force were introduced.

$$a_c = \frac{v^2}{r} \qquad F_c = m\frac{v^2}{r}$$

As seen in Figure 9.2, the acceleration of gravity and the force of gravity are always directed toward the center of the circular path followed by orbiting mass *m*. They are the centripetal acceleration and force on satellites in circular orbits.

$$a_c = g \qquad F_c = F_g$$

$$\frac{v^2}{r} = G\frac{m}{r^2} \qquad m\frac{v^2}{r} = G\frac{mM}{r^2}$$

Rearranging and simplifying two of these equations will solve for orbital speed *v*.

$$v_{\text{orbit}} = \sqrt{\frac{GM}{r}}$$

In the above equation, mass *M* is the mass of a planet or star at the center of an orbiting satellite. Note that the formula does not contain *m*, the mass of the orbiting satellite itself. The speed of a satellite does not depend on its own mass. The satellite's speed depends on the mass of the larger body it is orbiting and on the distance, *r*, from the center of the larger body. There apparently is an inverse relationship between speed and distance. Satellites closer to the central body orbit at higher speeds.

IF YOU SEE
orbital speed
or velocity

$$v_{\text{orbit}} = \sqrt{\frac{GM}{r}}$$

Speed depends on only the central mass, *M*, and the distance from it. Objects closer to the central mass have the greatest speeds.

EXAMPLE 9.3

Calculating a Satellite's Orbital Velocity

Calculate the speed of an orbiting satellite around Mars at a distance of 3.57×10^6 meters above the center of the planet. The mass of Mars is 6.42×10^{23} kilograms.

WHAT'S THE TRICK?

Use the general equation for finding the velocity of a satellite above a planet of known mass M at radius r above the planet's center.

$$v_{orbit} = \sqrt{\frac{GM}{r}}$$

$$v_{orbit} = \sqrt{\frac{\left(6.67 \times 10^{-11} \text{ N} \cdot \text{m}^2/\text{kg}^2\right)\left(6.42 \times 10^{23} \text{ kg}\right)}{\left(3.57 \times 10^6 \text{ m}\right)}}$$

$$v_{orbit} = 3.46 \times 10^3 \text{ m/s}$$

As with Example 9.2, you should note that the mathematical complexity of this example will most likely not be on the SAT Subject Exam in Physics. Rather, you will more likely be tested on your understanding of the equation and how the orbital speed will be affected by changing the variables. Look for such examples in the questions at the end of this chapter and on the practice exams.

Understanding the formulas associated with orbital motion will help you answer many conceptual questions concerning changing variables. Consider the key formulas discussed thus far:

$$F_G = G\frac{m_1 m_2}{r^2} \qquad \text{and} \qquad v_{orbit} = \sqrt{\frac{GM}{r}}$$

The SAT Subject Test in Physics will most likely ask questions based upon altering one or more of these variables and how that would affect orbital speed or the force of gravity. Actual numerical values will probably not be required since calculators are not allowed during the examination. For that reason, you should commit these fundamental gravity and orbital formulas to memory.

EXAMPLE 9.4

Determining the Effect of Changing Radius on an Orbiting Satellite

A satellite orbiting at a speed of *v* and a radius of *r* above the center of a planet climbs to a radius 2*r*. What is the satellite's new orbital speed?

(WHAT'S THE TRICK?)

Orbital speed is determined by the following formula.

$$v_{orbit} = \sqrt{\frac{GM}{r}}$$

Substituting 2*r* for *r* represents doubling the orbital distance when the satellite moves from *r* to 2*r*. The right side of the equation shown below is now multiplied by $\sqrt{1/2}$. In order to maintain the equality, the left side of the equation needs to be multiplied by the same factor.

$$\sqrt{\frac{1}{2}}\, v_{orbit} = \sqrt{\frac{GM}{2r}}$$

Doubling the orbital radius decreases the orbital velocity to $\sqrt{1/2}\, v$.

KEPLER'S LAWS

Seventeenth-century astronomer Johannes Kepler deduced three laws describing the motion of planets around the Sun.

1. Planets orbit the Sun along an elliptical path where the Sun is at one of the two foci of the ellipse, as shown in Figure 9.3.
2. A line drawn from the Sun to a planet would sweep out an equal area during an equal interval of time. In Figure 9.3, area$_1$ = area$_2$.

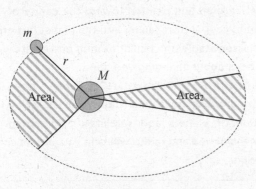

Figure 9.3 Kepler's laws

3. The square of the period of a planet's orbit is proportional to the cube of its radius in its orbit about the Sun.

These laws were deduced from Kepler's analysis of observations made by Tycho Brahe, a Danish astronomer and writer. They were considered controversial by several of Kepler's contemporary astronomers. The first law was particularly controversial. Most believed that planets orbited in perfect circles around the Sun.

IF YOU SEE an elliptical orbit

Kepler's laws

1. The central massive body is at one of the two foci.

2. The orbiting body moves faster when it is nearer the central body.

3. $T^2 \propto r^3$

Kepler's second law illustrates that when a planet is closer to the Sun (known as *perihelion*), it will move at a faster orbital speed than when it is at its farthest point from the Sun (known as *aphelion*). This will allow the imaginary line between Sun and planet to sweep out an equal area during an equal interval of time.

The third law relates a mathematical relationship between orbital period and orbital radius.

$$T^2 \propto r^3$$

Committing this relationship to memory will help you answer conceptual questions regarding the effect of changing the radius. This relationship works for all objects orbiting a common central mass.

SUMMARY

1. **NEWTON'S UNIVERSAL LAW OF GRAVITATION** describes the force of gravitational attraction between two objects based on their masses and the distance between them. The force is directly proportional to the masses and inversely proportional to the square of the distance between the masses.

2. **A SURFACE GRAVITATIONAL FIELD IS A RESULT OF THE MASS OF A PLANET AND THE DISTANCE OF THE SURFACE FROM ITS CENTER.** The direction of the field points toward the planet. A mass in that field would experience a force in the direction of the field. The gravity at any location in space near a celestial body (star, planet, and so on) can be determined using the formula $g = GM/r^2$.

3. **CIRCULAR ORBITS ARE GOVERNED BY NEWTON'S LAW OF UNIVERSAL GRAVITATION.** An orbiting satellite (or planet) has a tangential orbital speed as well as a constant acceleration pointed toward the center of its orbit. Its orbital speed is sufficient to keep it from falling into the planet it is orbiting. Memorizing the key formulas associated with circular motion and orbits will help you answer conceptual questions about changing variables.

4. **KEPLER'S LAWS DESCRIBE ORBITAL MOTION.** Kepler's three laws deduced from observation were controversial in their day. However, today they help describe the orbital motion of planets and satellites. Memorizing the mathematical relationship between period and radius will help you answer conceptual questions about orbital periods.

If You See	Try	Keep in Mind
Two planets separated by a distance	Newton's law of universal gravitation $$F_g = G\frac{mM}{r^2}$$	This calculates the forces acting on both planets, which are equal and opposite (Newton's third law).
A location where gravity, g, is requested or an object moves in a manner that changes the value of g	$$g = G\frac{M}{r^2}$$ Remember that m is the mass of any planet and r is the distance from the center of the planet to the point where gravity is to be determined.	Note the inverse-square relationship between the distance r and gravity g.
Factors affecting the orbital speed of orbiting satellites	$$v_{orbit} = \sqrt{\frac{GM}{r}}$$	This general formula can determine the velocity of any satellite at a distance r away from a planet of mass M.
An elliptical orbit	Kepler's laws, including $$T^2 \propto r^3$$	The speed of the body varies. It moves fastest when closest to the central body.

PRACTICE EXERCISES

1. A planet with twice the mass of Earth and a radius three times that of Earth would have a surface gravity of

    (A) $\quad g$

    (B) $\frac{2}{9}g$

    (C) $\frac{2}{3}g$

    (D) $\frac{3}{2}g$

    (E) $\frac{9}{2}g$

2. A satellite of mass m orbits Earth at an orbital radius r and with a speed v. What would the speed be for a satellite of mass $5m$ at the same orbital radius r?

    (A) $\quad v$

    (B) $\sqrt{5}\,v$

    (C) $\quad 5v$

    (D) $25v$

    (E) $\frac{1}{5}v$

3. A satellite orbits Earth at an orbital radius, r, with a period, T. If the satellite were to increase its orbital radius to $2r$, what would be its resulting period?

(A) $\frac{1}{2}T$

(B) $\sqrt{2}\,T$

(C) $2T$

(D) $\sqrt{8}\,T$

(E) $8T$

4. A satellite of mass m orbits Earth at an orbital radius r and a speed v. If the satellite were to drop down to an orbit of $\frac{1}{2}r$, what would be its resulting speed?

(A) $\frac{1}{2}v$

(B) $\frac{1}{\sqrt{2}}v$

(C) $\sqrt{2}\,v$

(D) $2v$

(E) $4v$

5. A satellite orbits Earth with an orbital radius of $2r_{Earth}$. What is the acceleration due to gravity on this satellite toward Earth, in terms of Earth's surface gravity, g?

(A) $\frac{1}{8}g$

(B) $\frac{1}{4}g$

(C) $\frac{1}{2}g$

(D) g

(E) $2g$

6. The gravitational force of attraction between two masses is 16 newtons. If the distance between the masses is quadrupled, what is the resulting force of attraction?

(A) 1 N

(B) 2 N

(C) 4 N

(D) 8 N

(E) 16 N

7. Which of Kepler's laws states that planets orbit the Sun in an ellipse?

(A) first law only

(B) second law only

(C) third law only

(D) first law and second law

(E) first law and third law

8. Which of the planets below would have the same free-fall acceleration on their surfaces?

	Mass	Radius
I.	M	$2R$
II.	$8M$	$2R$
III.	$2M$	R
IV.	M	R

(A) I and II
(B) II and III
(C) III and IV
(D) II and IV
(E) I and III

9. According to Kepler's second law, what quantity is swept out in an equal interval of time during a planet's orbit about the Sun?

(A) Length
(B) Volume
(C) Distance
(D) Area
(E) Displacement

10. A newly discovered planet, Planet X, has twice the mass of Earth. Planet X is orbited by a moon, which has twice the mass of Earth's moon. The orbital radius of the moon orbiting Planet X is the same as the orbital radius of Earth's moon. How does the orbital speed of the moon orbiting Planet X compare to the orbital speed of Earth's moon?

(A) It orbits at the same speed as Earth's moon.

(B) It orbits at $\frac{1}{2}$ the speed of Earth's moon.

(C) It orbits at 2 times the speed of Earth's moon.

(D) It orbits at $\sqrt{2}$ the speed of Earth's moon.

(E) It orbits at $\frac{\sqrt{2}}{2}$ the speed of Earth's moon.

ANSWERS EXPLAINED

	Key Words	Needed for Solution	Now Solve It
1. **(B)**	Twice the mass; a radius three times; gravity	Gravity formula $$g = G\frac{M}{r^2}$$	The gravity formula would solve for the gravity of Earth if the mass and radius of Earth were used. $$g_{\text{Earth}} = G\frac{M_{\text{Earth}}}{r_{\text{Earth}}^2}$$ Doubling the mass and tripling the radius will solve for gravity on a planet that is larger than Earth by those factors. $$\frac{2}{9}g = G\frac{2M_{\text{Earth}}}{\left(3r_{\text{Earth}}\right)^2}$$ The left side of the equation must be multiplied by $\frac{2}{9}$ in order to maintain the equality.
2. **(A)**	Mass; speed; orbital radius	Orbital-speed formula $$v_{\text{orbit}} = \sqrt{\frac{GM}{r}}$$	The mass of an orbiting body is not part of the orbital-speed formula. This means that the masses m and $5m$ are distracters and should be ignored. The formula indicates that orbital speed depends on the central mass M, in this case Earth, and the orbital radius r, as measured from Earth's center. Both satellites are orbiting the same mass M and both are at the same orbital radius r. This means both satellites will have the same speed v.
3. **(D)**	Orbital radius; period	Kepler's third law $$T^2 \propto r^3$$	Although the formula is incomplete, any changes to one side must be matched on the other side to maintain proportionality. $$(?T)^2 \propto (2r)^3$$ The right side has to be multiplied by 2^3 or 8. $$(?T)^2 \propto 8r^3$$ A number that when squared is equal to 8 is needed, which is $\sqrt{8}$.
4. **(C)**	Mass; orbital radius; speed	Orbital-speed formula $$v_{\text{orbit}} = \sqrt{\frac{GM}{r}}$$	If r is multiplied by $\frac{1}{2}$, then the left side of the equation must be multiplied by $\sqrt{2}$. $$\sqrt{\frac{1}{1/2}}\, v_{\text{orbit}} = \sqrt{\frac{GM}{\frac{1}{2}r}}$$ $$\sqrt{2}\, v_{\text{orbit}} = \sqrt{\frac{GM}{\frac{1}{2}r}}$$

	Key Words	Needed for Solution	Now Solve It
5. **(B)**	Orbital radius of $2r_{Earth}$	Inverse square law and/or gravity formula $$g = G\frac{M}{r^2}$$	The inverse square of 2 is $\frac{1}{4}$. The gravity at twice the orbital distance is $\frac{1}{4}$ of its value on Earth. Analyzing the equation yields the same answer. $$\frac{1}{4}g = G\frac{M}{(2r)^2}$$
6. **(A)**	Gravitational force; 16 newtons; distance between the masses is quadrupled	Inverse-square law and/or Newton's law of gravity $$F_g = G\frac{m_1 m_2}{r^2}$$	The inverse square of 4 is $\frac{1}{16}$, and $\frac{1}{16}$ of 16 newtons is 1 newton. Analyzing the equation yields the same answer. $$\left(\frac{1}{4}\right)16 = G\frac{m_1 m_2}{(4r)^2}$$
7. **(A)**	Kepler's laws	Knowledge/definitions	Kepler's first law describes the orbital path of planets as ellipses.
8. **(B)**	Planets; acceleration on their surfaces	Gravity formula $$g = G\frac{M}{r^2}$$	Both planet II and planet III have a g_{planet} of $2g$. $$\text{II: } 2g = G\frac{8M}{(2r)^2}$$ $$\text{III: } 2g = G\frac{2M}{r^2}$$
9. **(D)**	Kepler's second law; swept; equal interval of time	Kepler's laws	Kepler's second law states that a line connecting the Sun and a planet will sweep out equal areas in equal intervals of time.
10. **(D)**	Planet X has twice the mass of Earth; the orbital radius of the moon orbiting Planet X is the same as the orbital radius of Earth's moon	Orbital-speed formula $$v_{orbit} = \sqrt{\frac{GM}{r}}$$ The orbital speed is proportional to the square root of the central body's mass	Orbital speed depends on the central mass, M (Planet X), and the orbital radius, r. Planet X has twice the mass of Earth, but the orbital radius of its moon is the same as Earth's moon. $$\sqrt{2}\,v_{orbit} = \sqrt{\frac{G(2M)}{r}}$$ If mass M is doubled, then the left side of the equation must be multiplied by $\sqrt{2}$ to maintain equality.

Electric Fields

10

→ **CHARGE**

→ **ELECTRIC FIELDS**

→ **UNIFORM ELECTRIC FIELDS**

→ **ELECTRIC FIELDS OF POINT CHARGES**

Charge is a property primarily associated with electrons and protons. Charged objects create an electric field, which in many ways is similar to the gravity field surrounding masses. While gravity fields create a force on objects with mass, the electric field of one charged object creates a force that acts on other charged objects. Visual representations and the mathematical equations for gravity fields and electric fields are nearly identical. However, while gravity fields cause objects to attract each other, electric fields can cause charged objects to attract or repel one another. To gain a better understanding of electric fields, this chapter will do the following:

- Examine the properties of charge.
- Explore electric fields caused by charged plates and point sources.
- Draw, analyze, and solve problems for uniform electric fields.
- Draw, analyze, and solve problems for electric fields of point charges.

Table 10.1 lists the variables that will be discussed in this chapter.

Table 10.1 Variables Involved with Electric Fields

New Variables	Units
q = charge (small point charge) Q = charge (larger charge, charge on plates)	C (coulombs)
e = Charge of an electron	C (coulombs)
$\vec{E}$ = Electric field	N/C (newtons per coulomb) or V/m (volts per meter)
$\vec{F}_E$ = Electric force	N (newtons)
k = Coulomb's law constant or electrostatic constant	N • m^2/C^2 (newtons • meters squared per coulombs squared)

CHARGE

Charge is characteristic of electrons and protons. It is associated with a variety of electric properties. The magnitude of charge on an electron and on a proton is the fundamental value of charge, $e = 1.6 \times 10^{-19}$ coulombs, where coulombs (C) is the unit of charge. The charge on a proton is positive, while the charge on an electron is negative. When grouped together, the charge on an equal number of protons and electrons will cancel each other. However, if an object contains more protons than electrons or more electrons than protons, the object will have a net charge, q or Q. The variable q is typically used for a small charge, such as a point charge. The variable Q is typically used for a larger charge, such as the charge on plates. Charge, like matter, is conserved. Although the amount of charge remains constant it can move to another location or to another object. Although masses only attract one another, charges are capable of attracting and repelling each other. Opposite charges attract one another, while like charges repel.

Charged Objects

In addition to electrons and protons, there are other charged objects. If the number of electrons and protons in an atom are not equal, the atom has a net charge and is known as an **ion**. For example, the sodium ion, Na^+, has one less electron than a neutral sodium atom. The oxygen ion, O^{2-}, has two extra electrons compared to a neutral oxygen atom. Everyday objects can also contain excess charge.

Charged objects are usually split into two categories in beginning physics: point charges and charged plates. Point charges are spherical in nature and include electrons, protons, ions, and charged spheres. Regardless of the size of a charged sphere, the entire charge can be assumed to be located at a point in the center of the sphere. This includes hollow spheres and solid spheres. Charged plates consist of two metal plates, which are typically flat, parallel, of equal size, and separated by a distance. The two main types of charged objects create very different electrical effects and employ different equations. The sections that follow will compare and contrast these important charged objects.

Charge is always found in exact quantities. All charges are made up of whole numbers of either electrons or protons. Since the charge on each electron or proton is 1.6×10^{-19} coulombs, 6.25×10^{18} electrons or protons total to 1 coulomb of charge.

A neutral object does not mean the absence of charge. All objects are composed of atoms, which contain electrons and protons. Therefore, all objects contain charge. Neutral objects merely contain the exact same number of electrons and protons, and these opposite charges cancel each other.

Conservation of Charge

Although the individual charge on any one object in a problem may vary, the total charge of all the objects will remain constant. Charges can move from one object to another or can flow through a circuit. However, the total charge of a system (all the objects under examination) at the start of a problem will equal the total charge of the system at the end of a problem. In other words, total charge is conserved.

Charging

Charging an object involves moving extra charges onto or off of the object. How this is accomplished depends on whether the substance to be charged is a **conductor** or an **insulator**. A conductor is a substance that holds its electrons loosely. This allows the electrons to move freely throughout the conductor. The best examples of conductors are metals, which are used as wires in electrical circuits to transport electrons. Insulators are substances that hold their electrons tightly. As a result, insulators prevent the motion of charges. Plastics, which are often used to insulate people from electric shock, are a good example of insulators.

Both conductors and insulators can be charged. When a conductor is charged, the excess charges can move throughout the conductor and readily distribute over the entire outer surface of the conductor. When an insulator is charged, the charges stay in the spot where they have been placed.

All charging methods involve the transfer of electrons. Objects that gain electrons acquire a net negative charge, and objects that lose electrons acquire a net positive charge. **Charging by friction** involves rubbing two nonmetallic objects, such as a wool cloth and a plastic rod, against each other. This method can be used to deposit charge on an insulator. Conductors, such as metals, can be charged by conduction or induction. **Conduction** requires two conductors to touch each other. If one conductor has excess mobile electrons, then some of them will transfer to the other conductor. Charging by **induction** is done without physically touching the object that will be charged. This is best explained in Example 10.1.

EXAMPLE 10.1

Charging by Induction

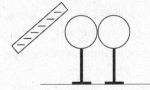

Two neutral, uncharged conducting spheres mounted on insulating stands are in contact with each other. A negatively charged rod is brought near the spheres but does not touch either sphere. How can the spheres both be charged, and what will the sign of the charge be on each sphere?

WHAT'S THE TRICK?

Although neutral, the two spheres still contain electrons and protons. Diagrams normally just show excess charges, like those in the rod. The negative charges in the rod will repel the negative charges in the spheres. This will cause the electrons in the spheres to move to the right, making the right sphere negative. The sphere to the left, which now has fewer electrons, will have a positive charge. If the two spheres are now physically separated, each becomes charged. These spheres become charged without coming into contact with the negative rod, which is induction.

 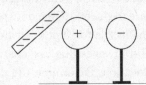

Note that this example also demonstrates conservation of charge. The original charge on the two spheres is zero since the spheres were initially neutral. Although the spheres each become individually charged, the resulting charges are equal and opposite. When added together, the total charge on the system (both spheres) is zero.

ELECTRIC FIELDS

Charged objects are surrounded by an electric field, E. The electric field is similar to the gravity field surrounding masses. Like gravity, the electric field is a vector quantity having both magnitude and direction. The electric field of a charged object creates an electric force, F_E, on other charged objects located in the field, just as the gravity field of a mass creates a force on other objects with mass.

Visual representations and the mathematical equations for electric fields and gravity fields are nearly identical. However, charges and their surrounding electric fields vary from mass and gravity fields in some unique ways. Gravity fields always point toward the mass responsible for the field. However, electric fields can point either toward or away from the charge depending on the sign of the charge. While gravity fields cause objects to only attract each other, electric fields can cause charged objects to attract or to repel one another. These differences make electric fields a little more complicated than gravity fields. The direction of the electric field and its effect on positive and negative charges is extremely important.

The SAT Subject Test in Physics involves two common electric-field configurations: uniform electric field between charged plates and the electric field surrounding spherical point charges. Each field type has its own set of equations and unique problem sets.

UNIFORM ELECTRIC FIELDS

Uniform electric fields exist between two parallel plates containing equal but opposite charges. These fields are considered to be uniform since both the magnitude and the direction of the field are the same at all points between the plates.

Visualizing Uniform Fields

Figure 10.1 compares a uniform electric field and a uniform gravity field.

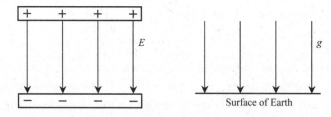

Figure 10.1 Electric field of charged plates (left) compared to gravity field (right)

Whereas gravity fields always point toward a mass, such as Earth, electric fields are a bit more complicated due to the existence of two types of charge. An electric field points in the same direction as an electric force points when it is acting on a positive charge. To find the electric field at a point in space due to a charge or to a group of charges, imagine a positive test charge at that location. Determine the direction of force on an imaginary test charge placed in that location. This will be the same as the direction of the electric field. This means that electric fields point away from positive charges and toward negative charges. A uniform field is drawn with parallel and equally spaced arrows. Often diagrams on exams consist of the arrows only, and the plates responsible for the electric field are not shown.

Magnitude of Uniform Electric Fields

Often the magnitude of the electric field, E, is given in a problem. Unlike the known value for the gravity field of Earth, $g = 9.8$ m/s^2, the electric field is unique to the plates used in each problem. Although the electric-field strength, E, may be given in a problem, you may also be required to calculate it using the equations that appear in the following sections. The units of the electric field are newtons per coulomb (N/C). The units of the gravity field will most likely be reported in meters per second squared (m/s^2). However, when analyzed as a field rather than as acceleration, the units for the gravity field can be reported as newtons per kilogram (N/kg).

Electric Force in Uniform Electric Fields

If a charge, q, such as a proton or an electron, is placed into a uniform electric field, it will experience an electric force, F_E. This is very similar to placing a mass, m, into the gravity field of Earth. See Figure 10.2.

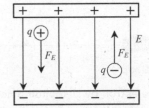 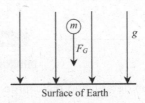

Figure 10.2 Force due to uniform fields

The magnitude of force can be determined using the following equations:

Electricity	Gravity
$F_E = qE$	$F_G = mg$

A force is an interaction between an object and an agent. The charged plates are the agent creating an electric field, E. Charge q is the object that experiences a force, F_E. The formula for the magnitude of the electric force is not dependent on whether charge q is positive or negative.

Figure 10.2 clearly shows that the direction of the electric force, F_E, is dependent on the sign of the charge located in the field. The force acting on a positive charge, such as a proton, will be in the direction of the electric field. However, the electric force acting on a negative charge is opposite the field.

Kinematics in Uniform Electric Fields

The sum of all forces acting on an object will determine its resulting motion. The electric force is merely another force. Solve problems involving electric force in the same manner as you solve all other problems involving forces.

1. Orient the problem.
2. Determine the type of motion.
3. Sum the force vectors in the relevant direction.
4. Substitute and solve.

IF YOU SEE
a charge
in a uniform
electric field

$F_E = qE$

The force on a positive charge is in the direction of the electric field, while the force on a negative charge is opposite the field.

One aspect of electric-force problems seems to vary from other force problems. Some problems include the force of gravity, while others ignore it entirely. Why and when is gravity important? Compared with electric force, the force of gravity is incredibly weak. Although the force of gravity is present, its effects are often negligible (too small to affect calculations). Electrons, protons, and ions have insignificant mass, and the force of gravity is usually ignored. In order for gravity to matter, the mass of the charged object must be fairly large. As a general rule of thumb, if the object is large enough to be seen, then the force of gravity is probably important.

Problems on exams frequently test the motion of electrons and protons. The direction of the motion depends on the sign of the charge, with like charges moving in opposite directions and unlike charges moving toward one another. When electrons are compared with protons, keep these facts in mind:

- Electrons and protons have the same magnitude of charge.
- When placed into the same electric field, they will both experience the same magnitude of electric force but in opposite directions.
- The electron has less mass, and the same force will cause it to have a greater acceleration than the proton.

EXAMPLE 10.2

Millikan Oil Drop

Robert Millikan determined the charge on an electron by suspending negatively charged oil drops in a uniform electric field created by two charged plates as shown in the diagram below. Determine the charge of an electron in terms of the mass of the oil drop (m), the electric field (E), and the gravity of Earth (g).

WHAT'S THE TRICK?

The oil drop has enough mass to include the force of gravity. In addition, there has to be an upward force countering gravity to keep the oil drop stationary.

Orient the problem: The electric field points from the positive plate to the negative plate. The electric force on the negative oil drop points upward, opposite the electric field. The force of gravity points downward.

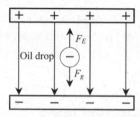

Determine the type of motion: This is static equilibrium.
Sum the force vectors in the relevant direction: Since the oil drop is stationary, the sum of the forces is zero. Simply set the two opposing forces equal to each other.

$$F_E = F_g$$

Substitute and solve:

$$qE = mg$$

Solve q in terms of the other variables.

$$q = \frac{mg}{E}$$

EXAMPLE 10.3

Motion of Charges in Uniform Fields

When an electron is released from rest in a uniform electric field E, it reaches a velocity of v after traveling a distance of x. In terms of v, what will be the electron's velocity if the magnitude of the electric field is doubled while traveling the same distance?

WHAT'S THE TRICK?

This problem combines force and kinematics for a particle that is accelerating. Gravity, while present, is negligible for particles as small as electrons.

Orient the problem: The electric force is the only force included in the calculations. As a result, this problem can be oriented in any manner.

Determine the type of motion: The particle is accelerating. Electrons move opposite the field.

Sum the force vectors in the relevant direction:

$$\Sigma F = F_E$$

Substitute and solve:

$$ma = qE$$
$$a = \frac{qE}{m}$$

The mass and charge of an electron remain constant. Acceleration is directly proportional to the electric field. If the electric field doubles, the acceleration will double.

$$(2a) = \frac{q(2E)}{m}$$

For the speed of an object accelerating from rest, $v_0 = 0$, and moving a set distance, use the following kinematic equation. Solve for v.

$$v^2 = v_0^2 + 2ax$$
$$v = \sqrt{2ax}$$

When distance, x, is held constant, velocity is proportional to the square root of acceleration. Doubling acceleration increases the right side of the equation by the square root of 2. To maintain the equality, velocity must also increase by this factor. The new velocity is $\sqrt{2}\,v$.

$$\sqrt{2}\,v = \sqrt{2(2a)x}$$

ELECTRIC FIELDS OF POINT CHARGES

Point charges are charges where the electric field is calculated as though it originates from a single point in space. Obvious point charges are individual electrons, protons, and ions. These tiny objects are essentially points in space. For a larger object to be a point charge, it must be spherical in shape and have the charge evenly distributed over its surface or throughout its volume.

Visualizing Uniform Fields

The field of a point charge radiates outward from the center of the charge and appears similar to the gravity field surrounding a planet viewed as a sphere in space. Figure 10.3 depicts the electric fields of individual positive and negative charges. They are not shown interacting with one another.

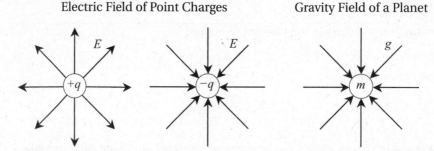

Figure 10.3 Electrical fields of positive and negative charges

Although the gravity field of a mass always points toward the mass, the direction of the electric field depends on the sign of charge q. Electric fields point away from positive charges and toward negative charges. The density of the lines indicates the strength of the field. The closer the lines are to each other, the stronger the field is.

When two charges are brought near each other, the electric fields intertwine. Figure 10.4 shows the interactions between unlike and like charges.

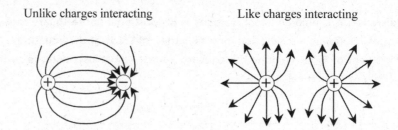

Figure 10.4 Interactions between unlike and like charges

Note that had the like charges on the right side of Figure 10.4 been negative, then the field lines would still have followed the same pattern. However, the arrows would have pointed toward the negative charges.

Magnitude of the Electric Field of Point Charges

The magnitude of the electric field, E, of a point charge is solved with an equation that bears a strong resemblance to the equation of the gravity field, g, at a point in space.

$$\text{Electric Field} \qquad\qquad \text{Gravity Field}$$

$$E = k\frac{q}{r^2} \qquad\qquad g = G\frac{m}{r^2}$$

The electrostatic constant, $k = 9 \times 10^9$ N • m^2/C^2, is analogous to the gravity constant, G. Charge q creates the electric field, while mass m creates the gravity field. The relationship between the charge, q, and the electric field, E, is directly proportional. Doubling the size of the charge, q, will double the magnitude of the electric field, E. The distance r is measured from the center of q (or m for gravity) to the point in space where the field is to be calculated. In both of these field formulas, the distance r is inverted and squared. Thus, changes in r are subject to the **inverse-square law**. Doubling the distance r will cause the magnitude of the field to become one-fourth of its original value.

Many problems do not require the entire field to be drawn. Instead, they require a vector direction of the electric field at a specific point in space. This can be accomplished using an **imaginary positive test charge**. Visualize the test charge at the location where field direction is needed. The direction the test charge would move if released is the same as the direction of the electric field. The force on a positive charge is always in the same direction as the field.

IF YOU SEE a spherical point charge

$$E = k\frac{q}{r^2}$$

Electric field is directly proportional to the charge, q, and inversely proportional to the square of the distance, $1/r^2$.

EXAMPLE 10.4

Electric Field of Point Charges

A conducting sphere contains a charge q. The electric field at a set distance from the center of the sphere is E. If the charge on the sphere and the distance from the sphere are both doubled, by what factor would the electric field change?

(WHAT'S THE TRICK?)

A sphere is a point charge, and its electric field is determined by the following equation.

$$E = k\frac{q}{r^2}$$

Double the charge, q, and the distance, r, to determine the effect on the electric field, E.

$$\frac{1}{2}E = k\frac{(2q)}{(2r)^2}$$

The electric field is reduced by a factor of half.

Superposition of Fields and Force

Problems often involve determining the value of the electric field due to more than one point charge. Each charge creates its own electric field that can be calculated using the following formula.

$$E = k\frac{q}{r^2}$$

The formula is solved once for each charge, where r is the distance from the charge being calculated to the point where the field is to be determined. This results in several values for the electric field, one for each charge present. Each calculated electric field is a vector that includes a specific direction. All of the electric-field vectors can be added together using vector mathematics to determine the total electric field.

EXAMPLE 10.5

Electric-Field Superposition

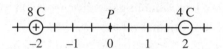

Distance in meters

A +8 coulomb charge is located 2 meters to the left of the origin. A –4 coulomb charge is located 2 meters to the right of the origin. Determine the magnitude and direction of the electric field at point P located at the origin.

WHAT'S THE TRICK?

Identify the +8 C charge as q_1 and the –4 C charge as q_2. Draw and label the electric-field vectors, E_1 for q_1 (away from positive charge) and E_2 for q_2 (toward negative charge) at point P.

Determine the magnitudes of E_1 and E_2. Magnitudes of vectors are always positive, so the sign on the charges can be ignored when determining electric-field magnitude.

$$|E| = k\frac{q}{r^2}$$

$$|E_1| = k\frac{q_1}{r^2} = (9 \times 10^9 \text{ N} \cdot \text{m}^2/\text{C}^2)\frac{(8 \text{ C})}{(2 \text{ m})^2} = 18 \times 10^9 \text{ N/C}$$

$$|E_2| = k\frac{q_2}{r^2} = (9 \times 10^9 \text{ N} \cdot \text{m}^2/\text{C}^2)\frac{(4 \text{ C})}{(2 \text{ m})^2} = 9 \times 10^9 \text{ N/C}$$

The resultant of adding these vectors is the total electric field due to the superposition of these charges. The direction of these vectors can be seen in the diagram. Since they both point to the right, they can both be regarded as positive vectors.

$$\vec{E} = \vec{E}_1 + \vec{E}_2 = (18 \times 10^9) + (9 \times 10^9) = 27 \times 10^9 \text{ N/C}$$

Superposition of Electric Charge

online.barronsbooks.com

EXAMPLE 10.6

Superposition

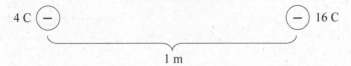

Two negative charges, 4 coulombs and 16 coulombs, are separated by a distance of 1 meter, as shown in the diagram above. Determine the location, as measured from the 4-coulomb charge, where the electric field is zero.

WHAT'S THE TRICK?

Both charges create individual electric fields that overlap. In order for their sum to equal zero, the magnitudes of the fields must be equal and their directions must be opposite. A positive test charge can be imagined to the left of, between, and to the right of the charges to locate a possible zero point. Since the charges are negative, the electric field of each charge will point toward that charge. The only place where the electric-field vectors point in opposite directions is between the charges, as shown in the diagram below. For the electric field of the smaller charge to equal and cancel that of the larger charge, the zero point must be closer to the smaller charge. An approximate location for the zero point is shown below.

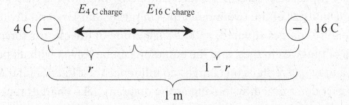

The problem asks for a distance from the 4-coulomb charge. Set this distance as r. The distance from the 16-coulomb charge is then $1 - r$. Set the magnitudes of the two fields equal to each other, and solve for the unknown distance r.

$$E_{2\text{-coulomb charge}} = E_{4\text{-coulomb charge}}$$

$$k\frac{4}{r^2} = k\frac{16}{(1-r)^2}$$

Cancel the constant k, and then take the square root of both sides of the equation.

$$\sqrt{\frac{4}{r^2}} = \sqrt{\frac{16}{(1-r)^2}}$$

$$\frac{2}{r} = \frac{4}{(1-r)}$$

$$r = \frac{1}{3} \text{ m}$$

Electric Force Due to Point Charges

The diagrams in Figure 10.5 show the forces acting on point charges.

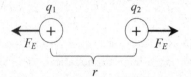

Like charges repel

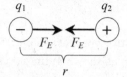

Opposite charges attract

Figure 10.5 Forces acting on point charges

Although the interaction between two negative charges is not shown, they would repel each other. The diagram would look very similar to that of the two positive charges shown on the left of Figure 10.5.

When two point charges interact, the resulting force acting on each charge can be determined using **Coulomb's law**. Coulomb's law is extremely similar to Newton's law of gravity.

Coulomb's Law $\qquad\qquad$ Newton's Law of Gravity

$$F_E = k\frac{q_1 q_2}{r^2} \qquad\qquad F_G = G\frac{m_1 m_2}{r^2}$$

The magnitude of the electric force, F_E, is determined by multiplying the electrostatic constant, k, by the magnitude of the two interacting charges, q_1 and q_2, and dividing this by the distance between the charges squared, r^2. The magnitude of the electric force is not dependent on the sign of the two charges, and the equation can be solved with all positive values.

The positive and negative signs on charges do influence the direction of the electric force, which can be easily determined by looking at the diagram. Like charges repel, while unlike charges attract. Note that the electric field of each charge is not included in Figure 10.5. Including the field arrows for both charges would have created too much clutter. Always remember that the electric force acting on a positive charge will match the direction of the electric field, while the force acting on a negative charge is opposite the field direction.

Newton's third law of motion is always in effect whenever two objects interact. Two force vectors are shown in each diagram in Figure 10.5. The force on charge 1 (object) is created by the electric field of charge 2 (agent). Similarly, the force on charge 2 (object) is created by the electric field of charge 1 (agent). Coulomb's law solves for the value of both of these force vectors as dictated by Newton's third law: Whenever two objects interact, there is an equal and opposite force between them.

Superposition of Force

When a system consisting of several point charges is present, force vectors add together in a manner similar to the superposition of electric-field vectors described earlier. If three or more charges are present and you must determine the force on one of them due to all the others, use Coulomb's law and superposition. Use Coulomb's law to find the force between the charge in question and every other charge acting on it. In addition, you can find the direction of each of these force vectors using the rules for attraction and repulsion. The result will be several force vectors that can be added together using vector addition to determine the net force acting on the charge in question.

IF YOU SEE
two charges
separated by
a distance

Coulomb's law

$$F_E = k\frac{q_1 q_2}{r^2}$$

Force is directly proportional to the charges and inversely proportional to the square of the distance of separation, $1/r^2$.

SUMMARY

1. **CHARGE IS A PROPERTY OF ELECTRONS AND PROTONS.** The fundamental charges—electrons and protons—are equal in magnitude, $e = 1.6 \times 10^{-19}$ coulombs, but have opposite signs. Neutral objects have an equal number of electrons and protons. Objects become charged when they have an imbalance in the number of electrons and protons. A coulomb of charge consists of 6.25×10^{18} electrons or protons. Although charges can move from place to place, the total charge in a system is always conserved.

2. **ALL CHARGES AND CHARGED OBJECTS ARE SURROUNDED BY A DISTORTION IN SPACE KNOWN AS AN ELECTRIC FIELD.** Electric fields and gravity fields are similar except that gravity fields always point toward a mass but electric fields point toward or away from a charge depending on the sign of the charge. The electric field of one charged object will create an electric force on another charged object.

3. **UNIFORM ELECTRIC FIELDS EXIST BETWEEN CHARGED PLATES.** Equally but oppositely charged parallel plates create an electric field between them that has the same value and direction at every point. The magnitude, E, of the field vector is different for every set of plates. The vector field arrows start on the positive plate and point toward the negative plate. When a charge is placed between the plates, it experiences an electric force, $F_E = qE$. The force on a positive charge is the same as the direction of the electric field. The force on a negative charge is opposite the field direction.

4. **THE ELECTRIC FIELD OF SPHERICAL POINT CHARGES VARIES.** Electrons, protons, ions, and charged spheres are all point charges. Mathematically, the electric field calculates from a point at the center of any size spherical charge distribution, whether solid or hollow. The magnitude of the field, $E = kq/r^2$, is directly proportional to the charge q and inversely proportional to the square of distance from charge, r (inverse-square law). The electric field of a point charge points away from a positive charge and points toward a negative charge. When more than one charge is present, the electric-field vectors of the individual charges can be calculated separately and then summed to determine the total electric field due to their overlap (superposition). The force between two charges is dictated by Coulomb's law, $F_E = kq_1q_2/r^2$. In addition, Newton's third law of motion applies to the interaction of charges.

5. **FORCE AND ACCELERATION ON PROTONS AND ELECTRONS.** Protons and electrons experience the same magnitude of electric force. They each have the same magnitude of charge but are opposite in sign. Electrons, however, experience a greater acceleration than protons because of the smaller mass of electrons.

If You See	Try	Keep in Mind
Parallel plates and/or parallel electric vectors	Uniform electric field $F_E = qE$	The force on positive charges matches the field, while the force on negative charges is opposite the field.
One or more spherical point charges	$E = k\dfrac{q}{r^2}$	Electric field is directly proportional to the charge, q, and inversely proportional to the square of the distance, r.
Two point charges separated by a distance	Coulomb's law $F_E = k\dfrac{q_1 q_2}{r^2}$	Electric force is directly proportional to each charge, q, and inversely proportional to the square of the distance, r.

PRACTICE EXERCISES

1. Adding extra electrons to a previously neutral atom results in

   (A) a new element
   (B) a positive ion
   (C) a negative ion
   (D) a positive isotope
   (E) a negative isotope

2. Two identical conducting spheres are initially separated. The left sphere has a –3-coulomb charge and the right sphere has a +2-coulomb charge. The spheres are allowed to touch each other briefly, and then they are separated. Determine the charge on the left sphere.

   (A) –1 C
   (B) $-\dfrac{1}{2}$ C
   (C)   0 C
   (D) $+\dfrac{1}{2}$ C
   (E) +1 C

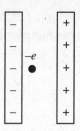

3. An electron is located between two charged plates, as shown in the diagram above. Which statement below is **not** true?

(A) The electric field of the plates is directed to the right.
(B) The electric field of the plates has the same magnitude at all points between the plates.
(C) The electric force on the electron is to the right.
(D) The acceleration of the electron will be constant until it impacts one of the plates.
(E) A uniform electric field exists between the plates.

4. A proton and an electron are released from rest in the same uniform electric field. Assume the proton and electron do not interact with one another. The acceleration of the proton, compared with that of the electron, is

(A) the same in both magnitude and in direction
(B) the same in magnitude and opposite in direction
(C) greater in magnitude and opposite in direction
(D) less in magnitude and the same in direction
(E) less in magnitude and opposite in direction

5. Two particles with charges Q and $2Q$ repel one another. How does the electric force acting on the particle with charge Q compare with the force acting on the particle with charge $2Q$?

(A) The force on charge Q is 1/4th of the force acting on charge $2Q$.
(B) The force on charge Q is 1/2 of the force on charge $2Q$.
(C) The force on charge Q is the same as the force on charge $2Q$.
(D) The force on charge Q is twice as large as the force on charge $2Q$.
(E) The force on charge Q is four times larger than the force on charge $2Q$.

6. An object has a charge of 3.0 coulombs and a mass of 2.0 kilograms. Determine the magnitude of the electric field that would create 12 newtons of electric force on this object.

(A) 0.25 N/C
(B) 0.50 N/C
(C) 2.0 N/C
(D) 4.0 N/C
(E) 8.0 N/C

7. The electric field of a charged sphere has a magnitude of E at a distance, d, from the center of the sphere. How does the magnitude of the electric field at a point that is $\frac{1}{2}d$ from the center of the sphere compare?

(A) It is $\frac{1}{4}E$.

(B) It is $\frac{1}{2}E$.

(C) It is E.

(D) It is $2E$.

(E) It is $4E$.

8. In the diagram above, two point charges, $-q$ and $+2q$, are held stationary. Determine the approximate location where the electric field is zero.

(A) A

(B) B

(C) C

(D) D

(E) E

Questions 9–10

In the diagram below, two negative charges, 1 coulomb and 2 coulombs, are initially separated by a distance of 3 meters. The electrostatic constant is $k = 9 \times 10^9$ N • m^2/C^2.

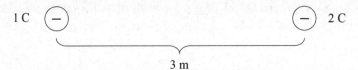

9. Determine the electric force acting on the 1-coulomb charge.

(A) 0.33×10^9 N

(B) 1×10^9 N

(C) 2×10^9 N

(D) 6×10^9 N

(E) 18×10^9 N

10. The charges are released and begin to move. Which statement regarding the force, acceleration, and velocity vectors is true for these charges?

(A) Force increases, acceleration increases, and velocity decreases.

(B) Force increases, acceleration increases, and velocity increases.

(C) Force decreases, acceleration decreases, and velocity decreases.

(D) Force decreases, acceleration decreases, and velocity increases.

(E) Force decreases, acceleration increases, and velocity decreases.

ANSWERS EXPLAINED

	Key Words	Needed for Solution	Now Solve It
1. **(C)**	Adding extra electrons; atom	Knowledge/definitions	Adding negatively charged electrons to an atom results in the creation of a negatively charged ion.
2. **(B)**	Identical conducting spheres; charge; touch; separated	Conservation of charge	Total charge is conserved and remains constant. $$\text{Total charge} = (-3 \text{ C}) + (+2 \text{ C}) = -1 \text{ C}$$ When the spheres touch, their excess charges neutralize, leaving only a –1 C charge on both spheres together. Since the spheres are equal in size, each sphere must contain half of this charge, –1/2 C.
3. **(A)**	Electron; between two charged plates	Knowledge/definitions	An electric field always points away from a positive charge and toward a negative charge. The vector field arrows of the charged plates point to the left.
4. **(E)**	Proton; electron; uniform electric field; acceleration	$F_E = qE$ The sum of forces yields acceleration	$$\Sigma F = F_E$$ $$ma = qE$$ $$a = \frac{qE}{m}$$ The charges on a proton and an electron are the same, but their masses are very different. The larger mass of the proton results in a smaller acceleration in the same electric field.
5. **(C)**	Electric force; charge Q; compare; charge $2Q$	Newton's third law	Newton's laws of motion apply to all forces. Any time two objects interact, the third law requires consideration. There is always an equal and opposite force between interacting objects.
6. **(D)**	Charge; magnitude of electric field; electric force	$F_E = qE$	Rearrange the equation to solve for the electric field: $$E = \frac{F}{q} = \frac{(12 \text{ N})}{(3.0 \text{ C})} = 4.0 \text{ N/C}$$ The mass was given as a distracter.
7. **(E)**	Charged sphere; electric field; $1/2d$	$E = k\dfrac{q}{r^2}$ Inverse-square law	For spherical charges, any change in distance causes the electric field to change by the inverse square. The inverse of 1/2 is 2, and its square is 4. $$(4)E = k\frac{q}{\left(\frac{1}{2}d\right)^2}$$

	Key Words	Needed for Solution	Now Solve It
8. **(A)**	Two point charges; where the electric field is zero	$E = k\dfrac{q}{r^2}$ Superposition	For the total field to be zero, the individual electric-field vectors of each charge must be equal in magnitude but opposite in direction. The zero point must be closer to the smaller charge so it can equal the field of the larger, more distant charge. This narrows the possibilities to answers A and B only. The answer must be A since the vectors of the electric field of each charge are opposite at point A. At point B, they point in the same direction and cannot cancel.
9. **(C)**	Two negative charges; electric force	Coulomb's law $F_E = k\dfrac{q_1 q_2}{r^2}$	Substitute the values into Coulomb's law: $F_E = (9 \times 10^9 \text{ N} \cdot \text{m}^2/\text{C}^2)\dfrac{(1 \text{ C})(2 \text{ C})}{(3 \text{ m})^2}$ $F_E = 2 \times 10^9 \text{ N}$
10. **(D)**	Two negative charges; released; force, acceleration, and velocity vectors	Attraction/repulsion Coulomb's law $F_E = k\dfrac{q_1 q_2}{r^2}$ Sum of forces $\Sigma F = F_E$ $ma = F_E$ Definition of acceleration and its relationship to the final velocity	Like charges repel. The charges will separate, increasing the distance r. According to Coulomb's law, the force of electricity is inversely proportional to the square of separation. As r increases, F_E decreases. Acceleration is directly proportional to the net force acting on an object, in this case the electric force. Since force decreases, acceleration also decreases. Many students will be fooled by answer C, believing a decreasing acceleration will slow an object. However, to slow an object, the force and resulting acceleration must be opposite the velocity of the object. However, as each charge moves, the force and acceleration are acting in the direction of motion. The charges are speeding up. Acceleration is the rate of change in velocity (change in velocity per second). When acceleration decreases, the change in velocity in the next second is smaller. However, the acceleration is still speeding up the object.

Electric Potential

11

→ **POTENTIAL OF UNIFORM FIELDS**

→ **POTENTIAL OF POINT CHARGES**

→ **ELECTRIC POTENTIAL ENERGY**

→ **MOTION OF CHARGES AND POTENTIAL**

→ **CAPACITORS**

A quantity known as potential, or **electric potential**, is unique to charged objects. Electric potential is associated with electric potential energy, work of the electric force, and conservation of energy. Electric potential, V, is a scalar quantity measured in volts (V). Electric potential, V, can be thought of as an electrical pressure at a point in space. Charges positioned at this point will have electric potential energy and the potential (pressure) to move.

Capacitance is the ability to store charge and electric potential energy. Charged plates are also known as capacitors, and they are useful components in electric circuits. The amount of charge and energy stored by a capacitor depends on its capacitance and on the potential difference between the charged plates. To gain a better understanding of electric potential, electric potential energy, and capacitors, this chapter will do the following:

- Explain the nature of electric potential for uniform electric fields.
- Explain the nature of electric potential for point charges.
- Examine the relationship between electric potential and potential energy.
- Explore the motion of charges based on potential.
- Introduce capacitors and capacitance.

Table 11.1 lists the variables that will be studied in this chapter.

Table 11.1 Variables Used for Electric Potential

New Variables	Units
V = Electric potential	V (volts)
ΔV = Potential difference	V (volts)
U_E = Electric potential energy	J (joules)
C = Capacitance	F (farads)
ϵ_0 = Permittivity of free space	8.85×10^{-12} C^2/N • m^2
U_C = Energy of a capacitor	J (joules)

POTENTIAL OF UNIFORM FIELDS

Uniform electric fields are a property of equal and oppositely charged parallel plates. Since the electric field between these plates is uniform, the electric potential is evenly distributed in the space between the plates. Figure 11.1 shows two commonly encountered ways to depict the electric potential, V, at various locations between charged plates that have been charged to a 6-volt potential difference, ΔV.

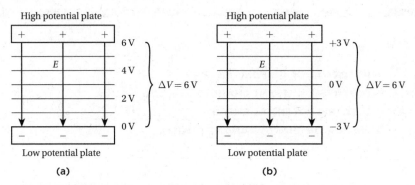

Figure 11.1 Electric potential for uniform fields

The magnitude of the electric potential, V, at a location in a uniform electric field can be determined with the following formula.

$$V = Ed$$

In this formula, distance, d, is measured from a location where the potential is equal to zero to the point in the field where potential is to be solved.

Electric potential is analogous to height in mechanics. For charged plates the value of potential depends on the location of a zero-potential reference line, much as the values of height depend on the location of a zero-height reference line. Commonly, the lowest position an object reaches during its motion is set as zero-height to make mechanics problems easier to solve. Similarly, the same method can be used with uniform electric fields. Potential is a scalar quantity, and the location with the lowest potential can be set as a zero-potential reference line, as seen in Figure 11.1(a). The higher-potential positive plates would then have a potential equal to the value of the potential difference between the plates. However, the zero-reference potential can be set anywhere. Figure 11.1(b) shows the zero-line set midway between the plates. No matter how the values of the potential lines have been quantified, the high and low potential lines must be separated by the value of the potential difference.

However, since the assignment of the zero-potential reference line is arbitrary, it is more important to consider the potential difference, ΔV, between key points when solving problems involving the motion of charges in uniform electric fields. It is the change in potential, ΔV, that will determine the work (change in energy) done on the charges by the electric field. If the electric field, E, and the distance, Δd, that a charge moves through are known, then the potential difference the charge experiences can be determined using the following equation.

$$\Delta V = E \Delta d$$

The electric potential can be viewed as lines that are perpendicular to the electric field and equally spaced. These lines are known as **equipotential lines**. Every point on a particular equipotential line has the same equal potential. In Figure 11.2, a point P, a positive charge,

**IF YOU SEE
the potential
difference
of charged
plates**

$\Delta V = E \, \Delta d$

Δd is the distance between the plates.

and a negative charge are all shown at a location where their electric potentials are 4 volts relative to the negative plate. They are all on the same equipotential line.

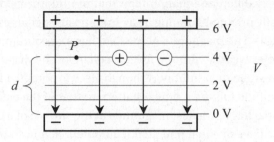

Figure 11.2 Electric potential

Some problems will ask for the electric potential of a point or charged object located between charged plates. This is calculated with the same equation used to find the potential of the plates themselves, $V = Ed$. However, the distance, d, is measured from the zero location (usually the negative plate) to the point or charged object, as shown in Figure 11.2.

EXAMPLE 11.1

Electric Potential of a Uniform Field

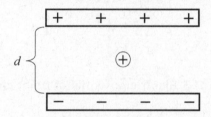

Two charged plates with a 20-newton-per-coulomb electric field are separated by a distance of 10 centimeters. A proton is located at the midpoint between the plates.

(A) Determine the potential difference between the charged plates.

WHAT'S THE TRICK?

To find the potential difference of the charged plates, use the following formula. Remember to change the distance to meters.

$$\Delta V = E\, \Delta d$$

$$\Delta V = (20 \text{ N/C})(0.10 \text{ m}) = 2 \text{ V}$$

(B) Determine the potential acting on the proton.

WHAT'S THE TRICK?

You need to find the potential at the point where the proton is located. Since the proton is at the midpoint between the plates, the distance (in meters) is half the distance between the plates.

$$V = Ed$$

$$V = (20 \text{ N/C})(0.05 \text{ m}) = 1 \text{ V}$$

POTENTIAL OF POINT CHARGES

The electric potential of point charges is visualized in an entirely different manner and is solved using an entirely different equation than that of uniform fields. The electric field of point charges is radially oriented pointing away from positive charges and radially oriented toward negative charges. The magnitude of the field has a maximum value at the surface of the charge and becomes weaker with the inverse square of the distance from the center of the charge. It has zero strength at infinity. Although electric potential is not a vector and has no direction, its magnitude follows a similar pattern to that of the electric field surrounding a point charge. Potential has its highest magnitude at the surface of a charge and diminishes to zero at infinity. The lines of equal potential (equipotential lines) are perpendicular to the field and form concentric circles around the charge. An example is shown in Figure 11.3 for a positive charge with a 6-volt potential at its surface.

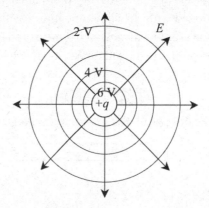

Figure 11.3 Electric potential of point charges

IF YOU SEE
a point
charge
and need
potential

$V = k\dfrac{q}{r}$

V is scalar. Include the sign on the charge *q*, but use the absolute value of distance *r*.

The equation for potential is very similar to the equation for the electric field. However, potential is a scalar quantity while the electric field is a vector.

Potential	Electric Field
$V = k\dfrac{q}{r}$	$E = k\dfrac{q}{r^2}$

Positive charges are considered to have a high positive potential, and negative charges have a low negative potential. Although the sign on the charge is important, the distance *r* from the center of the charge to any point *P* is set as positive regardless of its direction.

$$V = k\frac{(\pm q)}{|r|}$$

Electric Potential of Several Point Charges

If a problem consists of several point charges surrounding a point *P*, then merely add together the electric potentials of the individual charges. To find the total potential, simply sum the individual electric potentials.

$$V = V_1 + V_2 + V_3 + \cdots = k\frac{q_1}{r_1} + k\frac{q_2}{r_2} + k\frac{q_3}{r_3} + \cdots$$

Electric Potential of Point Charges

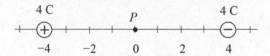

A +4-coulomb charge is located 4 meters to the left of the origin. A –4-coulomb charge is located 4 meters to the right of the origin. Determine the electric potential at point P, located at the origin.

WHAT'S THE TRICK?

The charges are spherical. Use the potential equation for point charges. Since there are two charges, add them. The sign on the charges is important. However, the sign on the distance from the origin is not.

$$V = k\frac{q_1}{r_1} + k\frac{q_2}{r_2}$$

$$V = (9 \times 10^9 \text{ N} \cdot \text{m}^2/\text{C}^2)\ \frac{(+4)}{|-4|} + (9 \times 10^9 \text{ N} \cdot \text{m}^2/\text{C}^2)\ \frac{(-4)}{|+4|}$$

$$V = 0 \text{ V}$$

ELECTRIC POTENTIAL ENERGY

Once the electric potential at a point in space is determined, any charge q can be inserted at that point and the **electric potential energy**, U_E, can be determined. Simply multiply the electric potential, V, at a point in space by the charge, q, located at that point.

$$U_E = qV$$

The relationship between electric potential and electric potential energy is similar to that between electric field and electric force. A point in space will have a specific electric-field value, E, and a specific electric-potential value, V. When a charge, q, is placed at a point in space, the electric field creates a force, F_E, on the charge. The electric potential can be used to determine its electric potential energy, U_E.

$$F_E = qE \qquad \text{and} \qquad U_E = qV$$

Whether the electric potential is due to charged plates, a single point charge, or several point charges, the equation to find electric potential energy is the same. In Figure 11.4(a), a point charge q is located in the uniform field of charged plates. In Figure 11.4(b), a point charge q_2 is located in the electric field created by point charge q_1.

IF YOU SEE
electric
potential
energy

$U_E = qV$

This equation solves for the energy of a charge q in any type of electric field.

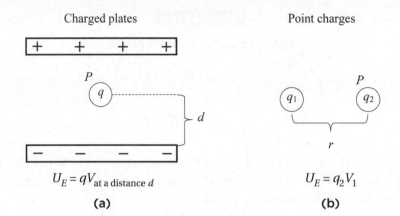

Figure 11.4 Electric potential energy

You can substitute the equations for electric potential, V, into the equations found in Figure 11.4(a) and Figure 11.4(b) to create alternate equations for electric potential energy.

$$V = Ed \qquad\qquad V_1 = k\frac{q_1}{r}$$

$$U_E = q(Ed) \qquad\qquad U_E = q_2\left(k\frac{q_1}{r}\right)$$

$$U_E = qEd \qquad\qquad U_E = k\frac{q_1 q_2}{r}$$

The resulting equations solve for electric potential energy without the need to determine electric potential. Electric potential energy is very similar to gravitational potential energy. These derived equations bear strong resemblances to the equations for gravitational potential energy.

$$U_G = mgh \qquad U_G = G\frac{m_1 m_2}{r}$$

Potential energy, U, is the energy of position. When objects are released, they lose their potential energy doing work. The lost energy is usually converted into kinetic energy. As a result, electric potential energy, U_E, can be thought of as the potential to move charges. Electric potential, V, is directly proportional to electric potential energy. In an electrical circuit or in problems involving the motion of charges, electric potential can be thought of as an electrical pressure that exists at a point in space or in a circuit. Any charge located at that point will have the energy needed to move. Electric potential is also commonly called voltage. High voltage is a high potential for charges to move and thus a greater chance of receiving an electric shock.

MOTION OF CHARGES AND POTENTIAL

When charges are released in electric fields, the charges experience a force, causing them to accelerate parallel to electric-field vectors. Positive charges accelerate in the direction of the electric field. Negative charges move opposite the electric field. While in motion, the charges experience a change in potential, known as a potential difference. Changes in speed are associated with changes in kinetic energy, which result from changes in potential energy. Energy is conserved in these processes.

Potential Difference of Moving Charges

When a charge is released in an electric field, the charge moves parallel to electric-field lines, causing a change in distance, Δd (uniform field) or Δr (field of a point charge). The change in distance results in a change in potential, ΔV, specifically known as a potential difference.

$$\Delta V = V_f - V_i$$

In addition, the term *potential difference* can refer to the difference in potential between separated charges, such as the potential between two charged plates.

Work of Electricity

When a charge accelerates, it changes speed and experiences a change in kinetic energy, ΔK. The work–kinetic energy theorem states that changes in kinetic energy are equal to work, $W = \Delta K$. If kinetic energy is changing, another energy in the system must be experiencing an equal but opposite change in order to ensure that energy is conserved. In problems involving a moving charge, the change in kinetic energy is offset by an equal change in electric potential energy. This is directly proportional to the potential difference through which the charge moves.

$$W_E = \Delta K = -\Delta U_E = -q\Delta V = -q(V_f - V_i)$$

Note that the above equation equates the values to one another but does not indicate the correct sign on each value. When a charge speeds up, work and the change in kinetic energy are both positive. In order for kinetic energy to increase, the electric potential energy must decrease. This means the sign on the change in electric potential energy is the opposite of the signs on work and the change in kinetic energy. When a charge slows down, work and the change in kinetic energy are negative while the sign on the change in electric potential energy is positive.

Solving for the correct sign on work using the change in electric potential energy equation, $W = -\Delta U_E = -(q\Delta V)$, is complicated since the equation contains a negative sign and all the variables may be either positive or negative. Most physics problems in beginning courses involve situations where the work of the electric force is positive. As a result, solving for the absolute value of work is actually easier.

$$|W| = |q| \bullet |(V_f - V_i)|$$

How can the correct sign be determined? One way is to assess the change in kinetic energy, which has the same sign as work. If a charge is speeding up, the change in kinetic energy is

IF YOU SEE
the work of
a moving
charge

$W = \Delta K = \Delta U_E$

$W = -q(V_f - V_i)$

Use $+W$ if the charge is speeding up, $-W$ if the charge is slowing down, and $W = 0$ if the charge is moving at constant velocity.

positive and the work is positive. If a charge is slowing down, the work is negative. Another way to determine the sign on work is to compare the directions of force and displacement. If force and displacement point in the same direction, work is positive. When they oppose each other, work is negative.

EXAMPLE 11.3

Work of Electricity

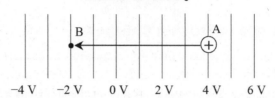

-4 V $\quad$ -2 V $\quad$ 0 V $\quad$ 2 V $\quad$ 4 V $\quad$ 6 V

The diagram above shows the equipotential lines in a region of space. Determine the work done on a 2-coulomb charge that is moved from point A to point B.

WHAT'S THE TRICK?

Work is equal to a change in energy. This problem involves equipotential lines. The energy that is changing is electric potential energy, ΔU_E. This is related to the potential difference, ΔV, through which the 2-coulomb charge moves.

$$W = -\Delta U_E = -q\,\Delta V = -q(V_f - V_i)$$
$$W = -\Delta U_E = -q\,\Delta V = -(2\text{ C})[(-2\text{ V}) - (+4\text{ V})]$$
$$W = +12\text{ J}$$

Work involves change, and the sign is often important. Be very careful to subtract the initial potential difference from the final potential difference and to include the given signs. Note that the positive 2-coulomb charge will be attracted towards the negative 2 V equipotential line and will therefore speed up. An increase in speed is positive for both work and change in kinetic energy.

IF YOU SEE
a charge
accelerating
through a
potential
difference

Conservation
of energy

$\frac{1}{2}mv^2 = q\Delta V$

Conservation of Energy

Conservation of energy dictates that the total energy (sum of the kinetic and potential energies) of a closed system must have the same value at any two points during a process.

$$K_1 + U_1 = K_2 + U_2$$

Electricity problems involve electric potential energy, $U_E = qV$.

$$K_1 + U_{E1} = K_2 + U_{E2}$$
$$\frac{1}{2}mv_1^2 + qV_1 = \frac{1}{2}mv_2^2 + qV_2$$

Most often, these problems involve an object at rest that accelerates through a potential difference, ΔV, to a point where the object's speed is at the maximum. For plates, maximum speed occurs when a charge moves from one plate with similar charge to the other plate with opposite charge. For a charge moving in the field of a point charge, the maximum speed occurs when the charge reaches infinity. For either of these cases, the maximum kinetic energy equals the change in electric potential energy. The conservation-of-energy formula abbreviates to:

$$\frac{1}{2}mv_{\max}^2 = q\,\Delta V$$

EXAMPLE 11.4

Conservation of Energy

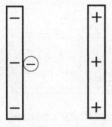

Two charged plates have a potential difference of V. An electron of mass m and charge e is initially at the negative plate. It accelerates through the potential differ- ence. Determine the electron's maximum speed, v, in terms of the given variables.

WHAT'S THE TRICK?

Remember conservation of energy. When the electron is located at the negative plate, it has only potential energy. When it reaches the positive plate, the electron has reached maximum speed. So, all of the potential energy has been entirely converted to kinetic energy.

$$\frac{1}{2}mv^2 = qV$$

$$v = \sqrt{\frac{2qV}{m}}$$

Although q usually represents a small charge, the variable e represents the funda- mental charge of an electron or a proton. In this problem, the charge was given as e, so q = e. The final answer should be in terms of the given quantities.

$$v = \sqrt{\frac{2eV}{m}}$$

CAPACITORS

Plates made of conducting material that are able to store excess charge and electric energy. This property has a useful application for electrical circuits. The capacity of charged plates to hold excess charges is an important factor.

Capacitance

Capacitance, C, which is measured in farads (F), is essentially the ability of a capacitor to store charge and energy. Two oppositely charged parallel plates are shown in Figure 11.5.

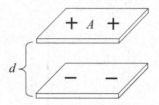

Figure 11.5 Oppositely charged parallel plates

The capacitance of the plates is directly proportional to the area, A, of one of the plates and inversely proportional to the distance of the plate separation, d.

$$C = \frac{\epsilon_0 A}{d}$$

IF YOU SEE a capacitor with a changing area or changing plate separation

$$C = \frac{\epsilon_0 A}{d}$$

Capacitance is directly proportional to area and inversely proportional to the distance of plate separation.

A constant, known as the permittivity of free space, $\epsilon_0 = 8.85 \times 10^{-12}$ C^2/N • m^2, is needed to turn the proportionality into an equation solving for capacitance, C. Multiplying by a value such as this constant would be difficult without a calculator. Instead, the SAT Subject Test in Physics will most likely focus on factors affecting capacitance, as demonstrated in Example 11.5.

Charging a Capacitor

A battery or a power supply is needed to provide the electric potential needed to charge capacitors or to run circuits. Batteries, like capacitors, are made up of plates of conducting material. The main difference is the presence of chemicals in a battery, which undergo a continuous reaction to keep the plates of the battery loaded with a fixed charge. The fixed charge on the plates of the battery creates a constant potential difference (voltage) across the plates and terminals (ends) of the battery. This potential provides the electrical pressure needed to charge capacitors and/or push charges through circuits. A power supply is a device that functions like a battery. It is plugged into an electrical outlet and adjusts the voltage delivered by the power company to a desired level.

Figure 11.6 shows two versions of the same capacitor connected to a 6-volt battery and uses circuit symbols for the battery and the capacitor.

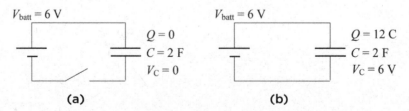

Figure 11.6 A capacitor connected to a battery

In Figure 11.6(a), the switch is open and the capacitor is initially uncharged, $Q = 0$. An uncharged capacitor has no potential, $V_C = 0$. In Figure 11.6(b), the switch has been closed for a long time, allowing the potential of the battery to push charges onto the capacitor. As the capacitor fills with charge, a potential (pressure) builds up on the plates of the capacitor. The capacitor will continue to fill as long as the potential of the battery is greater than the potential of the capacitor. This process happens quickly at first. As more and more charge builds up on the capacitor, though, forcing additional charges on the capacitor becomes more difficult. Eventually, the potentials become equal ($V_C = V_{\text{batt}}$), charging stops, and the capacitor is full. The amount of charge, Q, stored on a capacitor is a function of its capacitance, C, and its potential, V.

$$Q = CV$$

Energy of a Capacitor

Capacitors also store energy. Once a capacitor has been charged, the charges will remain on its plates as long as no pathway is provided for the charges to move from one plate to the other. The battery can even be removed from the circuit and the charges will remain in place on the capacitor. The charges are essentially held in a position to be used later. Therefore, the energy of a capacitor is potential energy, U_c.

$$U_c = \frac{1}{2} QV = \frac{1}{2} CV^2 = \frac{1}{2} \frac{Q^2}{C}$$

IF YOU SEE
a capacitor with changing conditions

Consider both

Q = CV
and

$$U_c = \frac{1}{2}QV$$

See Example 11.5.

EXAMPLE 11.5

Capacitors

A capacitor is connected to a variable power supply with potential V.

(A) Determine the effect on the charge stored on the capacitor if the potential of the power supply is cut in half.

(WHAT'S THE TRICK?)

The charge on a capacitor is tied to its capacitance, C, and the potential, V, across its plates. When a capacitor is fully charged, its potential will be equal to the potential of the power supply. Capacitance depends on the area, A, of the plates and the distance, d, between them. These quantities are not mentioned in the problem and are therefore assumed to remain constant. When capacitance is constant, the charge, Q, stored on the capacitor is directly proportional to the potential, V, of the capacitor. This potential is being cut in half.

$$Q = CV$$

$$\left(\frac{1}{2}Q\right) = C\left(\frac{1}{2}V\right)$$

Cutting the potential in half will cut the amount of charge stored in half.

(B) How does this change in power supply and potential affect the energy stored on the capacitor?

WHAT'S THE TRICK?

Care must be taken in analyzing energy. The first equation that comes to mind is

$$U_c = \frac{1}{2}QV$$

Some students focus on the previous answer only, and they cut the charge in half. Remember that both the charge and the potential are halved.

$$\left(\frac{1}{4}U_c\right)U_c = \frac{1}{2}\left(\frac{1}{2}Q\right)\left(\frac{1}{2}V\right)$$

The changes cause the energy stored to be one-quarter of its original value. This is confirmed when examining the second equation for the energy of capacitors.

$$U_c = \frac{1}{2}CV^2$$

$$\left(\frac{1}{4}U_c\right) = \frac{1}{2}C\left(\frac{1}{2}V\right)^2$$

Capacitance is constant, but potential is squared. This results in the same reduction in energy to one-quarter of its original value.

Discharging a Capacitor

A full capacitor stores both charge and energy. If a wire or a circuit is connected between the terminals of the capacitor, the potential difference and stored energy will cause charges to move from one plate of the capacitor to the other. This process occurs very rapidly at first and tapers off as the potential of the capacitor approaches zero. Eventually, the plates of the capacitor will reach a neutral charge.

One application involving capacitors is flash photography. A battery cannot deliver the surge of charge needed to create a quick and bright source of light. Instead, the battery is used to charge a capacitor slowly. Once the capacitor has been fully charged, a flash photo can be taken. Depressing the button on the camera to take the photo also closes a circuit between the plates of the capacitor. The charges stored on the capacitor surge from one plate of the capacitor to the other. Along the way, the charges pass through a specialized lightbulb, creating a quick and blinding flash.

SUMMARY

1. **POTENTIAL IS UNIQUE TO CHARGES AND ELECTRIC FIELDS.** Every point in space between charged plates or surrounding point charges has an electric potential. Electric potential can be thought of as electrical pressure that can move charges and operate circuits. The equations used for electric potential differ depending on the electric field. The potential of charged plates is $V = Ed$, while the potential of point charges is $V = kq/r$. Regions of positive charge are considered to be at high potential, while regions of negative charge are at low potential.

2. **ELECTRIC POTENTIAL ENERGY IS THE PRODUCT OF CHARGE AND POTENTIAL.** Although electric potential relies on only a location in an electric field, electric potential energy is also influenced by the size of the charge at that location. Think of electric potential energy as the potential to move a specific charge in a problem. Electric potential energy is similar to gravitational potential energy.

3. **A CHARGE RELEASED IN AN ELECTRIC FIELD WILL ACCELERATE THROUGH A POTENTIAL DIFFERENCE, DOING WORK.** When a charge is released in an electric field, it will accelerate along the electric field. As it does so, both the electric potential and electric potential energy will change. The change in electric potential is known as a potential difference. The change in electric potential energy is known as work. Positive charges released from rest move with the electric field from high potential to low potential. Negative charges released from rest move opposite to the electric field from low potential to high potential.

4. **WHEN A CHARGE ACCELERATES THROUGH A POTENTIAL DIFFERENCE, ENERGY IS CONSERVED.** When a charge is released from rest and moves parallel to electric field lines, the force of electricity does work on the charge. As a result, the charge will lose electric potential energy and gain kinetic energy. Energy is conserved in the process, and the gain in kinetic energy is equal to the change in potential energy.

If You See	Try	Keep in Mind
Potential difference of charged plates	$\Delta V = E \Delta d$	Δd is the distance between the plates.
Potential of an object located between charged plates	$V = Ed$	d is the distance from the negative plate to the specific location.
Potential at a location near a point charge	$V = k\dfrac{(\pm q)}{|r|}$	The sign on the charge is important, but the distance, r, is always positive. Voltage is a scalar quantity. It can be positive or negative, but it has no overall direction.
Electric potential energy	$U_E = qV$	The formula works for any type of electric field, but V must be calculated with the correct formula.
Work of electricity	$W = K = -\Delta U_E$ $W = \Delta K = -q(V_f - V_i)$	W and ΔK are positive when objects speed up and negative when objects slow. ΔU_E has the opposite sign.
A charge accelerated through a potential difference	$K_i + U_{Ei} = K_f + U_{Ef}$ $\dfrac{1}{2}mv_i^2 + qV_i = \dfrac{1}{2}mv_f^2 + qV_f$	If an object starts at rest one of the equations simplifies to: $\dfrac{1}{2}mv_{\max}^2 = q\,\Delta V$
Capacitor with changing properties	$C = \dfrac{\epsilon_0 A}{d}$ $Q = CV$ $U_c = \dfrac{1}{2}QV = \dfrac{1}{2}CV^2$	Changes to electric potential, V, may also cause the amount of charge stored, Q, to change.

PRACTICE EXERCISES

Questions 1-3

Two charged plates have a charge of 3.0 coulombs, are separated by a distance of 10 centimeters, and have a potential difference of 12 volts.

1. Determine the magnitude of the electric field between the plates.

    (A)   0.4 V/m
    (B)   1.2 V/m
    (C)   3.6 V/m
    (D) 40 V/m
    (E) 120 V/m

2. Determine the capacitance of the charged plates.

    (A)   0.25 F
    (B)   0.50 F
    (C) 18 F
    (D) 25 F
    (E) 36 F

3. Determine the energy stored by the charged plates.

    (A)   2 J
    (B)   6 J
    (C) 12 J
    (D) 18 J
    (E) 36 J

    ————————————————————————————

            12 V————————————
             0 V————————————
           −12 V————————————

4. The electric potential for a region in space is shown in the diagram above. Which statement is correct?

    (A) The electric field is directed to the right, $+x$ direction.
    (B) The electric field is directed to the left, $-x$ direction.
    (C) The electric field is directed up the page, $+y$ direction.
    (D) The electric field is directed down the page, $-y$ direction.
    (E) The diagram is incorrect. Potential is never negative.

Questions 5–6

A 1.5-coulomb charge with a mass of 0.50 kilograms is initially at rest. It is released and travels 2.0 meters in a uniform 3.0-volt-per-meter electric field.

5. How much work is done on the charge during this motion?

    (A) 2.25 J
    (B) 4.5 J
    (C) 9 J
    (D) 12 J
    (E) 18 J

6. What is the speed of the charge after traveling 2.0 meters?

    (A) 3.0 m/s
    (B) 6.0 m/s
    (C) $4\sqrt{3}$ m/s
    (D) $6\sqrt{2}$ m/s
    (E) 12 m/s

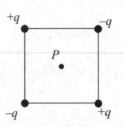

7. Four charges of equal magnitudes are arranged in a square, as shown in the diagram above. Which answer is true at point P, which is located at the exact center of the square?

    (A) $E \neq 0$ and $V < 0$
    (B) $E \neq 0$ and $V = 0$
    (C) $E \neq 0$ and $V > 0$
    (D) $E = 0$ and $V = 0$
    (E) $E = 0$ and $V > 0$

8. A conducting sphere with a mass of 2.0 kilograms and a charge of 1.0 coulomb is initially at rest. Determine its speed after being accelerated through a 16-volt potential difference.

    (A) 2.0 m/s
    (B) 4.0 m/s
    (C) 16 m/s
    (D) 24 m/s
    (E) 32 m/s

9. A capacitor is fully charged and the power supply is disconnected, isolating the capacitor completely. The plates are pulled apart. This results in the

(A) capacitance increasing and the potential increasing
(B) capacitance increasing and the potential decreasing
(C) capacitance decreasing and the potential increasing
(D) capacitance decreasing and the potential decreasing
(E) no change to either the capacitance or the potential

10. A capacitor is connected to a battery with potential V and is fully charged. The distance between the plates of the capacitor is doubled. As a result, the energy stored by the capacitor is

(A) $\frac{1}{4}$ of its original value

(B) $\frac{1}{2}$ of its original value

(C) the same as its original value
(D) 2 times greater than its original value
(E) 4 times greater than its original value

ANSWERS EXPLAINED

	Key Words	Needed for Solution	Now Solve It
1. **(E)**	Charged plates; distance; potential; electric field	$V = Ed$ Rearranged to solve for electric field	Convert 10 centimeters to 0.1 meters. $$E = \frac{V}{d} = \frac{(12\,\text{V})}{(0.1\,\text{m})} = 120\,\text{V/m}$$ For this question, the charge of 3.0 C and the plate area are not needed. Note that both N/C and V/m are valid electric-field units.
2. **(A)**	Charged plates; charge; potential; capacitance	$Q = CV$ Rearranged to solve for capacitance	$$C = \frac{Q}{V} = \frac{(3.0\,\text{C})}{(12\,\text{V})} = 0.25\,\text{F}$$ For this question, distance between the plates is not needed.
3. **(D)**	Charged plates; charge; potential; energy	$U_C = \frac{1}{2}QV = \frac{1}{2}CV^2$ Either equation works since the variables for both are now known.	$$U_C = \frac{1}{2}QV = \frac{1}{2}(3.0\,\text{C})(12\,\text{V}) = 18\,\text{J}$$ or $$U_C = \frac{1}{2}CV^2 = \frac{1}{2}(0.25\,\text{F})(12\,\text{V})^2 = 18\,\text{J}$$
4. **(D)**	Potential; electric field; direction	Knowledge/definitions	The electric field is a vector that is perpendicular to the lines of equal potential and that points from high potential (12 V) to low potential (−12 V). Therefore, the electric-field direction is down the page, −y.
5. **(C)**	Charge; at rest; released; travels 2.0 m; uniform 30 V/m electric field; work	Work–kinetic energy theorem $W = \Delta K = -\Delta U_E$ Potential difference in uniform electric fields $V = E\,\Delta d$	Work is related to changes in kinetic energy, which are equal in value to changes in potential energy. Changes in potential energy are related to changes in electric potential $$W = \Delta K = -\Delta U_E = -q\,\Delta V$$ $$W = -qE\,\Delta d$$ The variables given require using potential energy to solve for work, but there was no indication of the correct signs on the variables. Find the absolute value for work. Then assess the resulting motion of the object to confirm the correct sign on work. $$\lvert W \rvert = (1.5\,\text{C})(3.0\,\text{V/m})(2.0\,\text{m}) = 9.0\,\text{J}$$ Since the object started at rest, it must be speeding up. The change in kinetic energy is positive, and work is positive. In addition, force and displacement can be compared. When the directions of the force and displacement vectors are the same, work is positive. When these vectors oppose each other, work is negative.

	Key Words	Needed for Solution	Now Solve It
6. **(B)**	Charge; mass; initially at rest; travels 2.0 m; uniform 3.0-volt-per-meter electric field; speed	Conservation of energy starting from rest $$\frac{1}{2}mv^2 = q\,\Delta V$$ Potential difference in uniform electric fields $$V = E\,\Delta d$$	The change in distance is related to a change in electric potential and in potential energy. The change in potential energy equals the change in kinetic energy. $$\frac{1}{2}\,mv^2 = q\,\Delta V$$ $$v = \sqrt{\frac{2q\,\Delta V}{m}} = \sqrt{\frac{2qE\,\Delta d}{m}} = \sqrt{\frac{2(1.5\,\text{C})(3.0\,\text{V/m})(2.0\,\text{m})}{(0.5\,\text{kg})}}$$ $$v = 6.0\,\text{m/s}$$
7. **(D)**	Four charges; point P	Superposition of both electric field (see Chapter 10) and electric potential	Each charge has the same magnitude, $|q|$, and is the same distance, r, from point P. Therefore, the magnitude of the four resulting field vectors will all be the same. Vector direction is the critical factor in many problems. Drawing the four electric-field vectors, due to each charge, at point P reveals that all the vectors cancel each other, $E = 0$. However, electric potential is not a vector. The sign of the charge is included in calculations, but the sign on distance is not. The potential of the four charges is summed. $$V = k\frac{(+q)}{|r|} + k\frac{(+q)}{|r|} + k\frac{(-q)}{|r|} + k\frac{(-q)}{|r|} = 0\,\text{V}$$
8. **(B)**	Mass; charge; initially at rest; speed; potential difference	Conservation of energy $$\frac{1}{2}\,mv^2 = q\,\Delta V$$	$$\frac{1}{2}\,mv^2 = q\,\Delta V$$ $$v = \sqrt{\frac{2q\,\Delta V}{m}} = \sqrt{\frac{2(1.0\,\text{C})(16.0\,\text{V})}{(2.0\,\text{kg})}} = 4.0\,\text{m/s}$$
9. **(C)**	Capacitor; fully charged; isolating; plates are pulled apart	$$C = \frac{\epsilon_0 A}{d}$$ $$Q = CV$$	First equation: Pulling the plates apart increases the separation distance, d, which is inversely proportional to capacitance, C. As a result, capacitance decreases. Second equation: When a capacitor is isolated, charge remains constant since the charges cannot move. Capacitance is inversely proportional to electric potential. So, the decrease in capacitance, C, causes an increase in electric potential, V.

	Key Words	Needed for Solution	Now Solve It
10. **(B)**	Distance; doubled; energy stored; capacitor	$C = \dfrac{\epsilon_0 A}{d}$ $Q = CV$ $U_c = \dfrac{1}{2}QV = \dfrac{1}{2}CV^2$	Doubling distance, d, cuts capacitance, C, in half. $$\left(\dfrac{1}{2}C\right) = \dfrac{\epsilon_0 A}{(2d)}$$ Since the battery is still connected, the voltage remains constant. Using $U_c = \dfrac{1}{2}CV^2$ yields the quickest answer. $$\left(\dfrac{1}{2}U_C\right) = \dfrac{1}{2}\left(\dfrac{1}{2}C\right)V^2$$

Circuit Elements and DC Circuits

12

→ **PRINCIPAL COMPONENTS OF A DC CIRCUIT**

→ **DC CIRCUITS**

→ **HEAT AND POWER DISSIPATION**

Electric fields induced in a wire will allow current to flow in a circuit. As current flows, it can pass through resistors, illuminate lightbulbs, or accumulate along capacitors. This chapter will help to do the following:

- Define the principal components of a DC circuit.
- Solve problems involving simple DC circuits using Ohm's law.
- Determine heat and power dissipated from a DC circuit.

Table 12.1 lists the variables and units that will be encountered in this chapter.

Table 12.1 Variables and Units Used in Circuits

New Variables	Units
R = Resistance	Ω (ohms)
P = Power	W (watts) or joules per second
I = Current	A (ampere) or coulombs per second

PRINCIPAL COMPONENTS OF A DC CIRCUIT

Battery

The purpose of a **battery** is to create an electric potential difference. Think of a battery as a pump that creates the electrical pressure to push charges through an electric circuit. When a wire is connected between the positive and negative terminals of a battery, charges flow as a current through the wire on their way to the negative terminal of the battery. Potential (voltage) is analogous to height in mechanics. When a mass is released from rest, it moves from a high height to a low height while moving through a distance Δh. If a positive charge is released from rest, it moves from a high potential to a low potential while moving through a potential difference ΔV. Positive charges match the force and energy characteristics of masses in gravity fields. Thus, positive charges are the default charge in electricity. As a result, the convention is to visualize positive charges leaving the positive terminal of the battery (high potential) and flowing through the circuit on the way to the negative terminal (low potential). Although this is how circuits are analyzed, it is not in truth what is really happening. Only electrons are free to move in circuits, and the actual motion of charges is composed of electrons moving in the opposite direction. Mathematically it makes no difference, as protons and electrons have the same magnitude of charge. This only affects direction of charge flow. Even though it is not what is actually happening, the convention is to visualize the flow charges in a circuit as a positive flow.

The potential difference (voltage) between the positive and negative terminals of the battery provides the electrical pressure to the circuit. The overall resistance of the components in the circuit dictates the amount of charge allowed to flow from the battery. The flow of charge is known as the **current**, I, and is measured in units of amperes (A) or coulombs per second.

In circuit diagrams, a long line and a short line represent batteries, as shown in Figure 12.1(a). Sometimes a battery is represented by a series of long and short lines as seen in Figure 12.1(b). This represents multiple battery cells connected in series, much like a series of batteries lined up end to end in a flashlight.

(a)   (b)

Figure 12.1 Batteries

IF YOU SEE
current,
voltage, and
resistance

Ohm's law

$$I = \frac{V}{R}$$

While Ohm's Law if often written as, $V = IR$, the above version of the equation expresses causality correctly. The current flowing through a component in a circuit depends on the potential difference across that component and the resistance of the component. The magnitude of the current is directly proportional to potential and inversely proportional to resistance.

The side of the battery drawn with a long line is the positive terminal. The side drawn with a short line is the negative terminal. When batteries are connected into circuits, positive charges are thought of as departing from the positive terminal, traveling through the circuit, and arriving back at the negative terminal of the battery. Remember: Actual charge flow is electrons in the opposite direction.

Resistors

A **resistor** is a device with a known amount of resistance. As a result, resistors can be used to control current flow in the various branches of a circuit. The **resistance**, R, to current flow has many applications in how a circuit works. As current flow is resisted, some of the electrical energy is dissipated into heat energy in units of joules. Figure 12.2 shows the symbol for a resistor used in circuit diagrams.

Figure 12.2 Resistor symbol

Resistors are rated by how much they can resist the flow of current. The actual measurement of resistance is in units of ohms, Ω. Ohm's law can be used to determine the resistance of ohmic resistors:

$$V = IR$$

Using the voltage (V) applied to the circuit and the current (I) flowing through the circuit, you can determine the resistance (R). **Ohmic resistors** are resistors that obey Ohm's law. Resistors that do not obey Ohm's law are said to be nonohmic. Problems on the SAT Subject Test will most likely involve resistors and other devices having resistance that are ohmic. Unless otherwise stated in the exam, assume devices are ohmic.

Lightbulbs are essentially resistors that give off light. The principal difference between a plain resistor and a lightbulb is that a lightbulb uses some of the energy dissipated through resistance to produce light. In circuit diagrams, lightbulbs may be represented as resistors. However, the symbol for lightbulbs often varies, as shown in Figure 12.3.

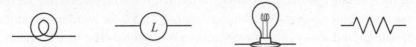

Figure 12.3 Lightbulb symbols

Switch

The purpose of a **switch** is to direct current flow around a circuit. In a schematic diagram of a circuit, a switch is represented as a line at an angle to the circuit. Figure 12.4 shows a switch in the open position. When open, a switch blocks the flow of current. When a switch is closed, the circuit becomes complete, and a current is able to flow.

Figure 12.4 Symbol for an open switch

DC CIRCUITS

The SAT Subject Test in Physics will most likely have series and parallel-resistor circuit problems for you to solve. Typically, these problems can be categorized as one of three types: series only, parallel only, or series-parallel. Each category has specific problem-solving strategies involved. Those strategies will be discussed in this section.

When multiple resistors are used in a circuit, they work together and have a combined resistance known as **equivalent resistance**. Adding resistors results in a single mathematical resistance value that describes a single resistor that is equivalent to the resistors working together. A single resistor possessing this equivalent resistance can replace all the resistors added together. The method of finding equivalent resistance differs for series and parallel circuits.

Series Circuits

In a series circuit, the electrical components are arranged so there is only one path through the circuit. Figure 12.5 shows three resistors arranged in series with a 12-volt battery.

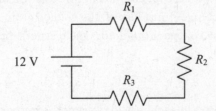

Figure 12.5 Series circuit

The equivalent resistance for resistors in series, R_s, is simply the sum of the individual resistances.

$$R_s = R_1 + R_2 + R_3 + \cdots$$

The equivalent resistance for an entire circuit is a single resistance value that represents the total resistance of all the resistors functioning together. This is the resistance that the voltage (electric pressure) produced by the battery must push against. The magnitude of the total current leaving the battery depends on the voltage (electric pressure) produced by the battery and the total equivalent resistance of the circuit, according to Ohm's law.

$$I = \frac{V}{R}$$

As more resistors are added in series, the total resistance increases. Current flow is inversely proportional to resistance. Increasing resistance decreases the total amount of current in a series circuit.

In the complete series circuit shown in Figure 12.5, there is only one pathway. This means the current must flow through every resistor in turn. Think of current as water and the wires and resistors as the pipes carrying the water from the top of a hill (the positive terminal of the battery) to the bottom of a hill (the negative terminal of the battery). Current is composed of charges, and charge is conserved in circuits. The charges leaving the battery as a current must return to the battery. In a series circuit, there is a single path that all the current must follow. With only one pathway available, the current leaving the battery has the same magnitude as the current passing through each of the resistors.

$$I_S = I_1 = I_2 = I_3 = \cdots$$

Although resistors in series all receive the same current, they do not have the same voltage. Voltage (electric potential) is similar to height in Newtonian mechanics. The 12-volt potential difference of the battery in Figure 12.5 can be thought of as a 12-volt hill. The positive terminal of the battery is at the top of a hill, with a value of 12 volts. The negative terminal of the battery is at the bottom of the hill, with a value of 0 volts. As charges flow through the battery energy from the chemical reaction in the battery increases the voltage of each of the charges from 0 to 12 volts. The action of the battery is analogous to a machine that pulls a rollercoaster from zero height to the top of a hill. Voltage is proportional to electric potential energy just as height is proportional to gravitational potential energy. When the charges leave the positive terminal of the battery they are thought of as flowing downhill toward the negative terminal of the battery, essentially falling through the 12-volt potential difference. On their way to the bottom, the charges pass through circuit components such as resistors. Energy is conserved in circuits, and the resistors in the circuit must use all the energy produced by the battery. As a result, the charges must lose all the voltage given to them by the battery. In a series circuit the sum of the voltage drops across the resistors must be equal to the voltage increase of the battery.

$$V_S = V_1 + V_2 + V_3 = \cdots$$

In addition, the voltage drop in each resistor in series is proportional to the resistance of the resistors.

EXAMPLE 12.1

Solving Series Circuits

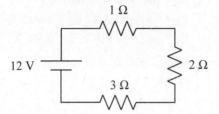

Determine the voltage drop across each resistor.

(WHAT'S THE TRICK?)

You can use a table like the one below to organize circuit problems. The column headings are the variables for Ohm's law, and any row in the table can be calculated using this law. The components used in the circuit are listed on the left. The top row contains the total values for the entire circuit and is labeled as the battery. The battery produces the total potential for the circuit, pushes against the total resistance in the circuit, and supplies the total current to the circuit. The values shown in bold type are the values given in the original problem. The values in regular type are the values determined during the course of the problem. They are preceded by a number enclosed in parentheses. This is the order in which the problem is solved. Below the table, the four steps are shown in detail.

	V	**I**	**R**
Battery	**12 V**	(2) 2 A	(1) 6 Ω
R_1	(4) 2 V	(3) 2 A	**1 Ω**
R_2	(4) 4 V	(3) 2 A	**2 Ω**
R_3	(4) 6 V	(3) 2 A	**3 Ω**

1. Add the resistors to find total equivalent resistance.

$$R_S = R_1 + R_2 + R_3$$
$$R_S = 1\,\Omega + 2\,\Omega + 3\,\Omega = 6\,\Omega$$

2. The battery supplies the total voltage, V_T, that that as the electric pressure pushing charges against the the total equivalent resistance of the entire circuit, R_T. Use Ohm's law to find the total current, I_T, leaving the battery.

$$I_T = \frac{V_T}{R_T}$$

$$I_T = \frac{12\text{ V}}{6\,\Omega} = 2\text{ A}$$

DC Circuits: Series

online.barronsbooks.com

3. The total current leaving the battery must move through each resistor, and is the same in every component is series.

$$I_T = I_1 = I_2 = I_3 = 2\text{ A}$$

4. Use Ohm's law for each resistor.

$$V_1 = I_1R_1 = (2\text{ A})(1\text{ }\Omega) = 2\text{ V}$$
$$V_2 = I_2R_2 = (2\text{ A})(2\text{ }\Omega) = 4\text{ V}$$
$$V_3 = I_3R_3 = (2\text{ A})(3\text{ }\Omega) = 6\text{ V}$$

Compare the ratio of resistance to the ratio of voltage

$$R_1 : R_2 : R_3 \qquad V_1 : V_2 : V_3$$
$$1 : 2 : 3 \qquad\quad 2 : 4 : 6$$

Energy is conserved, and in series resistors use energy proportional to their resistance. Voltage is directly proportional to electric energy and the voltage of resistors in series will be proportional to the resistances.
You can use the voltage values of the individual resistors to double-check your solution. In series, the voltage drops should sum to equal the voltage of the battery.

$$V_T = V_1 + V_2 + V_3$$
$$V_T = 2\text{ V} + 4\text{ V} + 6\text{ V} = 12\text{ V}$$

IF YOU SEE
a parallel circuit

$$V_T = V_1 = V_2 = \cdots$$
$$\frac{1}{R_T} = \frac{1}{R_1} + \frac{1}{R_2} + \cdots$$
$$I_T = I_1 + I_2 + \cdots$$

Parallel circuits have more than one pathway. Voltage is the same through each resistor, while resistance and current vary.

Using a table such as the one shown in Example 12.1 makes it easy to organize the values that may be required in a particular problem. This table illustrates the voltage, current, and resistance for individual resistors and the entire circuit. You can answer many general questions about a circuit simply by reading the completed table.

Parallel Circuits

In a parallel circuit, the electrical components are arranged so there is more than one path through the circuit. Figure 12.6 shows three resistors arranged in parallel with a battery.

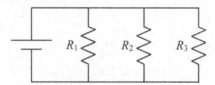

Figure 12.6 Parallel circuits

The equivalent resistance for resistors in parallel, R_P, is more complicated than for resistors in series. Parallel resistors are added together so that the sum of their individual reciprocals equals the reciprocal of the total resistance.

$$1/R_P = 1/R_1 + 1/R_2 + 1/R_3 + \cdots$$

Since the above formula solves for the reciprocal of parallel resistance, $1/R_P$, you must remember to invert the answer to determine the equivalent resistance in parallel, R_P. In Figure 12.6, all the resistors are in parallel. The equivalent resistance is also the total resistance that the battery must push against. When resistors are added in parallel the equivalent resistance is always less than the value of the smallest resistor that was summed. As more resistors are added in parallel, the total resistance decreases, causing the total current in the circuit to increase.

In a parallel circuit, the current moves through multiple pathways. The current flowing through each resistor is a portion of the total current that is produced by the battery. Think of the current as a fixed amount of water leaving the battery. When the water arrives at the junctions, where the paths in the circuit separate, the water will split. Some will flow through resistor 1, some through resistor 2, and the remainder through resistor 3. Then the water will reunite and continue back toward the battery. The total amount of water in the circuit is always the same. When it splits, it must still add up to the total amount that leaves the battery and that later returns to the battery. Therefore, the total current produced by the battery, I_T, is equal to the sum of the current in the parallel resistors. Again, this is based on conservation of charge. Whenever current arrives at a junction (intersection) in the circuit the current must split in a manner so that the amount of current entering the intersection must equal the amount of current leaving the intersection. The amount of current in the different pathways in a parallel circuit is inversely related to the ratio of resistance. As an example: If there are two parallel pathways and the second path has twice the resistance compared to the first path, then the second pathway will only carry half as much current as the first pathway.

$$I_T = I_1 + I_2 + I_3 + \cdots$$

The potential in a parallel circuit is the same for all resistors. Again, potential (voltage) can be thought of as height, and current can be represented by the flow of water. The battery pushes the water up the hill through a set height. In Figure 12.6, there are three paths, one through each resistor. If water leaves the top of the hill (positive terminal of the battery) and flows through the circuit, it may follow different paths. However, the water must always drop the same amount of height in order to return to the bottom of the hill (negative terminal of the battery). The drop in height must also be equal to the height created by the pumping action of the battery. In a parallel circuit, the voltage drops across each resistor are equal to each other and are equal to the total voltage produced by the battery.

$$V_T = V_1 = V_2 = V_3 = \cdots$$

Most electrical wiring for home use is wired in parallel. Each time a switch is closed, the same voltage (120 V for most homes) is applied across a resistor (lightbulb, television set, computer, and so on). The effect of having one appliance turned on or off does not affect the voltage applied to other appliances that are turned on. A home wired in series would require all appliances to be on all the time in order for the circuit to be completed.

EXAMPLE 12.2

Solving Parallel Circuits

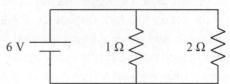

DC Circuits: Parallel

online.barronsbooks.com

Determine the current in each resistor.

WHAT'S THE TRICK?

As before, use a table to organize the problem.

	V	**I**	**R**
Battery	**6 V**	(2) 9 A	(1) $\frac{2}{3}\Omega$
R_1	(3) 6 V	(4) 6 A	$1\,\Omega$
R_2	(3) 6 V	(4) 3 A	$2\,\Omega$

1. To find the total resistance, add the inverse of each resistor.

$$\frac{1}{R_P} = \frac{1}{R_2} + \frac{1}{R_2}$$

$$\frac{1}{R_P} = \frac{1}{1\,\Omega} + \frac{1}{2\,\Omega} = \frac{3}{2}\ \Omega^{-1}$$

It is very important that you invert this value: $R_T = \frac{2}{3}\ \Omega$

2. Use Ohm's law to find the total current.

$$I_T = \frac{V_T}{R_T}$$

$$I_T = \frac{6\ \text{V}}{2/3\ \Omega} = 9\ \text{A}$$

3. Voltage remains the same in parallel.

$$V_T = V_1 = V_2 = V_3 = 6\ \text{V}$$

4. Use Ohm's law for each resistor.

$$I_1 = \frac{V_1}{R_1} = \frac{6\ \text{V}}{1\,\Omega} = 6\ \text{A}$$

$$I_2 = \frac{V_2}{R_2} = \frac{6\ \text{V}}{2\,\Omega} = 3\ \text{A}$$

Compare the ratio of resistance to the ratio of current in each resistor.

$$R_1 : R_2 \qquad I_1 : I_2$$
$$1 : 2 \qquad 6 : 3$$
$$2 : 1$$

When resistance is high current is low, and vice versa.

Use the current values of the individual resistors to double-check the solution. Charge is conserved. As a result, the current in each parallel pathway should sum to equal the current produced by the battery.

$$I_T = I_1 + I_2 = 6\ \text{A} + 3\ \text{A} = 9\ \text{A}$$

Series-Parallel Circuits

A series-parallel circuit is somewhat more complicated. However, you can employ the same table. Before using the table, however, it is helpful to find the total equivalent resistance, R_T, of the circuit.

EXAMPLE 12.3

Series-Parallel Circuits

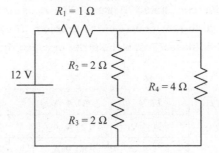

(A) Determine the equivalent resistance.

Group resistors and add them together to simplify the circuit progressively. In the diagram above, R_2 and R_3 are in series and can be consolidated as R_{23}.

$$R_{23} = R_2 + R_3 = 2\,\Omega + 2\,\Omega = 4\,\Omega$$

Now redraw the circuit.

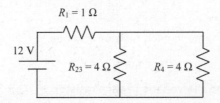

R_{23} and R_4 can now be added in parallel to obtain equivalent resistor R_{234}:

$$\frac{1}{R_{234}} = \frac{1}{R_{23}} + \frac{1}{R_4} = \frac{1}{4\,\Omega} + \frac{1}{4\,\Omega} = \frac{2}{4}\,\Omega^{-1}$$

Invert to solve for resistance in parallel.

$$R_{234} = 2\,\Omega$$

Redraw the circuit as shown below.

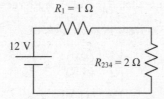

R_{234} can now be added to R_1 to find the total equivalent resistance in the circuit.

$$R_T = R_{1234} = R_1 + R_{234} = 1\,\Omega + 2\,\Omega = 3\,\Omega$$

DC Circuits: Series-Parallel

online.barronsbooks.com

CIRCUIT ELEMENTS AND DC CIRCUITS   255

(B) Determine the voltage drops and current through each resistor.

WHAT'S THE TRICK?

The values shown in bold type below are the values given in the original problem and the values of the equivalent resistances for R_{23}, R_{234}, and R_{1234}. The total resistance, R_{1234}, is the resistance the battery must push against. It is recorded in the table below on the row labeled "Battery." The values in regular type are the values determined during the course of the problem. They are preceded by a number enclosed in parentheses. This is the order in which the problem is solved.

	V	I	R
Battery	12 V	(1) 4 A	3 Ω
R_1	(3) 4 V	(2) 4 A	1 Ω
R_2	(7) 4 V	(6) 2 A	2 Ω
R_3	(7) 4 V	(6) 2 A	2 Ω
R_4	(4) 8 V	(5) 2 A	4 Ω
R_{23}	(4) 8 V	(5) 2 A	4 Ω
R_{234}	(3) 8 V	(2) 4 A	2 Ω

1. Use Ohm's law to solve for the total current.

$$I_T = \frac{V_T}{R_T} = \frac{12 \text{ V}}{3 \text{ Ω}} = 4 \text{ A}$$

2. Current remains the same in each resistor in series.

$$I_T = I_1 = I_{234} = 4 \text{ A}$$

3. Use Ohm's law to find the voltage in each resistor.

$$V_1 = I_1 R_1 = (4 \text{ A})(1 \text{ Ω}) = 4 \text{ V}$$
$$V_{234} = I_{234} R_{234} = (4 \text{ A})(2 \text{ Ω}) = 8 \text{ V}$$

Voltage adds in series. You can use this to double-check the values so far.

$$V_T = V_1 + V_{234} = 4 \text{ V} + 8 \text{ V} = 12 \text{ V}$$

4. There is a voltage drop of 4 V across the R_1 resistor. Since the battery has a total voltage of 12 V, this 4-V drop leaves 8 V to go across parallel resistors R_{23} and R_4. Since they are in parallel, R_{23} and R_4 both receive 8 V.

5. Use Ohm's law to find the current in each resistor.

$$I_4 = \frac{V_4}{R_4} = \frac{(8 \text{ V})}{(4 \text{ Ω})} = 2 \text{ A}$$

$$I_{23} = \frac{V_{23}}{R_{23}} = \frac{(8 \text{ V})}{(4 \text{ Ω})} = 2 \text{ A}$$

Current adds in parallel. You can use this to double-check the values.

$$I_{234} = I_{23} + I_4 = 2 \text{ A} + 2 \text{ A} = 4 \text{ A}$$

6. It has been established that 4 A flows through R_{234}. Of this total, 2 A flows through R_4. Therefore, the remaining 2 A must flow through R_2 and R_3.

7. Use Ohm's law to find the voltage in each resistor.

$$V_2 = I_2 R_2 = (2 \text{ A})(2 \text{ }\Omega) = 4 \text{ V}$$
$$V_3 = I_3 R_3 = (2 \text{ A})(2 \text{ }\Omega) = 4 \text{ V}$$

Voltage adds in series. You can use this to double-check the values.

$$V_{23} = V_2 + V_3 = 4 \text{ V} + 4 \text{ V} = 8 \text{ V}$$

HEAT AND POWER DISSIPATION

As current flows through resistors, the frictional resistance causes the wires to heat up. The amount of heat, Q, dissipated per second is known as power dissipation. Heat is similar to work in that heat is also a transfer of energy into or out of a system. Heat is the transfer of thermal energy. Be careful when working with heat in electricity problems. The variable Q is used both for heat and for charge. It is possible to mistake them for each other. Power is quantified in units of watts (joules per second).

Joule's Law

Joule's law states that the heat dissipated in a circuit is equal to the current squared multiplied by the resistance and the time that the current flows through the resistor.

$$\text{Heat} = Q = I^2 R t$$

In this case, the variable Q represents heat measured in joules. Joule's law shows that heat is directly proportional to the square of current. If current doubles, heat quadruples.

Power

Power is the rate of change in energy. It is measured in joules per second (J/s), which is known as watts (W).

$$P = \frac{\Delta E}{t}$$

In circuits, the change in energy may involve the electric energy generated (created) by power sources or dissipated (lost) in circuit components, $\Delta UE = QV$. Keep in mind that current is the amount of charge divided by time. Combining these equations creates a useful power equation for use in analyzing electric circuits.

$$P = \frac{\Delta E}{t} = \frac{QV}{t} = IV$$

The resulting equation, $P = IV$, can be combined with Ohm's law, $V = IR$, to create other useful power-equation variations.

$$P = IV = I^2 R = \frac{V^2}{R}$$

Most problems can be answered using both $P = IV$ and $V = IR$ as separate equations. However, being aware of all the power-equation variations does allow for faster problem solving.

IF YOU SEE
power

$P = IV$

$P = \dfrac{V^2}{R}$

$P = I^2 R$

Power is the rate of electric-energy dissipation.

IF YOU SEE
lightbulbs

Power determines brightness.

Series circuits have dimmer lights, while parallel circuits have brighter lights.

Brightness of Lightbulbs

The brightness of a lightbulb is related to the amount of power it dissipates. If given a circuit with different lightbulbs and asked to determine which lightbulb is brighter, simply treat the lightbulbs as resistors and solve for the current and voltage of each lightbulb, as shown in the previous circuit example problems. Then multiply the currents and voltages ($P = IV$) to determine the power dissipated by each bulb. The bulb dissipating the most power will be the brightest.

Lightbulbs are sold by the amount of power, in watts, that they dissipate. The greater the power printed on the bulb, the brighter it burns. However, the power that lightbulbs are rated as dissipating at is true only if the bulbs are used in typical household circuits. Homes in the United States are wired in parallel with voltages of 110 V to 120 V. In purely parallel circuits, voltage in all branches is the same as the voltage of the power source. Every lightbulb should then have the same voltage. Under these exact conditions, the printed power on a lightbulb is the power that will be dissipated. However, if the voltage supplied to a parallel circuit is not 110 V to 120 V or if the lightbulbs are moved to a series circuit, then the power dissipated will differ from the power rating printed on the lightbulb.

Remember that lightbulbs are essentially resistors that glow. Although the power rating on a bulb is not really constant, the resistance of a lightbulb is constant no matter which circuit it is used in. For lightbulbs, the relationship between power, potential, and resistance is important. This makes the following power equation very useful when working with lightbulbs.

$$P = \frac{V^2}{R}$$

In purely parallel circuits composed of lightbulbs, all branches and all bulbs have the same voltage as the power supply. This means they are all exposed to the maximum voltage, and they all dissipate the greatest amount of power. Therefore, lightbulbs in parallel circuits are at their brightest. Changing the voltage of the power supply changes the brightness of bulbs in parallel. According to the above equation, power is directly proportional to the square of voltage. Doubling voltage quadruples power dissipation.

If lightbulbs are wired in series, they use less power than they are rated for, and they burn dimmer. In series, voltage adds. Each bulb in series receives only a portion of the total voltage supplied by the power source. A lower voltage means less power and dimmer bulbs.

SUMMARY

1. **THE PRINCIPAL COMPONENTS OF A CIRCUIT** are batteries, resistors, and switches. Recognizing their symbols and functions is critical to solving circuit diagrams.

2. **STRATEGIES FOR SOLVING SERIES, PARALLEL, AND SERIES-PARALLEL CIRCUIT SCHEMATICS ARE BEST APPROACHED USING A TABLE.** The order of completing the table is helpful in understanding the values for voltage, current, and resistance in a circuit. Using the tables when confronted with a circuit diagram will help you answer the majority of questions pertaining to the circuit.

3. **JOULE'S LAW DESCRIBES THE HEAT ENERGY LOST THROUGH FRICTION IN UNITS OF JOULES.** Power is the amount of heat energy dissipated per second in units of joules per second (watts). Three equations based on Ohm's law can be used to solve for power. The easiest one to use in conjunction with the table is $P = IV$.

If You See	Try	Keep in Mind
Circuit schematic diagrams	Use the table and strategies for solving.	Remember the steps about how to obtain the values in the table.
Resistors in parallel-only circuits	Applying the shortcut method of solving for their individual currents based on the voltage of the battery and Ohm's law $I = V/R$	Parallel resistors receive the same voltage.
Questions about power dissipated in a circuit	$P = IV = I^2R = \dfrac{V^2}{R}$	Choose the formula for power that is the easiest to solve.
Voltage, current, and resistance	Ohm's law $V = IR$	Ohm's law works for individual resistors, groups of resistors, and the circuit as a whole.
A series circuit	Resistance adds normally $R_s = R_1 + R_2 + \cdots$ Current remains the same in series $I_s = I_1 = I_2 = \cdots$ Voltage adds in series $V_s = V_1 + V_2 + \cdots$	Trends occur in series: ■ High resistance ■ Low current ■ Low power ■ Dim lights
A parallel circuit	The inverses of resistance add $\dfrac{1}{R_P} = \dfrac{1}{R_1} + \dfrac{1}{R_2} + \cdots$ Voltage remains the same in parallel $V_P = V_1 = V_2 = \cdots$ Current adds in parallel $I_P = I_1 + I_2 + \cdots$	Trends occur in parallel: ■ Low resistance ■ High current ■ High power ■ Bright lights
Circuit diagrams	Organize the information in a table	Apply the rules for series and parallel circuits
Joule's law	$Q = I^2Rt$	When current flows through resistors, energy is dissipated as heat.
Power	$P = IV = I^2R = \dfrac{V^2}{R}$	Power may be referred to as the rate of electric-energy dissipation.
Brightness of bulbs	$P = IV = I^2R = \dfrac{V^2}{R}$	Bulbs consuming the most power are the brightest.

PRACTICE EXERCISES

Questions 1–3 refer to the following diagram.

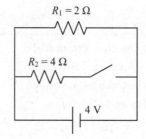

$R_1 = 2\,\Omega$

$R_2 = 4\,\Omega$

4 V

1. What current will flow through the 4-Ω resistor when the switch is closed?

    (A)   0.5 A
    (B)   1.0 A
    (C)   2.0 A
    (D)   4.0 A
    (E) 16.0 A

2. What is the total equivalent resistance of the circuit when the switch is closed?

    (A) $\dfrac{1}{2}\,\Omega$

    (B) $\dfrac{3}{4}\,\Omega$

    (C) $\dfrac{4}{3}\,\Omega$

    (D) 2 Ω

    (E) 4 Ω

3. What is the total power dissipated in the circuit when the switch is closed?

    (A)   1 W
    (B)   4 W
    (C) 10 W
    (D) 12 W
    (E) 16 W

Questions 4–8 refer to the following circuit diagram.

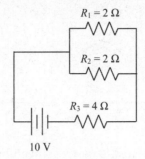

4. What current will flow through the 4-Ω resistor?

   (A) 0.5 A
   (B) 1.0 A
   (C) 2.0 A
   (D) 3.0 A
   (E) 4.0 A

5. What is the voltage drop across the 4-Ω resistor?

   (A) 2 V
   (B) 4 V
   (C) 6 V
   (D) 8 V
   (E) 10 V

6. How much heat is generated in the 4-Ω resistor in 2 seconds?

   (A) 8 J
   (B) 10 J
   (C) 16 J
   (D) 20 J
   (E) 32 J

7. How much power is dissipated in the 4-Ω resistor each second?

   (A) 8 W
   (B) 10 W
   (C) 16 W
   (D) 20 W
   (E) 32 W

8. What is the voltage drop across the 2-Ω resistor labeled R_1?

   (A) 2 V
   (B) 4 V
   (C) 6 V
   (D) 8 V
   (E) 10 V

Questions 9–10

The circuits shown below are connected to power sources that provide the same electric potential V.

I

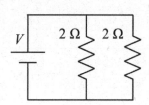

II

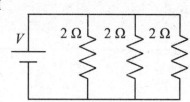

III

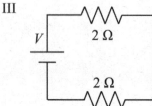

IV

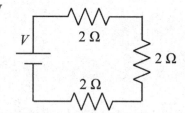

9. Which of the circuits will dissipate the most power?

   (A) I only
   (B) II only
   (C) III only
   (D) IV only
   (E) They will each dissipate the same amount of power.

10. In which circuit will the voltage of every resistor be the same as the voltage, V, of the battery?

   (A) I only
   (B) II only
   (C) III only
   (D) I and II
   (E) III and IV

ANSWERS EXPLAINED

	Key Words	Needed for Solution	Now Solve It
1. **(B)**	Current will flow through the 4-Ω resistor	Voltage remains the same in parallel $$V_{total} = V_1 = V_2 = \cdots$$ $$V = IR$$	Parallel resistors receive the same voltage. $$V_{total} = V_1 = V_2 = 4 \text{ V}$$ Use Ohm's law to solve for current. $$I = \frac{V}{R} = \frac{4 \text{ V}}{4 \ \Omega} = 1 \text{ A}$$
2. **(C)**	Total equivalent resistance of the circuit	Resistance in parallel $$\frac{1}{R_p} = \frac{1}{R_1} + \frac{1}{R_2}$$	$$\frac{1}{R_T} = \frac{1}{2} + \frac{1}{4} = \frac{3}{4}$$ Invert to solve for total resistance. $$R_T = 4/3 \ \Omega$$
3. **(D)**	Total power dissipated in the circuit	Three possible power equations $$P = IV = V^2/R = I^2R$$	Select the equation for power that matches the given variables. The total voltage at the battery is given and the previous problem solved for total resistance. $$P = \frac{V^2}{R} = \frac{(4 \text{ V})^2}{\left(\frac{4}{3}\Omega\right)} = 12 \text{ W}$$ Using alternate variations of the power formula will produce the same result, but they require you to solve for total current first.
4. **(C)**	Current will flow through the 4-Ω resistor	Resistors in parallel $$\frac{1}{R_p} = \frac{1}{R_1} + \frac{1}{R_2}$$ Resistors in series $$R_s = R_1 + R_2$$ Ohm's law $$V = IR$$ Current remains the same in series $$I_{total} = I_1 = I_2 = \cdots$$	Add resistors 1 and 2 in parallel. $$\frac{1}{R_{12}} = \frac{1}{R_1} + \frac{1}{R_2} = \frac{1}{2} + \frac{1}{2} = 1$$ $$R_{12} = 1 \ \Omega$$ Add resistors R_{12} and R_3 in series to find the total resistance $$R_T = R_{12} + R_3 = 1 + 4 = 5 \ \Omega$$ Use Ohm's law to find the total current in the circuit. $$I_T = \frac{V_T}{R_T} = \frac{(10 \text{ V})}{(5 \ \Omega)} = 2 \text{ A}$$ The 4-Ω resistor is in series, and the same 2-A current passes through it.
5. **(D)**	Voltage drop around the 4-Ω resistor	Ohm's law $$V = IR$$	Use the current found in the previous question. $$V = IR = (2 \text{ A})(4 \ \Omega) = 8 \text{ V}$$
6. **(E)**	Heat is generated; 2 seconds	Joule's law $$Q = I^2Rt$$	$$Q = I^2Rt$$ $$Q = (2 \text{ A})^2(4 \ \Omega)(2 \text{ s}) = 32 \text{ J}$$

	Key Words	Needed for Solution	Now Solve It
7. **(C)**	Power is dissipated; each second	$P = I^2R$	Power is the rate of energy use (joules per second). $$P = I^2R = (2\text{ A})^2(4\ \Omega) = 16\text{ W}$$
8. **(A)**	Voltage drop; R_1	Voltage adds in series $$V_s = V_1 + V_2 + \cdots$$ Voltage is the same in parallel $$V_p = V_1 = V_2 = \cdots$$	The battery has a voltage of 10 V. The 4-ohm resistor is in series with the battery and with resistors 1 and 2 combined. In question 5, the voltage of the 4-ohm resistor was found to be 8 volts. $$V_T = V_{12} + V_3$$ $$(10\text{ V}) = V_{12} + (8\text{ V})$$ $$V_{12} = 2\text{ V}$$ Resistors 1 and 2 are in parallel. $$V_{12} = V_1 = V_2 = 2\text{ V}$$
9. **(B)**	Dissipate the most power	Knowledge/definitions $$P = I^2R$$	The more resistors in parallel, the lower the overall resistance is. Lower overall resistance leads to more current flow.
10. **(D)**	Voltage of every resistor; same as voltage, V, of the battery	Knowledge/definitions $$V_p = V_1 = V_2 = \cdots$$	Resistors in parallel all receive a voltage that is equal to the total voltage supplied by the battery.

Magnetism

13

- → **PERMANENT OR FIXED MAGNETS**
- → **CURRENT-CARRYING WIRES**
- → **SOLENOIDS AND ELECTROMAGNETS**
- → **FORCE ON MOVING CHARGES**
- → **FORCE ON CURRENT-CARRYING WIRES**
- → **ELECTROMAGNETIC INDUCTION**

The source of magnetism and magnetic fields is moving charges. The magnetic field of one magnet (agent) will create a magnetic force on another magnet (object). Like gravity and electricity, magnetic force is a long-range force that acts at a distance even through empty space.

All charges are surrounded by an electric field. When the charges are in motion, they alter the space around them to create the effects known as magnetism. Three objects are surrounded by magnetic fields: fixed (permanent) magnets, current-carrying wires, and individual charges having velocity. Magnetism is also important in electric motors and in the generation of electricity, which is electromagnetic induction. This chapter will do the following:

- Examine the properties of magnetism and the types of magnets.
- Draw, analyze, and solve problems for uniform magnetic fields.
- Draw, analyze, and solve problems for the magnetic fields that surround current-carrying wires.
- Analyze how electricity can be generated by changing magnetic flux in order to create electrical pressure, to create emf, and to induce current flow.

Table 13.1 lists the variables discussed in this chapter.

Table 13.1 Variables Used in Magnetism

New Variables	Units
$\vec{B}$ = Magnetic field	T (tesla)
$\vec{F}_B$ = Force of magnetism	N (newtons)
μ_0 = permeability of free space	T • m/A (tesla • meters per ampere)
ϕ = magnetic flux	T • m^2 (tesla • meters squared)
$\mathcal{E}$ = emf	V (volts)

PERMANENT OR FIXED MAGNETS

Fixed magnets are the traditional magnets with which people are familiar. They include bar magnets and horseshoe magnets. Their magnetic properties are the result of electron spin in the orbitals of the magnet's atoms. Although the fixed magnet does not appear to be moving, in fact, the motion of the spinning electrons causes the magnetic field. In most substances, the net spin of all the electrons cancels. In some substances, such as iron, the net effect of the spinning electrons does not cancel. So, each atom acts like a tiny magnet. Groups of atoms having a similar magnetic orientation are known as **domains**. When all the domains of a magnetic substance are aligned, the substance becomes a fixed magnet with a set magnetic field.

The magnetic fields around and between fixed magnets have known properties. The magnetic field surrounding a fixed magnet appears similar to the electric field between an electron and a proton, as shown in Figure 13.1.

IF YOU SEE
fixed
magnets or
an external
uniform
magnetic
field

North to South

The direction
of the
magnetic
field between
magnetic
poles is north
to south.

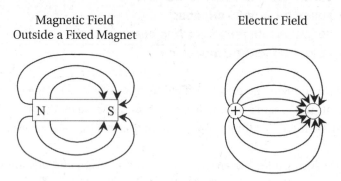

Figure 13.1 Magnetic and electric field lines

Fixed magnets have a north pole and a south pole. These poles are analogous to positive and negative aspects of an electric field, but they are not the same thing. Figure 13.1 shows only the magnetic field outside of a fixed magnet, which appears to extend from north to south just as the electric field extends from positive to negative. The electric field terminates on charges. However, the magnetic field forms continuous closed loops. The field lines in the diagram actually extend into and through the fixed magnet. Inside the magnet, the lines run from south to north. However, we are most concerned with the lines outside of fixed magnets, as this portion of the field interacts with other magnets. Fixed magnets can be used to create uniform magnetic fields, as shown in Figure 13.2. The letter $\vec{B}$ represents the vector for magnetic field, which is measured in teslas (T).

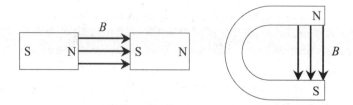

Figure 13.2 Uniform magnetic fields

In many problems, the magnets are not drawn. Only the magnetic field is given, or its direction is stated. Quite often, the magnets cannot be drawn as they are not in the plane of the page. The uniform magnetic fields shown in Figure 13.3 are oriented in the z-direction. The magnets that created them are located in front of and behind the plane of this page. Dots are used to indicate a field coming out of the page and ✕'s are used to indicate a field going into the page.

+z (out of page)   −z (into page)

Figure 13.3 Uniform magnetic fields in the z-direction

The even spacing of the symbols representing the field lines indicates that the magnetic fields are uniform. They have the same magnitude and direction at every point.

CURRENT-CARRYING WIRES

The magnetism of a current-carrying wire is essentially the sum of the magnetism of the moving charges that make up the current in the wire.

Visualizing the Field of Current-Carrying Wires

A wire is essentially a long cylinder. The magnetic field surrounding a current-carrying wire forms concentric circles around every part of the wire. Figure 13.4(a) shows a representation of the circling field. Figure 13.4(b) shows how this field can be rendered two-dimensionally.

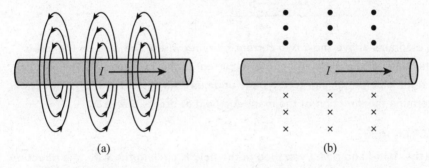

(a)   (b)

Figure 13.4 Magnetic fields around a current-carrying wire

IF YOU SEE a current-carrying wire and need direction

Right-hand rule

Point the thumb in the direction of the current. The direction of the curled fingers indicates field direction.

Figure 13.4(b) may seem confusing since it shows only a slice of the magnetic field passing through the plane of the page. Above the wire, the field is shown coming out of the page. Below the wire, it is shown entering the page. The directions of the fields above and below the wire are determined by the **right-hand rule**. The thumb of the right hand points in the direction of the current. The curled fingers point in the direction of the circular magnetic field created by the current. When the hand is oriented with the fingers above the wire, the tips of the fingers point out of the page in the +z-direction. The field is represented by dots. When the hand is oriented with the fingers below the wire, the tips of the fingers point into the page in the −z-direction, and the field is represented by ✕'s. What the field is doing in front of the wire and behind the wire cannot be shown in the plane of the page.

The diagrams in Figure 13.4 can be rotated 90 degrees so that the wire and its current appear to be coming out of or going into the page as shown in Figure 13.5. This diagram clearly shows the circling magnetic fields. Again, the right-hand rule can be used to determine the direction the field is circling. In the left diagram, point the right thumb into the page (✗) and the fingers will curl clockwise. In the right diagram, point the right thumb out of the page (dot) and the fingers will curl counterclockwise.

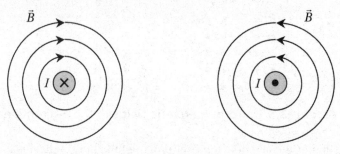

Figure 13.5 Using the right-hand rule to determine the direction of the magnetic field

Problems most often ask for the direction of the field at a specific point, as shown in Example 13.1.

Direction of the Magnetic Field Around a Wire

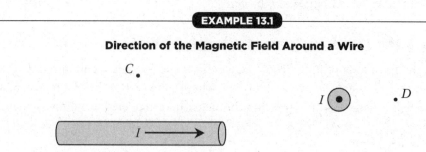

The diagrams above show two current-carrying wires. The wire on the left is in the plane of the page and carries a current in the +x-direction. The wire on the right is perpendicular to the page and carries a current in the +z-direction. Determine the direction of the magnetic field at points C and D.

WHAT'S THE TRICK?

Use the right-hand rule. Even though the field is circling the wire, the direction of the field at a single point is an instantaneous value. It will be tangent to the circling field at that point.

Point C: Out of the page, +z-direction.

Point D: Up, +y-direction

Magnitude of the Magnetic Field of a Wire

The magnitude of the magnetic field, B, of a long, straight, current-carrying wire is solved with the following equation.

$$B = \frac{\mu_0}{2\pi}\frac{I}{r}$$

The constant, μ_0, is known as the permeability of free space. It has the value $\mu_0 = 4\pi \times 10^{-7}$ T•m/A. The current in the wire is represented by I and is measured in amperes. The distance, *r*, is measured from the center of the wire to the point where the field is to be solved. The magnetic field is directly proportional to the current in the wire and inversely proportional to the distance from the wire.

EXAMPLE 13.2

Magnetic Field Due to a Current

A long, straight wire carries a current of 2.0 amperes. Determine the magnitude of the magnetic field at a point 10 centimeters from the wire.

WHAT'S THE TRICK?

Use the formula for long, straight wires. Always convert to acceptable units: from centimeters to meters.

$$B = \frac{\mu_0}{2\pi}\frac{I}{r}$$

$$B = \frac{\left(4\pi \times 10^{-7}\right)}{2\pi}\frac{(2.0)}{(0.10)} = 4 \times 10^{-6}\text{ T}$$

SOLENOIDS AND ELECTROMAGNETS

Coiling conducting wire around a hollow insulating tube creates a device known as a **solenoid**. When current passes through the coils of the solenoid it generates a magnetic field that resembles the field associated with permanent bar magnets, as shown in Figure 13.6.

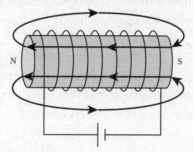

Figure 13.6 Magnetic field of a solenoid

Like the field of a permanent magnet, the magnetic field forms loops. Outside the solenoid the field curves from the north pole of the solenoid toward the south pole. Inside the hollow tube the field is nearly uniform and extends from the south pole toward the north pole of the solenoid. The solenoid has several advantages over a permanent fixed magnet. It can be turned off, its strength can be adjusted by changing the current passing through it, and the direction of the field can be changed by altering the direction of the current. When an iron core (iron rod) is inserted through the hollow tube it acts to greatly intensify the strength of the magnetic field, forming a device known as an **electromagnet**.

$$F_B = qvB$$

Uniform
circular
motion:

$$F_C = F_B$$

$$ma_C = qvB$$

$$m\frac{v^2}{r} = qvB$$

FORCE ON MOVING CHARGES
Uniform Magnetic Fields and Moving Charges

Tiny moving charges such as protons and electrons interact with magnetic fields. These moving charges are so small that the magnetic fields surrounding them are negligible. However, when moving charges move **perpendicularly** to external magnetic fields created by larger magnets, such as fixed magnets and current-carrying wires, the moving charges experience a force of magnetism. The following formula calculates the magnetic force on a charge moving in an external magnetic field.

$$\vec{F}_B = q\vec{v}\,\vec{B}\sin\theta$$

The force of magnetism on a moving charge is the product of its charge (q), its velocity ($\vec{v}$), and the magnetic field ($\vec{B}$), through which it is moving. The formula contains sin θ, where θ is the angle between the velocity and magnetic-field vectors. The value of sin θ is at its maximum and equal to 1 when $\theta = 90°$. Therefore, to receive a maximum force of magnetism, charges must move perpendicularly to the magnetic field. Most exam problems requiring the formula will send charges perpendicularly to the magnetic field and will solve only for vector magnitude. As a result, the formula often simplifies to the following.

$$F_B = qvB$$

All three vector quantities, $\vec{F}_B$, $\vec{v}$, and $\vec{B}$, in the above formula must be perpendicular to one another. They each lie on separate axes and require analysis in all the three dimensions (x, y, and z). This means that a charge moving parallel to the field will experience no magnetic forces at all. When charges move parallel to the magnetic field, $\theta = 0°$ and the magnetic force is zero. Zero quantities are often tricky conceptual questions. When charges move parallel to magnetic fields, they are subject to inertia because there is no force of magnetism. If the charges are stationary, they remain stationary. If they are moving at constant velocity, they continue moving at constant velocity.

Electric and gravity fields can cause objects to change speed and/or direction. Magnetic force acting on moving charges is centripetal, resulting in uniform circular motion. Charges moving in magnetic fields are accelerating, but they have constant speed. Figure 13.7 and Table 13.2 compare and contrast the three major uniform fields, the forces they create, and the motion they cause.

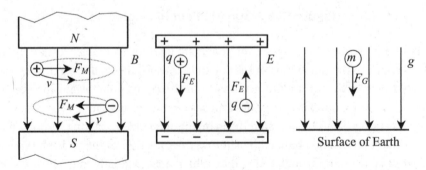

Figure 13.7 Three major uniform fields

Table 13.2 Forces and Motion Caused by Various Uniform Fields

	Magnetic Field, B	Electric Field, E	Gravity Field, g
Direction	North to south	Positive to negative	Toward Earth
Object	Moving charge, qv	Charge, q	Mass, m
Force	$F_B = qvB$	$F_E = qE$	$F_G = mg$
Resulting motion	Circular motion perpendicular to the field. Positive and negative charges circle in opposite directions as determined by the right-hand rule.	Linear acceleration parallel to the field. Positive charges follow the field, while negative charges move opposite the field.	Linear acceleration in the direction of the field.
Sum of forces	F_C replaces ΣF $F_C = F_B$ $ma_C = qvB$	$\Sigma F = F_E$ $ma_E = qE$	$\Sigma F = F_G$ $ma_g = mg$ $a_g = g$
Velocity	Constant speed $a_C = \dfrac{v^2}{r}$	$v_f^2 = v_i^2 + 2a_E x$ $v_f = v_i + a_E t$	$v_f^2 = v_i^2 + 2gy$ $v_f = v_i + gt$

The circular motion of moving charges is easier to visualize if the magnetic field is rotated so that it is oriented in the z-direction (out of or into the page), as shown in Figure 13.8.

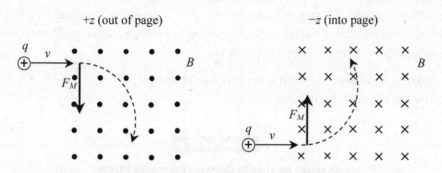

Figure 13.8 The magnetic field rotated in the z-direction

Only positive charges are shown in Figure 13.8. If negative charges were shown, they would circle in the opposite direction. The direction, whether clockwise or counterclockwise, that a charge will circle is determined by a slightly different version of the right-hand rule. This version of the right-hand rule is for determining the direction of force on a moving, charged particle by a magnetic field. The previous version of the rule was used to determine the direction of a magnetic field created by the current in a long, straight wire. When magnetic force is involved, the fingers are not curled and should be extended straight, with the thumb and fingers separated by 90° as shown in Figure 13.9.

IF YOU SEE
a positive
charge
moving in
a magnetic
field

**Right-hand
rule**

Thumb: $\vec{v}$

Fingers: $\vec{B}$

(Keep your
fingers straight
when force is
involved.)

The direction
the palm
pushes: $\vec{F}_B$

Use your
left hand
for negative
charges.

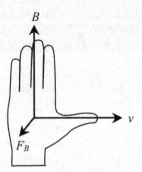

Figure 13.9 Using the right-hand rule
with magnetic force

To find the direction of force, point the thumb in the direction of the velocity, v, of the moving charge. Point the extended fingers in the direction of the magnetic field lines. The direction that the palm of the hand pushes (a force is a push) is the direction of the force of magnetism. Note that on the left side of Figure 13.9, the palm is pushing out of the page, and the force of magnetism shown is in the positive z-direction.

The right-hand rule gives the direction of force, which can be used to determine whether the object will circle clockwise or counterclockwise. When charges move in a magnetic field the only force acting on them is the force of magnetism. This force causes uniform circular motion since the magnetic force acts perpendicularly to motion. Therefore, the force of magnetism must point toward the center of the circle. In the left diagram in Figure 13.8, the force of magnetism points downward. Thus, the center of the circle must be below the point where the charge enters the field. The charge must circle clockwise. The opposite is true for the charge entering the $-z$ field in the right diagram in Figure 13.8.

Negative (opposite) charges will always move in a direction opposite that of positive charges. If a positive charge circles clockwise, a negative charge circles counterclockwise. You can handle negative charges in one of two ways.

1. Use the right-hand rule and give the opposite answer.
2. Use the left hand for negative charges.

EXAMPLE 13.3

Charges Moving in Uniform Magnetic Fields

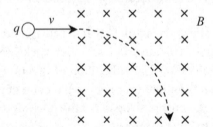

A charge, q, moving at a speed of v enters a uniform magnetic field, B, as shown in the diagram above.

(A) Determine the radius of the circular path in terms of the given variables.

In circular motion, centripetal force (F_C) is used instead of ΣF. The force causing the circular motion and pointing toward the center of the circle is the force of magnetism, F_B.

$$F_C = F_B$$

$$m\frac{v^2}{r} = qvB$$

$$r = \frac{mv}{qB}$$

(B) Determine whether the charge shown in the diagram is positive or negative.

The center of the clockwise circular motion is below the point where the charge enters the field. The palm of the hand must be oriented downward to push toward the center. Only the left hand is capable of aligning with all three vectors: v, B, and F_B. Therefore, the charge is negative.

Work Done by the Magnetic Force

The force of magnetism does no work on moving charges. The force of magnetism always acts perpendicularly to the motion of charges. In order to do work, a force must be parallel to an object's motion. Since no work is done, the force of magnetism cannot change the kinetic energy and velocity of a moving charge. However, the perpendicular force of magnetism can change the direction of moving charges, causing them to circle at constant speed.

Current-Carrying Wires and Moving Charges

A charge may also be moving through the magnetic field created by a current-carrying wire. The equation for the magnitude of the force of magnetism acting on a charge near a current-carrying wire is the same as when a charge moves in a uniform magnetic field. Although the magnitude of a uniform magnetic field is usually a given, the magnitude of the field surrounding a wire can be determined by using an equation.

$$F_B = qvB \qquad\qquad B = \frac{\mu_0}{2\pi}\frac{I}{r}$$

Determining direction of the force on the charge will require you to use both versions of the right-hand rule. First, you must determine the direction of the magnetic field of the wire. Use the curled-finger version to find the direction of the magnetic field on the side of the wire where the charge is moving. Next, you must determine the force on the moving charge using the straight-finger version and seeing which way the palm is pushing.

FORCE ON CURRENT-CARRYING WIRES

Any combination of magnets will result in the field of one magnet (agent) creating a force on the second magnet (object). In the preceding section, fixed magnets and currents created forces on moving charges. What about other interactions? The field of a fixed magnet can create a force on a current-carrying wire, causing the wire to move. The field of one current-carrying wire can create a force on another current-carrying wire.

Fixed Magnets and Current-Carrying Wires

If a current-carrying wire passes through an external magnetic field, the field will create a force of magnetism on the wire, causing the wire to move. An external magnetic field is not the field of the object mentioned in the problem. The external magnetic field (✕'s) shown in Figure 13.10 is caused by a set of fixed magnets that are positioned in front of and behind the page and cannot be shown in the diagram. This field acts as the agent that puts a force on the wire, the object. The field circling the wire is not shown since the wire does not act on itself.

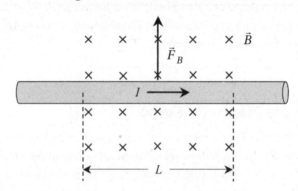

Figure 13.10 Current-carrying wire in an external magnetic field

The force acting on the wire is the product of the current in the wire (I), the length of the wire inside the field ($\vec{L}$), and the external magnetic field ($\vec{B}$). Like the force acting on moving charges, the formula contains sin θ.

$$\vec{F}_B = I\vec{L}\,\vec{B}\,\sin\theta$$

As long as the wire is perpendicular to the field (the most likely scenario on exams), the formula simplifies to the following when solving for vector magnitude.

$$F_B = ILB$$

The direction of this force is determined with the right-hand rule. Currents by definition are considered positive, so the right hand is needed. Use the left hand only if a problem specifically addresses a negative current or the actual electron flow. Since this problem involves force, use the straight-finger version of the right-hand rule. The thumb is the direction of the moving charges, which in this case is the direction of the current, I.

The force of magnetism acting on a current-carrying wire is the basis of all electrical motors. In a motor, the wire is wrapped so that it forms many coils, which extends the length, L, and increases the magnetic force, F_B, of the wire. The coils of wire are arranged so the magnetic force can rotate the coils of wire continuously. The magnetic force in an electric motor converts electrical energy into mechanical energy.

Two Current-Carrying Wires

The magnetic field surrounding one current-carrying wire will create a force on a second nearby current-carrying wire. Consider two wires side by side, as shown in Figure 13.11. Wire 1 (agent) is surrounded by a magnetic field that creates a force of magnetism on wire 2 (object). This force will either attract wire 2 to wire 1 or repel it. You can also analyze the problem in reverse. Wire 2 is surrounded by a magnetic field. This field creates a force on wire 1. Newton's third law dictates that the forces acting on both wires are equal and opposite.

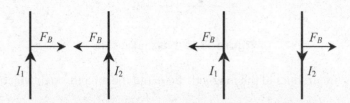

Figure 13.11 Focus on current-carrying wires

Interacting wires involve two equations: one for the field of the first wire and one for the force acting on the second wire.

$$B_1 = \frac{\mu_0}{2\pi} \frac{I}{r} \qquad F_{B2} = I_2 L_2 B_1$$

The magnetic field (B_1) of the first wire depends on the current in the first wire (I_1) and the distance between the two wires (r). The force created on the second wire is the product of its current (I_2), the length of wire 2 (L_2), and the field of the first wire (B_1).

Both versions of the right-hand rule are needed to solve for direction. The curled-finger version is used to find the field direction of wire 1. The straight-finger version is used to find the force on wire 2. To answer questions of this type quickly, memorize the fact that currents in the same direction attract, and currents in the opposite direction repel. Be careful with this, as it is the opposite of other rules for attraction and repulsion. Normally, opposites attract and likes repels. However, with wires, currents in opposite directions repel and currents in the same direction attract.

> **Current-Carrying
> Parallel Wires**
>
> *online.barronsbooks.com*

ELECTROMAGNETIC INDUCTION

The process of generating electricity is known as electromagnetic induction. This is the opposite of the electric motor. In a motor, a current-carrying wire passing through a magnetic field receives a force, causing it to move. In electromagnetic induction, a force is applied, causing a loop of wire or a magnet to move in relation to each other. This induces a current to flow in the wire loop. Induction occurs when the amount of magnetic field, known as **magnetic flux**, ϕ, experiences a change. The change in magnetic flux, $\Delta\phi$, creates an electric potential that induces charges to flow though the loop. When an electric potential is induced, it is known as an **emf**, $\mathcal{E}$, rather than a voltage, and emf has the units of volts. Induction converts mechanical energy into electrical energy.

Magnetic Flux

Magnetic flux, ϕ, can be thought of as the amount of magnetic field, B, passing straight through an area of space, A, as shown in Figure 13.12.

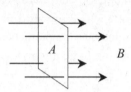

Figure 13.12 Magnetic flux

Magnetic flux is the product of the magnetic field and the area through which it passes.

$$\phi = BA$$

In electromagnetic induction, the area used is where the magnetic field and the coil of wire overlap, as shown by the gray areas in Figure 13.13.

Figure 13.13 Electromagnetic induction

The coil of wire can have any shape, but is often rectangular or circular. It can consist of a single loop of wire or of many loops of wire. When many loops are shown, the wire coil may appear as a spiral helix, where the area, A, is the area of one of the coils.

Inducing a current requires a change in flux, $\Delta\phi$, and this fact is often the basis of conceptual questions. The change in flux can be determined as follows:

$$\Delta\phi = \phi_f - \phi_i$$
$$\Delta\phi = (BA)_f - (BA)_i$$

From the equations, it is apparent that a change in flux results when either the magnetic field or the area change. This change can be accomplished in several ways.

- Moving the magnet and/or the coil toward each other increases the magnetic field, $\Delta\vec{B}$. See Figure 13.14.

Figure 13.14 Coil and magnet moving toward each other

IF YOU SEE
a magnet moving into a loop of wire or a loop of wire moving into a magnetic field

Electro-magnetic induction

Induction requires a change in flux (a change in either the magnetic field or the area).

- Moving the magnet and/or the coil away from each other decreases the magnetic field, $\Delta\vec{B}$. See Figure 13.15.

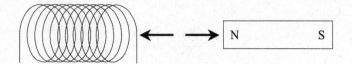

Figure 13.15 Coil and magnet moving away from each other

- Spinning a loop of wire in a magnetic field changes the area, ΔA. See Figure 13.16.

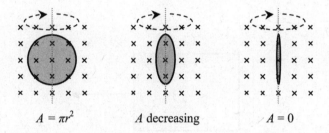

$A = \pi r^2$ A decreasing $A = 0$

Figure 13.16 Rotating loop of wire in a magnetic field

- Moving the coil into or out of a magnetic field can change the area, ΔA. See Figure 13.17.

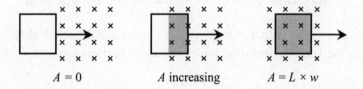

$A = 0$ A increasing $A = L \times w$

Figure 13.17 Loop of wire moving into a magnetic field

Emf

Michael Faraday found that if the flux in a loop or coil of wire changes, then a current is created and flows through the loop or coil of wire. This process is known as electromagnetic induction. Changing the flux, $\Delta\phi$, in a closed loop induces (creates) a potential (voltage) in the loop. This special induced voltage is known as an emf and is represented by the variable symbol, $\mathcal{E}$. It has the units of volts (V). Although it is known by another name, emf acts as any voltage does. The induced emf is the pressure that induces charges to move as a current in the loop of wire. Remember that a change in magnetic flux (a change in magnetic field or area) through a closed loop of conducting material is the necessary event that will induce an emf, $\mathcal{E}$, which in turn will induce a current to flow in the conducting loop.

The mathematical relationship between changing flux and the induced emf is described by Faraday's law.

$$\mathcal{E} = \frac{\Delta\phi}{t} \qquad \text{or} \qquad \mathcal{E} = \frac{\phi_f - \phi_i}{t}$$

The equation of Faraday's law does include a minus sign. However, the minus sign is not needed to determine the magnitude of the induced emf and has been omitted from these equations. The importance and usage of the minus sign is not a factor in being successful on the SAT Subject Exam.

$$\mathcal{E} = \frac{(BA)_f - (BA)_i}{t}$$

If a rectangular loop of wire moves into or out of a uniform magnetic field at a constant speed, the above formula simplifies greatly. Figure 13.18 depicts a loop moving with a speed of v into a magnetic field of B.

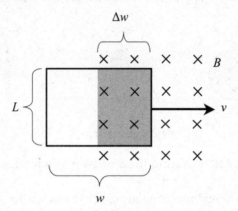

Figure 13.18 Loop of wire entering a magnetic field

As the loop moves into the field, the width of the loop exposed to the field is changing, Δw. The speed of the loop equals the rate at which the width is changing.

$$v = \Delta w / t$$
$$\mathcal{E} = \frac{\Delta \phi}{t} = \frac{B \, \Delta A}{t} = \frac{BL \, \Delta w}{t} = BLv$$
$$\mathcal{E} = BLv$$

The resulting simplified formula is referred to as the **motional emf** formula. It is commonly used for rectangular loops entering or leaving uniform magnetic fields at constant speed.

Induced Currents

Inducing an emf creates an electrical pressure that induces charges to move as a current. The magnitude of the induced current can be determined using Ohm's law, where emf is used instead of potential.

$$V = IR \qquad \text{becomes} \qquad \mathcal{E} = IR$$
$$P = IV \qquad \text{becomes} \qquad P = I\mathcal{E}$$

Lenz's Law

Problems may deal with only the direction of the induced current, which is dictated by Lenz's law and the right-hand rule. Generating electricity involves two magnetic fields. The first is the external magnetic field that passes through the loop or coil of wire. The second magnetic field is created when the current begins to flow and turns the wire into a second magnet. Lenz determined that the current in the loop flows in a direction so that the magnetic field of the current opposes the change in the original external magnetic field. There are two possibilities.

1. If the magnetic flux is increasing (magnet and loop moving toward each other or the area of the loop is increasing), the current must flow in a manner so that its magnetic field counters the increase. To counter the increase, the magnetic field of the current in the loop must cancel out the original external magnetic field. In order to accomplish this, the magnetic field of the current must point in the opposite direction of the field created by the external magnet.

2. If the magnetic flux is decreasing (magnet and loop moving away from each other or the area of the loop is decreasing), the current must flow in a manner that increases the magnetic field. In order to accomplish this, the magnetic field of the current in the loop must point in the same direction as the original external magnetic field.

Lenz's law is complicated and may simply be tested in a commonsense manner. Questions just ask how the direction of the current is affected when:

- the poles of the magnet moving into a loop are switched;
- the magnet reverses direction; and
- the magnet leaves the field as opposed to entering the field.

In each of the above cases, the direction of the current reverses.

Transformers

Transformers are electric devices used to change (transform) voltage and current. Transformers typically consist of two separate circuits that each contain coils of wire. While the coils are in separate circuits, they are parallel to each other and in close proximity.

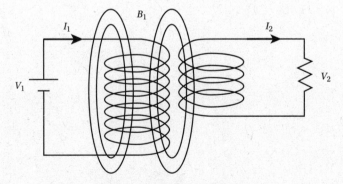

Figure 13.19 Transformer

In Figure 13.19 the primary circuit, on the left, contains a voltage source, V_1, that sends a current, I_1, through the primary coil, which results in a magnetic field, B_1, that passes through and surrounds the primary coil. The key to a transformer's operation is a changing current in the primary coil. This is accomplished by using alternating current (AC), which is an oscillating current that continually changes. When the current, I_1, in the primary coil is changing, the magnetic field, B_1, is also changing. The changing magnetic field passes the area of the loops of the secondary coil. This change in flux induces an emf, V_2, in the secondary coil. As a result, a current, I_2, is induced in the secondary coil without the two circuits making physical contact.

The ratio of voltage in the secondary coil, V_2, to the voltage in the primary coil, V_1, is the same as the ratio of the number of loops in the secondary coil, N_2, to the number of loops in the primary coil, N_1.

$$\frac{V_2}{N_2} = \frac{V_1}{N_1}$$

In Figure 13.19 the secondary coil has half as many loops as the primary coil. As a result, the induced voltage, V_2, will be half of the primary voltage, V_1. This type of transformer is known as a step-down transformer since it decreases voltage in the secondary coil. A step-up transformer would have more loops in the secondary coil, causing voltage to increase in the secondary coil.

In an ideal transformer, energy would be transferred from one coil to the other with no loss in energy to the environment. As a result, the rate of energy transfer would be constant and both circuits would have the same power. Since power is the product of current and voltage ($P = IV$), any change to voltage would be countered by an inverse change in current. If voltage increases, then current decreases by the same factor, and vice versa, in order for power to remain constant.

SUMMARY

1. **FIXED MAGNETS ARE USED TO CREATE UNIFORM MAGNETIC FIELDS.** Fixed magnets, such as an iron rod or a horseshoe magnet, owe their magnetic properties to the spin of electrons in the atomic orbitals. The magnetic field of a fixed magnet runs from north to south outside of the magnet. When two fixed magnets are oriented correctly, a uniform magnetic field is created between them.

2. **CURRENT-CARRYING WIRES ARE SURROUNDED BY MAGNETIC FIELDS.** The magnetic field around a current-carrying wire forms concentric circles. The magnitude of the field is directly proportional to the current in the wire and inversely proportional to the distance from the wire. The direction of the magnetic field is determined with the right-hand rule (thumb pointed in the direction of the current, curled fingers pointed in the direction of the circular magnetic field).

3. **MOVING ELECTRONS, PROTONS, AND IONS ARE TINY MAGNETS.** A charged particle (q), moving with a speed (v), in a magnetic field ($\vec{B}$), will experience a force of magnetism, $\vec{F}_B = qv\vec{B}$. The magnetic field can be due to either a fixed magnet or a current-carrying wire. The direction of the force of magnetism for a positive particle is determined by the right-hand rule (thumb pointed in the direction of the velocity of the particle, extended fingers pointed in the direction of the external magnetic field, palm pointed toward the direction of force). The opposite direction of force is used for a negative particle. The force of magnetism acting on charged particles causes them to circle at constant speed. The mathematics of uniform circular motion can be applied, $\vec{F}_C = \vec{F}_B$. Remember that a magnetic field cannot change the speed of a charged particle. Since the force of magnetism acts perpendicularly to the direction of motion of the particle, magnetic fields do not do any work on charged particles.

4. **A CURRENT-CARRYING WIRE PASSING THROUGH AN EXTERNAL MAGNETIC FIELD WILL JUMP.** The force on a current-carrying wire can be determined with the equation $\vec{F}_B = IL\vec{B}$. The direction of the force is found using the right-hand rule. This is the essence of an electric motor, which is a device that converts electric energy into mechanical energy.

5. **TWO CURRENT-CARRYING WIRES WILL EITHER REPEL OR ATTRACT ONE ANOTHER.** Current-carrying wires are magnets that attract and repel. When currents are parallel and in the same direction, the wires attract. When the currents are parallel and in the opposite directions, the wires repel.

6. **ELECTROMAGNETIC INDUCTION IS THE PROCESS OF GENERATING A CURRENT.** To generate a current, the magnetic flux through a loop or coil of wire must be changing, $\Delta\phi$. Faraday's law shows that the induced emf (generated electrical pressure) is equal to the rate of change in flux: $\mathcal{E} = \Delta\phi/t$. Emf is a generated voltage. It causes a current to flow through a closed conducting path, such as a loop or a coil of wire. The magnitude of the resulting current can be found with an adaptation of Ohm's law, $\mathcal{E} = IR$. The direction can be determined using Lenz's law and the right-hand rule.

If You See	Try	Keep in Mind
A fixed magnet or two magnets with north and south poles facing	Magnetic field runs north to south. If two magnets are parallel with opposite poles facing, the field is uniform (same value everywhere).	When the magnetic field runs in the z-direction ($\times$'s or dots), the magnets are in front of and behind the page and are not shown.
A long, straight wire	$$B = \frac{\mu_0}{2\pi}\frac{I}{r}$$ To find the magnitude of the field and right-hand rule (curled fingers) to find the direction of the field	The field circles the wire and is proportional to the current in the wire and inversely proportional to the distance from the wire.
A charge moving in a uniform magnetic field	Force acting on a moving charge $$F_B = qvB$$ and right-hand rule (extended fingers) for positive charges and left hand for negative charges	The charges move at constant speed in uniform circular motion. $$F_C = F_B$$ $$ma_C = qvB$$ $$m\frac{v^2}{r} = qvB$$
A wire in an external magnetic field (a field created by a set of fixed magnets)	Force acting on a wire $$F_B = ILB$$ and right-hand rule (extended fingers)	The force causes the wire to jump. This is the basis of all electric motors, which convert electric energy into mechanical energy.
A magnet and a loop or coil of wire moving toward or away from one another, or a loop of wire moving into or out of a magnetic field.	Electromagnetic induction; according to Faraday's law, a change in flux induces an emf $$\mathcal{E} = \frac{\Delta\phi}{t}$$ An emf is a created voltage (pressure) that induces a current $$\mathcal{E} = IR$$ The direction of the current is determined by Lenz's law and the right-hand rule	There must be a change in flux. Either the magnetic field or the area of the wire loop exposed to the field must be changing. Reversing the motion or the poles of the magnets reverses the current.

PRACTICE EXERCISES

1. A wire carries a current out of the page. Which diagram below correctly describes the magnetic field of the wire?

(A)

(B)

(C)

(D)

(E) None of these

2. A wire carries a current, I. At a distance of r, the magnitude of the magnetic field is $\bar{B}$. If both the current in the wire and the distance from the wire are doubled, the magnitude of the magnetic field changed by a factor of

(A) $\frac{1}{4}$

(B) $\frac{1}{2}$

(C) 1

(D) 2

(E) 4

3. Which of the following fields CANNOT change the speed of the object that is acted upon by each field?

    I. Uniform gravity fields
   II. Uniform electric fields
  III. Uniform magnetic fields

(A) I only
(B) II only
(C) III only
(D) Both I and II
(E) Both II and III

Questions 4–6

Negative charge q moving with speed v in the $+x$-direction enters a uniform $+z$ magnetic field, $\vec{B}$, as shown in the diagram below.

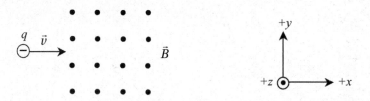

4. What is the direction of the force of magnetism acting on the charge at the instant it first enters the magnetic field?

(A) $+x$
(B) $+y$
(C) $-y$
(D) $+z$
(E) $-z$

5. How is the magnitude of the force of magnetism acting on the charge affected by doubling the magnetic field?

(A) $\frac{1}{2}$ its original value
(B) the same as its original value
(C) $\sqrt{2}$ times larger than its original value
(D) 2 times larger than its original value
(E) 4 times larger than its original value

6. How is the acceleration of the charge affected by doubling its initial velocity?

   (A) $\frac{1}{2}$ its original value

   (B) the same as its original value

   (C) $\sqrt{2}$ times larger than its original value

   (D) 2 times larger than its original value

   (E) 4 times larger than its original value

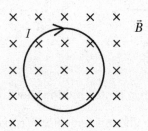

7. In the diagram above, a loop of conducting material is placed into a uniform external magnetic field. A current runs clockwise through the loop, and the magnetic field is in the $-z$-direction. The force on the loop is

   (A) toward the center of the loop and acts to shrink the loop

   (B) away from the center of the loop and acts to expand the loop

   (C) in the $+z$-direction

   (D) in the $-z$-direction

   (E) in the direction of the current

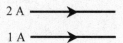

8. Two parallel wires carry currents in the same direction, as shown in the diagram above. One wire carries a 2-ampere current, and the other carries a 1-ampere current. The force of magnetism acting on the 1 A wire is

   (A) half of the force acting on the 2 A wire, and the wires attract

   (B) half of the force acting on the 2 A wire, and the wires repel

   (C) twice the force acting on the 2 A wire, and the wires attract

   (D) twice the force acting on the 2 A wire, and the wires repel

   (E) the same as the force acting on the 2 A wire, and the wires attract

Questions 9–10

A rectangular loop of wire is moving toward and enters a uniform magnetic field at constant velocity, as shown in the diagram below. The loop is shown in two positions: initial conditions in solid lines and final conditions in dashed lines.

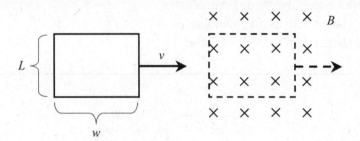

9. Which statement is NOT correct for the motion shown in the diagram?

   (A) During the entire motion, the magnetic flux is constant.
   (B) Magnetic flux begins to change at the instant the loop enters the field.
   (C) A change in flux is needed in order to induce an emf.
   (D) In order to create a current, an emf must be induced in a closed loop of conducting material.
   (E) The resulting current in the loop is $\frac{\mathcal{E}}{R}$.

10. During the interval where a current is induced, its magnitude is

   (A) BLv

   (B) $\dfrac{BLv}{R}$

   (C) $BLvR$

   (D) $\dfrac{R}{BLv}$

   (E) $\dfrac{(BLv)^2}{R}$

ANSWERS EXPLAINED

	Key Words	Needed for Solution	Now Solve It
1. **(D)**	Wire; current out of page: magnetic field	Knowledge/definitions Right-hand rule (curled-fingers version)	The magnetic field forms circles around current-carrying wires, which means C and D are possible. The right-hand rule indicates that the current is counterclockwise, which is answer D.
2. **(C)**	Wire; current; distance; magnitude of the magnetic field	$B = \dfrac{\mu_0}{2\pi} \dfrac{I}{r}$	The magnetic field is directly proportional to current and inversely proportional to distance. $$B = \dfrac{\mu_0}{2\pi} \dfrac{(2I)}{(2r)}$$ Doubling both current and distance results in the same strength in the magnetic field: a factor of 1.
3. **(C)**	Fields; *cannot* change the speed	Knowledge/definitions	Gravity and electric fields can speed up objects and slow them down when objects move parallel to these fields. However, uniform magnetic fields cause objects to circle at constant speed.
4. **(B)**	Negative charge; enters a uniform +z magnetic field; direction of force	This charge is negative Left-hand rule	Point the thumb of the left hand in the direction of the velocity vector. While keeping the fingers straight, point them out of the page. The palm is now oriented so that it pushes up the page in the +y-direction.
5. **(D)**	Force of magnetism; doubling the magnetic field	$F_B = qvB$	The force of magnetism is directly proportional to the strength of the magnetic field. $$(2F_B) = qv(2B)$$ The magnitude of force is doubled.
6. **(D)**	Acceleration; doubling its initial velocity	Knowledge/definitions $$F_C = F_B$$	Charges moving in a uniform magnetic field experience uniform circular motion with centripetal acceleration. $$F_C = F_B$$ $$ma_C = qv_B$$ $$2a_C = \dfrac{q(2v)B}{m}$$ Doubling the velocity doubles the acceleration.

	Key Words	Needed for Solution	Now Solve It
7. **(B)**	Diagram; current runs clockwise through the loop; magnetic field; force	Right-hand rule (straight-finger version)	The right hand is used for currents, which are considered positive. When force is involved, the fingers are kept straight and point into the page, $-z$-direction. The thumb points in the direction of the current, which is circling. However, when the hand is placed at any point on the loop, the palm always points outward away from the center of the loop.
8. **(E)**	Parallel wires; currents in the same direction; 2-ampere current; 1-ampere current; force	Knowledge/definitions Right-hand rule	Two parallel wires with current in the same direction will attract one another. However, the key to this problem is actually Newton's third law. When two objects interact, there is an equal and opposite force between them. The only answer involving equal force is E. The two different currents act as distracters.
9. **(A)**	Loop; entering a uniform magnetic field	Knowledge/definitions	Magnetic flux is the intersection of the magnetic field and the area of the loop. Initially, the magnetic flux is zero. In the end, it is at a maximum.
10. **(B)**	Loop; entering a uniform magnetic field: current is induced	$\mathcal{E} = IR$ $\mathcal{E} = BLv$	Rearrange Ohm's law, adapted for emf, to solve for current. $$\mathcal{E} = IR \qquad \text{becomes} \qquad I = \frac{\mathcal{E}}{R}$$ Substitute BLv from the motional emf formula. $$I = \frac{BLv}{R}$$

Simple Harmonic Motion

14

→ **TERMS RELATED TO SHM**
→ **OSCILLATIONS OF SPRINGS**
→ **OSCILLATIONS OF PENDULUMS**
→ **GRAPHICAL REPRESENTATIONS OF SHM**
→ **TRENDS IN OSCILLATIONS**

Simple harmonic motion (SHM) is the periodic and repetitive motion of an object, such as a pendulum, a mass on a spring, or a plucked guitar string. In each of these examples of SHM, the object is displaced from its initial point of equilibrium and oscillates in a predictable manner about the equilibrium position. In this chapter, we will explore the following:

- The types and terminology of SHM.
- Solving problems for springs and pendulums.
- Interpreting graphical representations of SHM.
- Summarizing key values in energy and force during SHM.

TERMS RELATED TO SHM

Period

Period, T, is the time in seconds for an object in SHM to complete one full oscillation. An example of a full oscillation would be the movement of a pendulum from its initial starting position back to that position again. When more than one oscillation is given, the period can be determined by dividing the total time by the number of oscillations:

$$T = \frac{\text{total time}}{\text{\# of oscillations}}$$

Note that the number of oscillations has no units. It is simply the count of an event.

Frequency

Frequency, f, is the number of full oscillations that occur in 1 second. Divide the number of oscillations by the time interval to determine the frequency.

$$f = \frac{\text{\# of oscillations}}{\text{total time}}$$

IF YOU SEE period

$T = \dfrac{\text{time}}{\text{\#oscillations}}$

Period is the time of one complete oscillation, measured in seconds.

$$f = \frac{\text{\# oscillations}}{\text{time}}$$

Frequency is the number of oscillations in 1 second and has the units of 1/s, s^{-1}, or Hz.

Frequency is the reciprocal of period and therefore has units of 1/seconds, which is also known as Hertz (Hz). The relationship between period and frequency can be expressed as follows:

$$T = \frac{1}{f} \qquad \text{and} \qquad f = \frac{1}{T}$$

Period and frequency are inversely proportional. As one increases, the other decreases.

EXAMPLE 14.1

Determining the Period from the Frequency

A plucked guitar string vibrates at a frequency of 100 Hertz. What is the period of vibration of the string?

WHAT'S THE TRICK?

Period is the reciprocal of frequency.

$$T = \frac{1}{f}$$

$$T = \frac{1}{100\,Hz}$$

$$T = 0.01\ s$$

Amplitude

Amplitude, A, is the magnitude of the maximum displacement, $A = x_{max}$, of an oscillating particle or wave relative to its rest position. Amplitude is a measure of the intensity of the oscillation and is directly proportional to the energy imparted into the oscillating system. The amplitude does not affect the period or the frequency.

The amplitude can be assigned as either a positive or negative value, depending on which side of the equilibrium position is being described. For example, the amplitude of a plucked guitar string can either be above or below the equilibrium position of the string at rest. These two maximum amplitudes are individually assigned values of $+A$ and $-A$, respectively.

OSCILLATIONS OF SPRINGS
Determining the Spring Constant

Imagine a mass suspended by a spring, as represented in Figure 14.1. The force of gravity pulls the mass toward Earth, while the restorative force of the spring pulls the mass upward in an effort to restore the spring to its original, unstretched position. At equilibrium, the mass will be at rest and the spring will be stretched by an amount proportional to the force of gravity upon the mass.

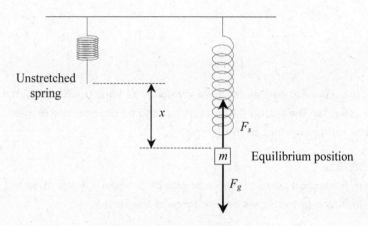

Unstretched spring

x

F_s

m Equilibrium position

F_g

Figure 14.1 Hooke's law

The **restorative force** of a spring, F_s, is represented by the following equation, known as **Hooke's law**:

$$F_s = -kx$$

In Hooke's law, x represents the displacement of the spring from its unstretched position, in meters, and k represents the **spring constant**, in newtons per meter. The spring constant, k, is specific to each spring, regardless of the mass suspended from the spring.

The **equilibrium position** is reached when the restorative force of the spring is equal in magnitude but opposite in direction to the gravity force acting on the mass. This can be described by setting the force of gravity equal to but opposite in sign to the restorative force of the spring.

$$F_s = F_g$$
$$kx = mg$$

IF YOU SEE
a stretched
or a
compressed
spring

Hooke's law

$F_s = -kx$

The direction of the force is opposite the direction of the spring's displacement.

EXAMPLE 14.2

Determining the Spring Constant

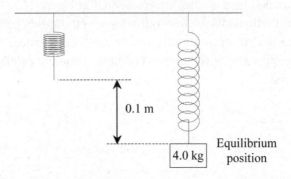

A 4.0-kilogram mass is suspended by a spring. In its equilibrium position, the mass has extended the spring 0.10 meters beyond its unstretched length. What is the spring constant of this spring?

WHAT'S THE TRICK?

Equilibrium is reached when the force of gravity is equal in magnitude but opposite in direction to the restorative force of the spring.

$$F_g = F_s$$
$$mg = kx$$
$$k = \frac{mg}{x}$$
$$k = (4.0 \text{ kg})(10 \text{ m/s}^2)/(0.10 \text{ m})$$
$$k = 400 \text{ N/m}$$

Remember that this value for the spring constant will be the same for this particular spring regardless of any other masses suspended by the spring. This concept will become important in an example later in the chapter.

Determining the Period of a Spring in SHM

The period of a spring in SHM is the amount of time, in seconds, in which the mass on a spring moves from an initial position back to that same initial position. The actual motion of a spring-mass system follows the same linear path up and down or back and forth. However, if the motion of an oscillating spring is graphed versus time, the resulting function is a sine wave. Figure 14.2 is a graph of position versus time. The positions and stretch of a vertically oriented spring-mass system have been superimposed on the graph at five key locations.

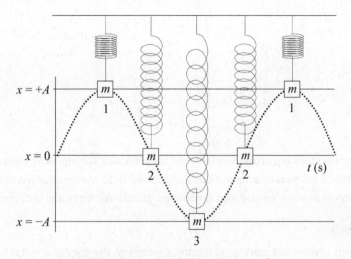

Figure 14.2 Position versus time

IF YOU SEE
period of
a spring in
SHM

$$T_s = 2\pi\sqrt{\frac{m}{k}}$$

Period depends on only the mass attached to the spring and the spring constant.

The sine wave represents the predictable, periodic nature of an oscillating spring-mass system during a time interval. If the mass hanging from a spring were released at position 1 in Figure 14.2, the mass would descend vertically, passing through equilibrium at position 2. The mass would reach maximum displacement at position 3, where it would reverse direction. On the return trip, the mass would pass through equilibrium at position 2 a second time and would return to its starting location at position 1. One complete cycle occurs when the oscillator returns to its initial position (position 1) for the first time. The period is the time of one complete cycle. For a spring-mass system, the period, T_s, can be determined as follows:

$$T_s = 2\pi\sqrt{\frac{m}{k}}$$

You should note that the period depends on only the mass attached to the spring, m, and the spring constant, k. The period DOES NOT depend on the displacement of the spring or the gravity in which the spring is oscillating.

EXAMPLE 14.3

Determining Period and Frequency of an Oscillating Spring

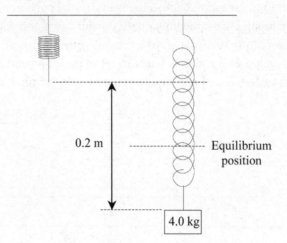

A 4.0-kilogram mass is suspended from an unstretched spring. When released from rest, the mass moves a maximum distance of 0.20 meters before reversing direction. What are the period and frequency of this spring-mass oscillator?

WHAT'S THE TRICK?

For problems involving a spring, you must determine the spring constant first. The mass moves a distance of 0.20 meters before it reverses direction. This is the entire up-down motion. The equilibrium position is in the middle of this motion. So, it occurs at 0.10 m.

$$F_s = F_g$$
$$kx = mg$$

$$k = \frac{mg}{x}$$

$$= \frac{(4.0 \text{ kg})(10 \text{ m/s}^2)}{(0.10 \text{ m})}$$

$$= 400 \text{ N/m}$$

Apply the spring constant to the formula for the period of a spring-mass oscillator.

$$T = 2\pi\sqrt{\frac{m}{k}}$$

$$= 2\pi\sqrt{\frac{(4.0 \text{ kg})}{(400 \text{ N/m})}}$$

$$= 0.20\pi \text{ s}$$

Frequency is simply the reciprocal of period.

$$f = \frac{1}{T}$$

$$= \frac{1}{0.2\pi}$$

$$= \frac{5}{\pi} \text{ Hz}$$

OSCILLATIONS OF PENDULUMS

Pendulums are simple harmonic oscillators that take the form of a mass suspended at the end of a string. The period of a pendulum is the time for the mass at the end of the pendulum to oscillate from an initial position back to that initial position again, as shown in Figure 14.3.

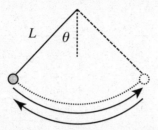

Figure 14.3 The period of a pendulum

Pendulums are not perfect simple harmonic oscillators. They approximate simple harmonic oscillators as long as the displacement angle, θ, is small ($\theta \leq 10°$). Keep in mind that diagrams are not drawn to scale on exams. Most questions on the SAT Subject Test will assume that pendulums are operating as simple harmonic oscillators regardless of how they are drawn. For pendulums with small displacements, the length of the string, L, and the acceleration of gravity, g, are the only variables contributing to the period of the pendulum. The period of a pendulum can be determined using the following formula.

$$T_p = 2\pi\sqrt{\frac{L}{g}}$$

As the length of the string increases, the period increases, causing the frequency to decrease. The amount of mass at the end of the string and the displacement from equilibrium (distance the mass is moved sideways) DO NOT affect the period of the pendulum at all.

IF YOU SEE period of a pendulum

$$T_p = 2\pi\sqrt{\frac{L}{g}}$$

Period depends on the length of the pendulum's string and gravity if not on Earth.

GRAPHICAL REPRESENTATIONS OF SHM

When the displacement of an object in SHM is plotted on a graph of displacement versus time, the result is a sine wave, as shown in Figure 14.4.

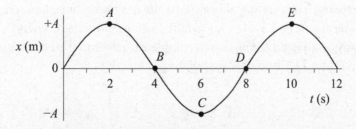

Figure 14.4 Displacement versus time

One complete oscillation is represented by the time for the graph to go from point A to point E. The period is the time of one complete cycle.

$$T = t_E - t_A = 10 - 2 = 8 \text{ seconds}$$

The frequency is the inverse of the period and is therefore $\frac{1}{8}$ Hz. This means that one-eighth of an oscillation occurs every second.

TRENDS IN OSCILLATIONS

In addition to being able to determine the period and frequency of a simple harmonic oscillator, the SAT Subject Test in Physics may ask questions involving the trends in key variables, such as speed, kinetic energy, displacement, potential energy, force, and acceleration. These trends are summarized in Figure 14.5.

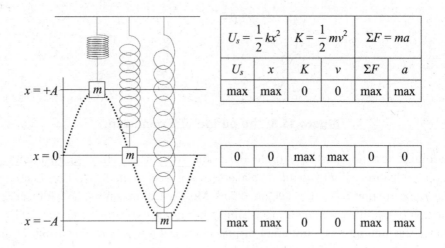

$U_s = \frac{1}{2}kx^2$		$K = \frac{1}{2}mv^2$		$\Sigma F = ma$	
U_s	x	K	v	ΣF	a
max	max	0	0	max	max

0	0	max	max	0	0

max	max	0	0	max	max

Figure 14.5 Trends in oscillations

Energy in Oscillations

Chapter 7, "Energy, Work, and Power," explains Hooke's law and simple harmonic motion as it relates to energy. In an oscillation, energy continually changes form from potential energy to kinetic energy. Potential energy depends on displacement. When the oscillator reaches its maximum displacement, the potential energy will also reach its maximum value. At the equilibrium position, the oscillator has a displacement of zero and a potential energy of zero. Kinetic energy is the complete opposite. At maximum displacement, an oscillator has an instantaneous speed of zero and a kinetic energy of zero. When an oscillator passes through the equilibrium point, it attains its greatest speed and has maximum kinetic energy. However, the total mechanical energy, $\Sigma E = U + K$ (potential energy plus kinetic energy), remains constant and is always conserved throughout a complete oscillation. These energy relationships are indicated in Figure 14.5, and they are graphed in Figure 14.6.

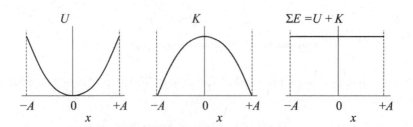

Figure 14.6 Energy graphs of oscillations

Conservation of energy is often needed to solve oscillation problems. Since the total energy is constant at every point in an oscillation, the total energy at any point 1 can be set equal to the total energy at any point 2.

$$\Sigma E_1 = E_2$$
$$U_1 + K_1 = U_2 + K_2$$

For a spring, substitute the potential energy of a spring.

$$\frac{1}{2}kx_1^2 + \frac{1}{2}mv_1^2 = \frac{1}{2}kx_2^2 + \frac{1}{2}mv_2^2$$

For a pendulum, substitute the potential energy of gravity.

$$mgh_1 + \frac{1}{2}mv_1^2 = mgh_2 + \frac{1}{2}mv_2^2$$

Force and Acceleration

Acceleration is the result of the sum of the forces acting on an object. Figure 14.7 diagrams the key force and energy relationships for a mass-spring system that is oscillating on a horizontal frictionless surface. In an oscillation, the sum of forces is zero when the object is at the equilibrium position. When the sum of forces is zero, the acceleration is also zero, $\Sigma F = ma$. This surprises many students since the acceleration is zero at the exact instant where velocity is the greatest. However, if an oscillator is held at equilibrium and then released, it will not move at all. This confirms that acceleration is zero in this position. When the oscillator is moved out of equilibrium (the spring is stretched/compressed or the pendulum is moved to either side), the sum of the forces increases with displacement, reaching its highest value at maximum displacement. This is clearly evident when the oscillator is released. It moves rapidly toward equilibrium, gaining more and more speed. When it reaches equilibrium, the oscillator has so much speed that it coasts right through the equilibrium position, even though the acceleration and sum of forces are zero at this point. Then the oscillator begins building force and acceleration on the other side of equilibrium until these values max out at the point where the oscillator stops instantaneously and reverses direction. This concept often fools students. For an oscillator, the acceleration and sum of forces are greatest at maximum displacement and are zero at equilibrium.

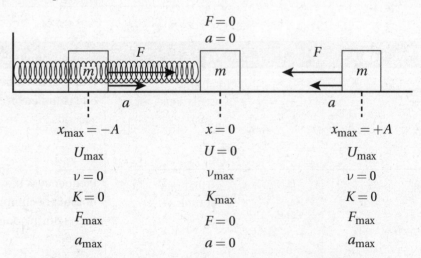

Figure 14.7 Force and energy trends in oscillations

Conservation of energy

The total energy at any point 1 is equal to the total energy at point 2.

$$\Sigma E_1 = \Sigma E_2$$

$$U_1 + K_1$$
$$= U_2 + K_2$$

SUMMARY

1. **DEFINITION OF SHM.** An object in simple harmonic motion (SHM) oscillates up and down, or back and forth, about an equilibrium position. The farther from equilibrium the greater the restoring force that acts to bring the object back to equilibrium. Examples include spring-mass systems, pendulums, and guitar strings.

2. **TERMS RELATED TO SHM.** The period of oscillation is the amount of time, in seconds, for one complete oscillation. Frequency is the reciprocal of period and is defined as the number of oscillations occurring each second. The units for period are seconds. The units for frequency are 1/seconds or Hertz.

3. **FINDING THE SPRING CONSTANT.** Every spring has a unique spring constant, k. Determining the spring constant is a priority when working with springs and SHM. For springs that suspend a mass vertically, the spring constant can be determined by setting the restorative force, $F_s = kx$, of the spring equal to the force of gravity, $kx = mg$.

4. **PERIOD AND FREQUENCY FOR A SPRING DEPEND ON THE MASS ATTACHED TO THE SPRING AND THE SPRING CONSTANT.** The period and frequency of a spring *do not* depend on the displacement (amount of stretch or compression) of the spring. They also *do not* depend on the gravity in which the spring is oscillating.

5. **PERIOD AND FREQUENCY FOR PENDULUMS DEPEND ON THE LENGTH OF THE STRING AND THE GRAVITY.** The period and frequency of a pendulum *do not* depend on displacement (the distance the pendulum is moved right or left). They also *do not* depend on the amount of mass hanging from the pendulum.

6. **DETERMINING PERIOD AND FREQUENCY FROM A GRAPH.** Graphs of displacement versus time for SHM oscillators result in sine waves. The time for one complete oscillation to occur is the period of the SHM oscillator. The reciprocal of that value is the frequency.

7. **TOTAL ENERGY IS CONSERVED IN AN OSCILLATION.** Kinetic and potential energies are continually changing during the oscillation, but total energy is conserved. The total energy at any point 1 during an oscillation can be set equal to the total energy at any point 2 during the oscillation, $U_1 + K_1 = U_2 + K_2$.

If You See	Try	Keep in Mind
Period	$T = \dfrac{\text{total time}}{\text{\# of oscillations}}$ and $T = \dfrac{1}{f}$	Period is the time of one complete cycle in seconds.
Frequency	$f = \dfrac{\text{\# of oscillations}}{\text{total time}}$ and $f = \dfrac{1}{T}$	Frequency is the number of complete cycles during one second.

A stretched spring	Hooke's law $$F_s = kx$$ Determining the spring constant, k, for vertical springs displaced by a mass is done as follows: $$F_s = F_g$$ $$kx = mg$$	The force of the spring, F_s, acts opposite the displacement, x. Finding the spring constant is often a critical first step.
Periods or frequencies of oscillators	Springs: $$T_s = 2\pi\sqrt{\frac{m}{k}}$$ Pendulums: $$T_p = 2\pi\sqrt{\frac{L}{g}}$$ Frequencies can be found by substituting these periods into $$f = \frac{1}{T}$$	The period of a spring depends on the mass suspended and the spring constant. The period would not be affected if the spring were to oscillate in an environment with different gravity. The period of a pendulum depends on the length of the string and gravity when not on Earth. In addition, the magnitude of the displacement for a mass-spring system or a pendulum bob does affect the period of oscillation.
Energy in an oscillator	Conservation of energy: $$E_1 = E_2$$ Springs: $$U_{s1} + K_1 = U_{s2} + K_2$$ $$\frac{1}{2}kx_1^2 + \frac{1}{2}mv_1^2 = \frac{1}{2}kx_2^2 + \frac{1}{2}mv_2^2$$ Pendulums: $$U_{g1} + K_1 = U_{g2} + K_2$$ $$mgh_1 + \frac{1}{2}mv_1^2 = mgh_2 + \frac{1}{2}mv_2^2$$	Total energy is constant. This means that the total energy at any point 1 in an oscillation is equal to the total energy at any other point 2.

PRACTICE EXERCISES

Questions 1–2
An oscillator completes 20 cycles in 50 seconds.

1. Determine the period of the oscillation.

   (A)   0.4 s
   (B)   2.5 s
   (C)   20 s
   (D)   50 s
   (E)   100 s

2. Determine the frequency of the oscillation.

   (A)   0.4 Hz
   (B)   2.5 Hz
   (C)   20 Hz
   (D)   50 Hz
   (E)   100 Hz

---

3. A 4.0-kilogram mass is attached to a spring and lowered to equilibrium. During this process, the spring stretches 2.0 meters. Determine the spring constant.

   (A)   0.0125 N/m
   (B)   1.25 N/m
   (C)   20 N/m
   (D)   40 N/m
   (E)   80 N/m

4. Which of these factors affects the period and frequency of an oscillating spring?

   I.   Spring constant
   II. Spring stretch
   III. Mass attached to the spring

   (A) I only
   (B) II only
   (C) III only
   (D) I and II only
   (E) I and III only

5. Which of these factors affects the period and frequency of a pendulum?

   I. Length of the string
  II. Displacement of the mass from equilibrium
 III. Mass attached to the string

  (A) I only
  (B) II only
  (C) III only
  (D) I and II only
  (E) I and III only

6. A pendulum oscillates with a period of T. If the length of the string were quadrupled, what would be the resulting period?

  (A) $\sqrt{2}T$
  (B) $2T$
  (C) $4T$
  (D) $8T$
  (E) $16T$

7. Which of the following is Hooke's law?

  (A) $PV = nRT$
  (B) $E = mc^2$
  (C) $F = ma$
  (D) $F = -kx$
  (E) $T = 2\pi\sqrt{m/K}$

Questions 8–9

The diagram below shows a mass-spring system in oscillation and identifies three key points in the motion. Positions I and III are the maximum displacements, and position II is the equilibrium position.

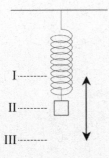

8. At which position(s) is the kinetic energy greatest?

  (A) I only
  (B) II only
  (C) III only
  (D) I and II only
  (E) I and III only

9. At which position(s) is the acceleration the greatest?

(A) I only

(B) II only

(C) III only

(D) I and II only

(E) I and III only

---

10. Which graph correctly depicts the total energy during an oscillation?

(A)

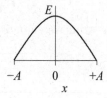

(B)

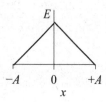

(C)

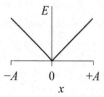

(D)

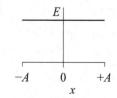

(E)

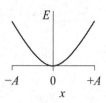

ANSWERS EXPLAINED

	Key Words	Needed for Solution	Now Solve It
1. **(B)**	20 cycles . . . 50 seconds . . . determine the period	$T = \dfrac{\text{total time}}{\text{\# of oscillations}}$	$T = \dfrac{(50\text{ s})}{(20\text{ cycles})} = 2.5\text{ s}$ **Note:** A cycle does not have any units. It is merely the count of a specific event.
2. **(A)**	20 cycles . . . 50 seconds . . . determine the frequency	Either $f = \dfrac{\text{\# of oscillations}}{\text{total time}}$ or $f = \dfrac{1}{T}$	Either $f = \dfrac{(20\text{ cycles})}{(50\text{ s})} = 0.4\text{ Hz}$ Or use the answer from question 1. $f = \dfrac{1}{2.5\text{ s}} = 0.4\text{ Hz}$
3. **(C)**	Spring . . . lowered to equilibrium . . . stretches 2.0 meters	At equilibrium, forces are equal and opposite $F_s = F_g$ $kx = mg$	Finding the spring constant is often a critical first step. $k = \dfrac{mg}{x} = \dfrac{(4.0\text{ kg})(10\text{ m/s}^2)}{(2.0\text{ m})} = 20\text{ N/m}$
4. **(E)**	Factors affect the period and frequency of an oscillating spring	Period of a spring $T_s = 2\pi\sqrt{\dfrac{m}{k}}$	Only the mass and spring constant are contained in the formula for the period of a spring, and only these factors affect its period.
5. **(A)**	Factors affect the period and frequency of a pendulum	$T_p = 2\pi\sqrt{\dfrac{L}{g}}$	Only length and gravity are contained in the formula for the period of a pendulum, and only these factors affect its period. Gravity is most often Earth's gravity, which remains constant. However, if the pendulum were moved into a new gravity, then gravity would matter.
6. **(B)**	Pendulum . . . length of the string were quadrupled . . . resulting period	$T_p = 2\pi\sqrt{\dfrac{L}{g}}$	The quadrupled length is under a square root. $(2)T_p = 2\pi\sqrt{\dfrac{(4L)}{g}}$ This will increase the period to $2T_p$.
7. **(D)**	Hooke's law	$F = -kx$	You should memorize equations and relevant names associated with key laws and major principles. The minus sign in the equation indicates that the vector representing the restoring force, F, is opposite the direction of x, the vector representing the spring stretch or compression. When solving for the magnitude of the restoring force, the minus sign is ignored, $F = kx$.

	Key Words	Needed for Solution	Now Solve It
8. **(B)**	Diagram . . . spring-mass systems in oscillation . . . kinetic energy greatest	Knowledge/definitions	The kinetic energy is greatest when speed is greatest. This occurs at equilibrium.
9. **(E)**	Diagram . . . spring-mass systems in oscillation . . . acceleration the greatest	Knowledge/definitions	This occurs at maximum displacement when the spring is stretched the farthest from equilibrium. At this position, the sum of force and acceleration are the most extreme. Strangely, this is the position where motion stops instantaneously. Therefore, students are tricked into thinking acceleration is zero when it is really at its maximum. At equilibrium, acceleration is zero, but this is where the speed is greatest. Watch out for this!
10. **(E)**	Graph . . . total energy	Knowledge/definitions	Total energy (potential plus kinetic) is conserved during an oscillation and must remain constant. Choice A is the graph of kinetic energy, and choice B is the graph of potential energy. Add these two graphs together. The result is a straight horizontal line at the maximum (total) energy during the oscillation.

Waves

15

→ **TRAVELING WAVES**

→ **MECHANICAL WAVES**

→ **ELECTROMAGNETIC WAVES**

→ **DOPPLER EFFECT**

→ **SUPERPOSITION AND STANDING WAVES**

The previous chapter on simple harmonic motion (SHM) introduced various terms, concepts, and formulas dealing with oscillations. Oscillations involve a single vibrating object. When energy from one oscillator is transferred to another oscillator, it is known as a wave. The principal function of waves is to transmit energy. Wave behavior can take the form of mechanical waves or electromagnetic (light) waves. This chapter will expand upon those ideas by first reviewing some general wave properties and then demonstrating how these apply to mechanical and electromagnetic waves. Table 15.1 shows the variables that will be discussed. In particular, this chapter will do the following:

- Define transverse and longitudinal traveling waves and their behavior in a medium.
- Define the properties of mechanical and electromagnetic waves.
- Explain the Doppler effect.
- Describe superposition and how it leads to standing waves and beats.

Table 15.1 Variables That Describe Waves

New Variables	Units
λ = wavelength	m (meters)
c = speed of light	3×10^8 m/s (meters per second)

TRAVELING WAVES

A **wave** is an organized disturbance consisting of many individual oscillators. Although each individual oscillator barely moves, the resulting wave can travel great distances at a constant speed in a specific medium. While traveling, this disturbance transmits energy from one place to another without moving any physical objects over the same distance. The energy propagates outwardly from the source of the oscillations.

There are two types of traveling waves: **transverse** and **longitudinal**. Figure 15.1 illustrates the two types.

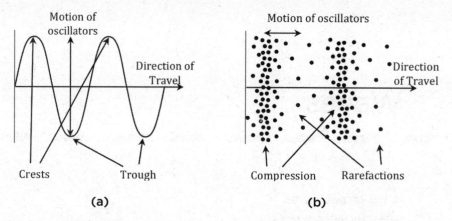

Figure 15.1 Transverse (a) and longitudinal (b) waves

In transverse waves, the oscillators vibrate perpendicularly to the direction of wave propagation (they transverse/cross equilibrium). In longitudinal waves, the oscillators vibrate parallel (along or longitudinal) to the direction of wave propagation.

Mathematically, both transverse and longitudinal waves can be graphically represented as sine-wave functions. The longitudinal wave has a unique appearance that is difficult to depict in diagrams. As a result, diagrams used to analyze longitudinal waves depict a sinusoidal form that resembles the appearance of transverse waves. Figure 15.2 identifies the principal components of both transverse and longitudinal waves. The figure depicts an instantaneous view of a wave. The amplitude of the oscillations is shown on the y-axis, while the horizontal position of the wave is frozen on the x-axis.

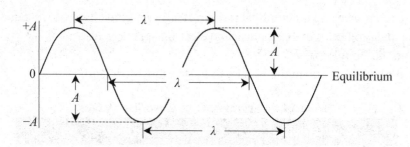

Figure 15.2 Principal wave components

The **amplitude**, A, is the maximum displacement, $A = x_{max}$. It can be measured from equilibrium ($x = 0$) to either $+A$ or $-A$. The wave illustrated in Figure 15.2 is a sine wave because the wave begins at the origin and then proceeds upward toward the positive maximum displacement, $+A$. However, the wave might start at the positive maximum displacement (making it a cosine wave), at the origin and proceed downward, or at the negative maximum displacement and proceed upward. Regardless, the wave will exhibit a repeated sinusoidal pattern between the two maximum displacements.

When the wave pattern repeats, it is known as a **wavelength**, λ. Wavelength is the distance measured between two successive identical portions of a wave. It is often easiest to see the wavelength between two successive crests or two successive troughs. However, the wavelength can also be determined as the distance between three crossings of the equilibrium line.

Medium and Wave Speed

The **medium** is the substance through which a wave propagates. For example, sound waves most often move in the medium air. Motion of a wave through one, and only one, medium is at constant speed. When a wave changes medium, the wave speed most often changes to a new constant speed in the new medium. The **speed of the wave**, v, in a specific medium is the product of the wave's frequency, f, and its wavelength, λ. In addition, the constant-speed formula is also applicable when waves travel in a constant medium.

$$v = f\lambda \quad \text{and} \quad v = \frac{d}{t}$$

As long as the wave travels in only one medium, the wave speed will remain constant and the frequency and wavelength will vary inversely.

An example of this phenomenon is sound waves. Frequency of sound is perceived as **pitch**. As the frequency increases, the sound wave itself does not travel any faster. If increasing frequency did affect the wave speed, then high-pitched notes would reach the ears of an observer before low-pitched notes. This, however, is not the case. An observer detects multiple frequencies produced by a single source simultaneously. This means that high- and low-pitched sounds must have different wavelengths. Therefore, frequency and wavelength must vary inversely. High-pitched (high-frequency) sounds have short wavelengths. Low-pitched (low-frequency) sounds have long wavelengths.

Another example of this phenomenon is light waves. The frequency of light is perceived as color. Low-frequency light is perceived as red, and high-frequency light is perceived as blue. If increasing frequency did affect the wave speed, then different colors would arrive at the eye of an observer at different times. This is not the case. Red and blue frequencies, for example, produced by a single source are viewed simultaneously by an observer and appear magenta. Red frequencies simply have longer wavelengths, while blue frequencies have shorter wavelengths.

You should also note that if a wave changes mediums as it travels, its speed is affected. However, when a wave changes medium, the frequency remains the same. As an example, a yellow swimming suit will appear yellow both above and below water. If changing mediums changes wave speed while keeping frequency constant, then wavelength must be changing. The relationship between wave speed and wavelength is directly proportional. If a wave speeds up when entering a new medium, the wavelength will become longer. If a wave slows down when entering a new medium, the wavelength will become shorter.

IF YOU SEE a wave traveling in only one medium

Constant wave speed

Frequency and wavelength vary inversely, and wave speed is constant.

IF YOU SEE a wave changing mediums

Constant frequency

The velocity and wavelength of a wave are directly proportional.

EXAMPLE 15.1

Wave Equation

A wave travels through a medium with velocity v, frequency f, and wavelength λ. If the wave then enters another medium that increases the velocity to $2v$, what will be the corresponding frequency and wavelength?

WHAT'S THE TRICK?

The velocity of a wave is dependent upon the medium in which it travels. When waves change mediums, only the velocity and wavelength are affected. Frequency remains unchanged. Therefore, according to the wave equation, $v = f\lambda$, if velocity doubles, wavelength must also double and frequency will remain the same.

Effect of Amplitude on a Wave

The amplitude of a wave is the maximum displacement of the oscillating particles composing the wave as measured from their equilibrium positions. The amplitude of a wave does not affect the wave speed, wavelength, or frequency of the wave. The amplitude affects only the energy of the wave. A good example of this is the effect of amplitude on a sound wave. In terms of sound, amplitude is the volume of a sound. If, for example, a note of a certain frequency is being played through a loudspeaker, the note will not change if the volume is increased. Similarly, if the amplitude (volume) is increased, the speed of the sound coming out of the loudspeaker will not travel any faster to its intended observer.

MECHANICAL WAVES

Mechanical waves involve the displacement of molecules in a medium from a position of equilibrium. Examples of mechanical waves and their mediums include vibrations on a string, ripples on a pond, and sound moving through air. A medium can therefore be a solid, liquid, or gas. These waves are disturbances created by a source that travels outwardly from the source at a constant velocity for that medium.

IF YOU SEE
speed of
sound

Medium
dependent

In general sound travels faster in denser mediums. This means it moves slowest in gases.

Sound Waves

Sound is a form of mechanical waves that travels as longitudinal (compression) waves. Longitudinal sound waves are difficult to illustrate. However, they mathematically graph as sinusoidal functions. As a result, graphs of sound waves often make them appear similar to transverse waves for illustrative purposes. Mechanical waves, such as sound, require a medium in order to propagate and travel. Sound can travel through solids, liquids, and gases. Each medium affects the speed and wavelength of the sound but not the frequency. The speed of a wave in a medium depends on many factors (density, elasticity, temperature, etc.). However, as a general rule, sound travels faster in denser mediums (see Table 15.2). This means sound moves fastest in solids and slowest in gases.

Table 15.2 Speed of Sound in Different Mediums

Medium	Speed of Sound
Aluminum	5,100 m/s
Water	1,480 m/s
Air (25°C)	345 m/s

ELECTROMAGNETIC WAVES

Electromagnetic waves are **light waves**. These include radio waves, microwaves, infrared light, visible light, ultraviolet light, X-rays, and gamma rays. Researchers determined in the early twentieth century that these waves do not require a medium in which to travel and can therefore move through the vacuum of space. However, if they do travel in a medium, such as air, water, or glass, their wave behavior will be affected by that medium.

All forms of electromagnetic waves travel at the same constant speed in a vacuum. This value is known as the speed of light, c. Its value is 3×10^8 meters per second. For light moving in a vacuum, the speed formula can be modified.

$$c = f\lambda$$

The speed of electromagnetic waves is the complete opposite of mechanical waves. Electromagnetic waves have their highest speed in a vacuum, where mechanical waves cannot even exist. While mechanical waves typically speed up in denser mediums, electromagnetic waves are slowed as a medium's optical density increases. Table 15.3 shows commonly encountered speeds for light waves.

Table 15.3 Speed of Light in Different Mediums

Medium	Speed of Light (approximate)
Air/Vacuum	3.00×10^8 m/s
Water	2.25×10^8 m/s
Glass	2.00×10^8 m/s

A practical example of this phenomenon is seen when a beam of light from a laser pointer strikes a glass block at an angle. The light beam will move through the glass at a new angle due to the change in speed of light when it enters the glass block. This phenomenon, known as refraction, will be covered in Chapter 16, "Geometric Optics."

EXAMPLE 15.2

Speed of Sound vs. Speed of Light

During a thunderstorm, a flash of lightning is seen. Then, 5.0 seconds later, a thunderous crack is heard. How far away was the lightning?

WHAT'S THE TRICK?

This problem illustrates the disparity between the speed of light and the speed of sound. They are both traveling simultaneously through the medium of air, but each has vastly different speeds. The speed of light in air is nearly the same as in a vacuum. This speed is so fast that the light from the lightning arrives nearly instantaneously. The travel time is so short that it is negligible (near zero) and not worth including in calculations.

The speed of sound in air is about 340 meters per second. The distance to the lightning strike can be determined using the constant speed of sound in air.

$$v = d/t$$
$$d = vt$$
$$d = (340 \text{ m/s})(5.0 \text{ s}) = 1{,}700 \text{ m}$$

IF YOU SEE
speed of
light

Medium dependent

The speed of light is greatest in a vacuum. Light travels slower in denser mediums, which is the opposite of mechanical waves.

Electromagnetic Spectrum

The oscillations of an electromagnetic field create a spectrum of electromagnetic waves. These waves transmit energy in direct proportion to their frequency. At the low end of the spectrum are radio waves. At the high end are X-rays and gamma rays. In between these ends of the spectrum is visible light. The human eye is sensitive to the frequencies in this range, which are perceived as colors. The electromagnetic spectrum is shown in Figure 15.3. It starts at the left with long-wavelength, low-frequency, and low-energy radio waves. The spectrum progresses in order to the short-wavelength, high-frequency, high-energy gamma rays. Visible light is broken into the colors of the rainbow in their relative order as well.

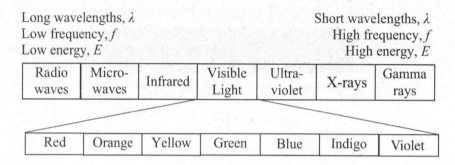

Figure 15.3 The electromagnetic spectrum

Knowing the exact frequencies and wavelengths of each portion of the electromagnetic spectrum will not be important. However, knowing the relative wavelength, frequency, and energy relationships is helpful. Remembering the order of the visible-light spectrum, from low frequency to high frequency, with the acronym ROYGBIV will also help. Note that the energy of the blue end of the spectrum is higher than the red end. This is often mistaken by students who wrongly think that red is the high-energy color, when in reality blue and violet light possess greater energy.

EXAMPLE 15.3

Electromagnetic Spectrum

Sort the following electromagnetic waves in order from lowest energy to highest energy:

X-rays, radio waves, ultraviolet, visible light, and microwaves

WHAT'S THE TRICK?

This trick requires you to know the order of the spectrum and which end has the highest energy. Remember that visible light is right in the middle of the spectrum, with infrared next to red and ultraviolet next to violet. Ultraviolet causes skin cancer, so it possesses higher energy and will be on the right. X-rays and gamma rays sound dangerous and possess even higher energy, so they are to the right of ultraviolet. This leaves radio waves and microwaves on the low-energy end, to the left of infrared. The correct order for the electromagnetic waves given is:

Radio waves, microwaves, visible light, ultraviolet, and X-rays

---

DOPPLER EFFECT

Wave Front Model of Sound

Diagrams of sound waves in exams often show the source of sound as a dot and the waves of sound as expanding circles, similar to the result seen when dropping a rock into a pond. When a rock is dropped into a still pond, waves move outward in every direction from the point of impact. The crests of the waves appear as expanding circles when viewed from above. This view of waves is known as the **wave front model**. The expanding circles represent the expanding wave crests. The distance between the circles (crests) is the wavelength. Although drawn as circles on paper, sound-wave fronts actually form expanding three-dimensional spheres. A stationary sound source would emit spherical wave fronts as depicted in Figure 15.4.

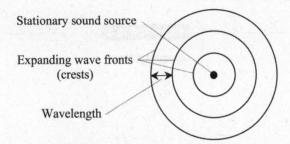

Figure 15.4 Wave front model

The Doppler Effect and Sound

When a wave is produced by a source in a medium, the wave propagates through the medium at a constant speed, v. Once produced by the source, the speed of the wave in that particular medium will not change, regardless of the speed of the source.

An interesting phenomenon occurs when the source of a sound is moving with respect to a stationary observer. Figure 15.5 shows a car moving to the right at constant speed. If the driver presses the horn continuously, then sound waves will leave the car and travel outward in expanding spheres.

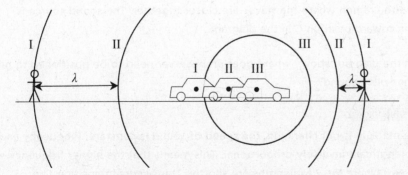

Figure 15.5 The Doppler effect

To make it simple, assume one sound wave is emitted every second. Each sound-wave front moves outward from the location where the car was at the time the wave was emitted. The first sound wave, I, was emitted when the car was at position I. This wave has been expanding for three seconds. The second wave, II, was emitted a second later when the

**IF YOU SEE
a moving
sound
source**

Doppler effect

If the sound
source
is moving
toward the
observer,
the observer
hears a higher
frequency. If it
is moving
away, the
observer
hears a lower
frequency.

car was at position II. This wave has been moving for two seconds. The third wave, III, was emitted when the car was at position III. This wave has been traveling for only one second.

The sound of the horn will differ depending on the location of an observer, and the effects observed are known as the **Doppler effect**. For an observer in front of the car, the wavelengths appear to be shorter than they actually are. Since the speed of sound is constant, the shorter wavelengths create a higher frequency, and the horn will sound as though it has a higher pitch than it actually does. For an observer behind the car, the wavelengths appear to be longer than they actually are. The observer hears a lower frequency with a lower pitch than the horn actually makes. Note that the sound of the horn is not changing at all. It merely appears to have a higher frequency and shorter wavelengths when the sound source is moving toward the observer, It also apparently has a lower frequency and longer wavelength when the sound source is moving away from the observer.

EXAMPLE 15.4

Doppler Effect

B

A (((•))) C

D

(A) A sound source emits the wave fronts shown in the diagram above. In what direction is the sound source traveling compared to a stationary observer?

WHAT'S THE TRICK?

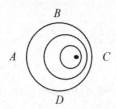

When sources move away:
• Perceive longer wavelengths
• Hear lower pitch

v_{source}

When sources approach:
• Perceive shorter wavelengths
• Hear a higher pitch

The wave fronts pile up in front of the sound source, so the sound source moves toward the region where the waves are closer together. The sound source is moving toward position C in the diagram.

(B) In the diagram above, where does an observer need to be positioned to hear a higher-pitch sound?

WHAT'S THE TRICK?

If the medium is not changing, the speed of sound is constant. Frequency and wavelength are inversely proportional. This means that the higher frequency will be heard where the wavelengths are shorter. This occurs in position C in the diagram.

The previous explanations and examples address the most common problems involving moving sound sources and stationary observers. In some problems, the observer may be moving, or both the source and observers may be moving. The key to any Doppler-shift problem is the relative motion between the source and the observer. Whether the source moves toward the observer, the observer moves toward the source, or they both move toward each other, the observed effect is the same. When a sound source and observer approach each other, the perceived wavelength of sound decreases and the observed frequency increases (high pitch). If the distance between the source and observer is increasing, then the perceived wavelength increases and the observed frequency decreases (low pitch).

Speed of the Source and the Speed of Sound

Several scenarios are possible depending on the speed of the source of sound. Figure 15.6 depicts four possible Doppler-effect diagrams for moving sound sources.

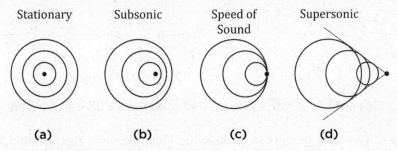

Figure 15.6 The Doppler effect for moving sound sources

A **stationary object** will create concentric circular wave fronts as seen in Figure 15.6(a). A **subsonic object**, as seen in Figure 15.6(b), is an object moving with a speed less than the speed of sound (about 340 m/s in air). The best examples are a car continuously honking its horn or a siren on an emergency vehicle. When the sound source reaches the **speed of sound**, as shown in Figure 15.6(c), the wave fronts pile up on one another. These sound waves constructively interfere (add up) with each other to create a phenomenon known as the **sound barrier**. If a sound source exceeds the speed of sound, it is said to be **supersonic**, which is shown in Figure 15.6(d). When this occurs, the wave fronts constructively interfere in a manner that creates a wake of sound that is similar to the wake of a boat. When this wake of sound passes by a person, the compression of waves sounds like a boom. This is known as a **sonic boom**.

The Doppler Effect and Light

The Doppler effect for light waves is best known for the redshift seen in the color spectrum of most stars. If a star is moving toward Earth, the wavelengths of light will appear shorter than they actually are. This will cause the light from the star to shift very slightly toward the blue end of the visible spectrum. When a star moves away from Earth the wavelengths of light will appear longer than they actually are, and the visible light emitted by the star will shift very slightly toward the red end of the spectrum. The spin of galaxes and galactic clusters causes some stars to appear blue-shifted and others to appear red-shifted. However, the majority of stars appear red-shifted compared to their expected color spectrums. This overall redshift is evidence that the universe is expanding.

SUPERPOSITION AND STANDING WAVES
Superposition

Superposition is when two or more waves occupy the same point in space, at a given moment, and their combination displaces the medium to reflect the sum of their individual displacements. In Figure 15.7, two wave pulses, *A* and *B*, generated in a string travel toward one another. In Figure 15.7(a), the pulses are the same size and are on the same side of the string. In Figure 15.7(b), the pulses are the same size, but pulse *B* is inverted.

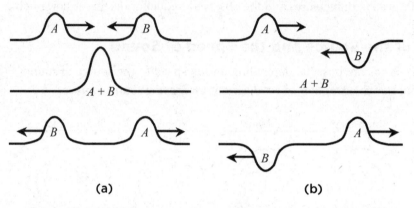

Figure 15.7 Constructive (a) and destructive (b) wave pulses

When pulses *A* and *B* superimpose (occupy the same location), their combined displacements add to create a composite pulse. After superimposing for an instant, the wave pulses continue on their way in their original directions. In Figure 15.7(a), the wave pulses add to create a larger wave. This is an example of **constructive interference**. The waves overlap. The medium displacement is larger than it was for the waves individually. Figure 15.7(b) is an example of **destructive interference**, which occurs when waves overlap and cause a smaller displacement of the medium. When opposite waves are exactly the same size, they will cancel entirely, as shown in Figure 15.7(b).

Standing Waves

When a wave is trapped between two boundaries, the individual points move up and down but do not travel. They appear to stay in one place. This is known as a **standing wave**. An example can be seen on a guitar in which the guitar strings are attached at one end to the headstock and at the other end to the bridge. Plucking the string will create a standing wave. As one wave strikes a boundary, it bounces back and interferes with the next incoming wave. This creates an alternating pattern where part of the standing wave moves back and forth while several points do not appear to move at all. Figure 15.8 shows the incoming wave as a solid line and the reflected wave as a dashed line.

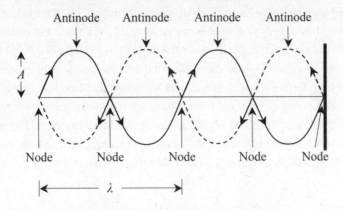

Figure 15.8 A standing-wave pattern

The **nodes** are where the superposition of two waves creates destructive interference. The **antinodes** are the locations of greatest constructive interference. The amplitude, A, is the distance from the equilibrium line to the maximum displacement of the string. The maximum amplitude occurs at the antinodes. The wavelength of the wave is the distance between three successive nodes or three successive antinodes. The nodes are spaced exactly $\lambda/2$ away from each other, and so are the antinodes. The wavelength of a standing wave is 2 times the distance between two successive nodes or antinodes.

Fundamental Frequency and Harmonics

When a simple instrument such as a guitar string or a flute is played, standing waves occur in the string or in the air inside the flute. The simplest standing waveform that can be produced in the string or between the ends of the flute produces a frequency known as the **fundamental frequency**, f_1. The fundamental frequency is also known as the **first harmonic**. In a string and in an open tube (open at both ends), the fundamental frequency is associated with a waveform consisting of half of a wavelength. For closed tubes (closed at one end), the waveform consists of a quarter wavelength. Figure 15.9 shows the standing waveform associated with each of these simple instruments.

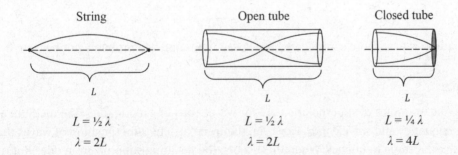

Figure 15.9 First-harmonic waveforms

The values for wavelengths determined in Figure 15.9 can now be substituted into the wave-speed equation in order to determine the fundamental frequency, f_1.

- Strings and open tubes: $v = f_1 (2L)$
- Closed tubes: $v = f_1 (4L)$

Other standing waves can be created in strings and in tubes. All of these patterns are known as the **harmonics**. They have a variety of frequencies (f_1, f_2, etc.) associated with the number of wavelengths fitting along the string or between the ends of the tubes. Usually questions regarding the harmonics are associated with the difference in frequency between the fundamental frequency, f_1 (the first harmonic), and the second harmonic, f_2. More than likely, these questions will focus on either a string or an open tube, which are mathematically the same. To create the second harmonic, the next complete standing wave must fit along the string or inside an open tube. Figure 15.10 shows the next wavelength capable of creating a standing wave in strings and tubes. The wavelengths depicted are associated with the frequencies of the second harmonic, f_2.

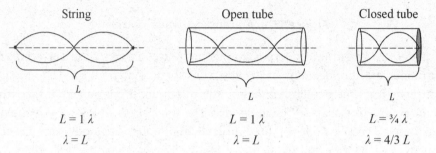

String | Open tube | Closed tube

L | L | L

$L = 1\ \lambda$ | $L = 1\ \lambda$ | $L = \frac{3}{4}\ \lambda$

$\lambda = L$ | $\lambda = L$ | $\lambda = 4/3\ L$

Figure 15.10 Second-harmonic waveforms

For strings and open tubes, the wavelength for the second harmonic is half of the wavelength associated with the fundamental. The wavelength of the harmonics can be easily determined if the wavelength, λ_1, associated with the fundamental frequency is known. Simply multiply this key wavelength by the inverse of the harmonics number, n.

$$\lambda_n = (1/n)\lambda_1$$

The wavelength of the second harmonic can be calculated as follows:

$$\lambda_2 = (1/2)\lambda_1$$

Frequency and wavelength are inversely proportional. As a result, the frequencies can be found by multiplying the fundamental frequency by n.

$$f_n = nf_1$$

The pattern is a bit more complicated for closed tubes and will probably not be tested.

Beats

So far, the examples of superposition are for waves traveling along the same medium at the same frequency and wavelength. However, there can also be superposition of waves that do not have the same frequency. When this occurs, the resulting superposition does not reflect a perfect sinusoidal pattern. A good example of this is the superposition of sound waves produced by two musical instruments at slightly different pitches (musical notes).

Imagine two violinists playing the same note. A sound wave is produced from each violin. As the waves propagate through the medium of air, they will undergo superposition. If the waves are perfectly identical, they will complement each other and add constructively. If, however, the waves are slightly different, there will be both constructive and destructive portions. The destructive portions decrease the amplitude (volume) at a regular rate and cause what are known as **beats**.

In Figure 15.11, two completely different waves are added to create a third wave pattern, which displays the characteristic known as beats. Figures 15.11(a) and 15.11(b) display the amplitude versus time graphs for two sound waves, which have different frequencies and wavelengths. Figure 15.11(c) illustrates the superposition of these two waves.

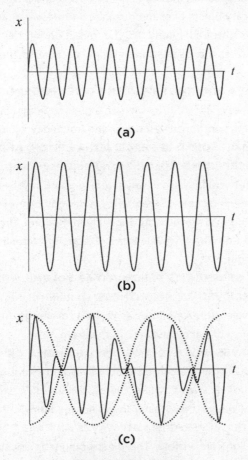

(a)

(b)

(c)

Figure 15.11 Beats

Three distinct beats occur during the graphed time interval. This is known as the **beat frequency**. The beat frequency created by any two waves can be quickly determined by the absolute value of the difference between their frequencies.

$$f_{\text{beat}} = |f_1 - f_2|$$

Beats are detected by the human ear as quick outbursts of loud and quiet because amplitude (volume) is being affected. Most people can detect seven or fewer beats per second. For two waves that are very dissimilar in frequency, there are too many beats per second to be detected by the human ear.

SUMMARY

1. **A WAVE IS AN ORGANIZED DISTURBANCE CONSISTING OF MANY INDIVIDUAL OSCILLATORS.** There are two types of waves: longitudinal and transverse. In both types, the wave disturbance transmits energy from one place to another without moving any physical objects over the distance. In transverse waves, the individual oscillators vibrate perpendicularly to the direction of the wave disturbance. In longitudinal waves, the oscillations vibrate parallel to the direction of wave propagation.

2. **WAVES TRAVEL AT A CONSTANT SPEED IN A GIVEN MEDIUM.** Speed is the product of frequency and wavelength. If a wave changes mediums, the speed and wavelength will change proportionally. However, the frequency will remain constant.

3. **THE FREQUENCY OF SOUND IS PERCEIVED AS PITCH, AND THE FREQUENCY OF LIGHT IS PERCEIVED AS COLOR.** High-frequency sound is heard as a high-pitch sound. High-frequency light waves are gamma rays and X-rays. Blue and violet are high-frequency colors. Frequency is directly proportional to energy but inversely proportional to wavelength for a given medium. High-frequency waves have high energy and short wavelengths. Frequency remains constant when a wave changes medium.

4. **THE AMPLITUDE OF SOUND IS PERCEIVED AS VOLUME, AND THE AMPLITUDE OF LIGHT IS PERCEIVED AS BRIGHTNESS.** Amplitude does not affect speed, frequency, or wavelength. Amplitude is related to energy. Increasing amplitude increases the energy of each oscillator in the wave.

5. **MECHANICAL WAVES INVOLVE THE DISPLACEMENT OF MOLECULES IN A MEDIUM.** Mechanical waves include sound waves and vibration of strings. Sound waves are longitudinal waves, while the vibrations of a string are transverse waves. Mechanical waves travel faster in denser mediums.

6. **ELECTROMAGNETIC WAVES ARE LIGHT WAVES AND ARE TRANSVERSE WAVES.** Not all forms of light are visible. The electromagnetic spectrum includes radio waves, visible light, ultraviolet light, microwaves, X-rays, and gamma rays. Electromagnetic waves do not need a medium in which to travel. They travel fastest in a vacuum and are slowed by denser mediums, the opposite of mechanical waves.

7. **DOPPLER EFFECT OF SOUND AND LIGHT.** When a source emits a constant frequency of sound or light, an observer will perceive a different frequency if the source is moving toward or away from the observer than if the source is stationary. If the source moves toward the observer, the observer will perceive an increased frequency and decreased wavelength. If the source moves away from the observer, the observer will perceive a decreased frequency and an increased wavelength.

8. **SUPERPOSITION IS WHEN TWO OR MORE WAVES MEET AND ADD TOGETHER.** When wave crests are in the same direction and align, superimposing waves add constructively, creating larger waves. If the crests arrive out of sync, then waves can interfere destructively, resulting in smaller waves. Destructive intereference can even cause the waves to cancel each other completely. In a standing wave, superposition of waves is contained between two fixed points. Constructive and destructive interference create a pattern that seems to vibrate in place. When waves are not aligned, an irregular pattern is formed, creating a phenomenon known as beats.

If You See	Try	Keep in Mind
A wave traveling in only one medium	Constant speed $v = f\lambda$	Frequency and wavelength vary inversely, and wave speed is constant.
A wave changing mediums	Constant frequency $v = f\lambda$	The velocity and wavelength of a wave are directly proportional.
Speed of sound	Medium dependent $v = f\lambda$	In general, sound travels faster in denser mediums.
Speed of light	Medium dependent $c = f\lambda$	The speed of light is greatest in a vacuum. Light travels slower in a denser medium (the opposite of sound).
A moving sound source	Doppler effect	If the sound source is moving toward an observer, then a higher pitch is heard. If the source is moving away, then a lower pitch is heard.

PRACTICE EXERCISES

1. Which of the following are longitudinal waves?

    I. Sound waves
    II. Light waves
    III. Waves propagating in a string

    (A) I only
    (B) II only
    (C) III only
    (D) I and II only
    (E) I and III only

2. A sound wave from a source has a frequency of f, a velocity of v, and a wavelength of λ. If the frequency were doubled, how would the speed and wavelength be affected?

    (A) v, λ
    (B) $2v, \lambda$
    (C) $v, 2\lambda$
    (D) $\frac{1}{2}v, \lambda$
    (E) $v, \frac{1}{2}\lambda$

3. When light moving in air enters a block of glass, which of the following properties do NOT change?

   (A) Wave speed
   (B) Wavelength
   (C) Frequency
   (D) Both B and C
   (E) Both A and B

4. As an ambulance siren rapidly approaches an observer, the observer will perceive a

   (A) frequency increase due to a wave speed increase
   (B) frequency increase due to a wave speed decrease
   (C) frequency increase due to a wavelength increase
   (D) frequency increase due to a wavelength decrease
   (E) frequency decrease due to a wavelength increase

5. What characteristic of a wave is affected during the superposition of waves?

   (A) Amplitude
   (B) Frequency
   (C) Pitch
   (D) Wavelength
   (E) Wave speed

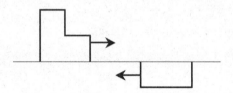

6. Two wave pulses travel toward each other as shown in the diagram above. Which of the following diagrams represents the superposition of the pulses when they meet?

(A)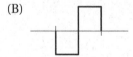

(D)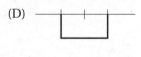

(B)

(E)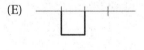

(C)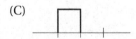

7. In a standing wave, a node is the position where

   (A) constructive interference occurs
   (B) destructive interference occurs
   (C) amplitude is maximum
   (D) both A and C
   (E) both B and C

Questions 8–9

Consider the following properties of a wave and its behavior.

   (A) Amplitude
   (B) Frequency
   (C) Wavelength
   (D) Wave speed
   (E) Superposition

8. The brightness of light and the volume of sound are associated with which wave characteristic?

9. The pitch of a musical instrument and the color of light are associated with which wave characteristic?

_____

10. Which sequence correctly lists the electromagnetic waves from shortest wavelength to longest wavelength?

   (A) Radio waves, visible light, X-rays
   (B) Radio waves, X-rays, visible light
   (C) Visible light, radio waves, X-rays
   (D) X-rays, visible light, radio waves
   (E) X-rays, radio waves, visible light

ANSWERS EXPLAINED

	Key Words	Needed for Solution	Now Solve It
1. **(A)**	Longitudinal waves	Knowledge/definitions	Sound waves are always longitudinal. The other choices form transverse waves.
2. **(E)**	Frequency were doubled . . . how would the speed and wavelength be affected	Knowledge/definitions $v = f\lambda$	Wave speed in a given medium is constant, so speed is unaffected. As a result, frequency and wavelength are inversely proportional. $$v = (2f)\left(\frac{1}{2}\lambda\right)$$ Wavelength is halved when frequency doubles.
3. **(C)**	Light moving in air . . . enters a block of glass . . . properties do NOT change	Knowledge/definitions $v = f\lambda$	When a wave changes mediums, the frequency remains the same.
4. **(D)**	Siren rapidly approaches an observer . . . perceive	Knowledge/definitions Doppler effect	When a sound source approaches an observer, the sound's wavelength appears to be less than it really is. This causes the perceived frequency to increase.
5. **(A)**	Superposition	Knowledge/definitions Superposition	When two waves superimpose, their amplitudes are added together.
6. **(C)**	Wave pulses travel toward each other . . . superposition	Superposition	Sketch the waves as they meet each other: The area below the equilibrium line cancels out the rectangular area above the equilibrium line. This leaves the small square on the top left as the sum of the two pulses.
7. **(B)**	Node	Knowledge/definitions	Destructive superposition of two waves causes a node.
8. **(A)**	Brightness . . . volume	Knowledge/definitions	The amplitude of light waves is brightness. The amplitude of sound waves is volume.
9. **(B)**	Pitch . . . color	Knowledge/definitions	The frequency of sound waves is pitch. The frequency of light waves is color.
10. **(D)**	Shortest wavelength to longest wavelength	Knowledge/definitions Electromagnetic spectrum	The shortest wavelengths are the electromagnetic waves with the highest energy and frequency. X-rays is first and then visible light. Radio waves are the longest wavelengths with the lowest frequencies.

Geometric Optics

16

→ **RAY MODEL OF LIGHT**
→ **REFLECTION**
→ **REFRACTION**
→ **PINHOLE CAMERA**
→ **THIN LENSES**
→ **SPHERICAL MIRRORS**

When light strikes a new medium, such as a reflective mirror, transparent glass, or water, a variety of things can occur. Light can be absorbed, reflected, or refracted (bent) when it strikes the new medium. Usually all of these occur to some degree. For simplicity, however, problems often focus on one of these possibilities at a time. The absorption of light will be addressed in the next chapter. In this chapter, the reflection and refraction of light will be explored in order to do the following:

- Predict the location and appearance of images due to reflection.
- Analyze refraction, the bending of light rays when they move from one medium into another.
- Analyze the properties of a pinhole camera.
- Determine the location of and identify the types of images formed by spherical lenses and mirrors.

Table 16.1 lists the variables that will be used.

Table 16.1 Variables for Geometric Optics

New Variables	Units
n = index of refraction	No units
θ_c = critical angle	Degrees
R = radius of curvature	m (meters)
f = focal distance	m (meters)
d_o = object distance	m (meters)
d_i = image distance	m (meters)
h_o = object height	m (meters)
h_i = image height	m (meters)
M = magnification	No units but may be followed by a $\times$

RAY MODEL OF LIGHT

Light can be viewed in many ways. Each model of light has its advantages in solving specific problems. In Chapter 15, "Waves," light was encountered as a sinusoidal form. In certain problems, it is advantageous to view light as advancing wave fronts (crests). In this chapter, the angles at which light strikes a surface will be important. So, a different model of light is needed. The three models of light are represented visually in Figure 16.1.

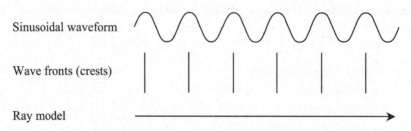

Sinusoidal waveform

Wave fronts (crests)

Ray model

Figure 16.1 Three models of light

The **ray model** of light shows the path of light as a straight line with an arrow indicating the direction of the light. The advantage of using the ray model of light is it allows angles of reflection and refraction to be measured.

REFLECTION

IF YOU SEE
a mirror

Law of
reflection

$\theta_i = \theta_r$

Angles are measured with respect to a normal line, which is perpendicular to the surface of the mirror.

When light strikes a flat, polished surface, it is reflected in a manner similar to a ball's bouncing off of a wall or the floor. In Figure 16.2, an **incident ray** (inbound ray) of light is shown striking a surface at an **incident angle** of θ_i. Whenever surfaces are involved, angles are always measured from a **normal** line. The normal is a line drawn perpendicular to the surface area. The resulting **reflected ray** bounces off the surface with an **angle of reflection** of θ_r.

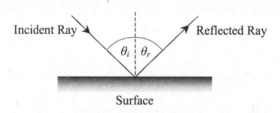

Incident Ray   Reflected Ray

θ_i | θ_r

Surface

Figure 16.2 Reflection

The angle of reflection is the same as the incident angle. This relationship is known as the **law of reflection**.

$$\theta_i = \theta_r$$

Specular reflection occurs when a surface is flat, smooth, and polished; all the reflected rays leave the surface parallel to one another, and the image of the reflection appears similar to the original object. Rough surfaces create **diffuse reflection**, where the irregularities of these surfaces cause the reflected rays to move off in random directions. Reflection allows us to see these rough surfaces, but clear images of objects cannot be reflected by rough surfaces.

Plane Mirror

The simplest mirror is a flat mirror, known as a **plane mirror**. When an object, such as a person, is viewed in a plane mirror, several rays of light can be visualized as starting from the object, moving toward the mirror, and reflecting off of the mirror's surface. Each ray follows the law of reflection. Images are formed where rays intersect. However, the rays starting from the head of the person in Figure 16.3 diverge (spread out). If the reflected rays are traced backward (back trace), shown as dashed lines in Figure 16.3, then an intersection can be found. The image will appear at the intersection of reflected rays.

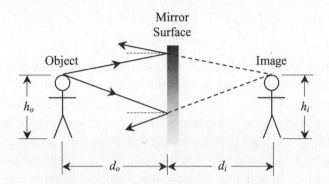

Figure 16.3 Reflection off a plane mirror

The image formed by a plane mirror has several characteristics.

1. The image is upright. It appears right side up.
2. The image is the same size as the object, $h_i = h_o$, resulting in a magnification equal to one, $M = 1$.
3. The image distance and object distance are equal, $d_i = d_o$.
4. The image is a **virtual image**. No light passes through the mirror. Therefore, intersecting light does not form the image. This means the image cannot be projected onto a screen, which is an important characteristic of a virtual image. Yet virtual images can be seen. An image is easily recognized as a virtual image since it is always upright.
5. Plane mirrors reverse front and back.

REFRACTION

When light moves into a new medium with a different density, its speed and wavelength change. If the light strikes this medium at an angle, the ray will bend at the surface of the new medium. The bending of light due to a change in density is known as **refraction**.

Index of Refraction

The **index of refraction**, n, is a ratio of the speed of light in a vacuum, c, to the speed of light in the medium, v, in which the light is moving:

$$n = \frac{c}{v}$$

Since both c and v are measured in meters per second, the units cancel, and the index is simply a value with no units. Light moves the fastest in a vacuum and is slower in a medium. This means the numerator, c, will either be equal to or greater than the denominator, v. As a result, the index of refraction can never be less than 1. When light is traveling in a vacuum, v and c have the same value, and the index of refraction is equal to 1, $n_{\text{vacuum}} = 1$. Air has a very low density. The speed of light in air is very nearly equal to the speed of light in a vacuum. This means that the accepted value for the index of refraction of air is also 1, $n_{\text{air}} = 1$. Knowing

$$n = \frac{c}{v}$$

The index of refraction compares the speed of light in a vacuum to the speed of light in a medium. It has no units and is proportional to the density of the medium.

that the indexes of refraction for a vacuum and for air are both equal to 1 is often important when solving problems. The index of refraction increases as optical density increases.

Snell's Law

The amount that light refracts (bends) when it enters a new medium can be determined using **Snell's law**.

$$n_1 \sin \theta_1 = n_2 \sin \theta_2$$

Snell's law shows the relationship between the indexes of refraction of mediums 1 and 2 and the angles of the light rays when light moves from medium 1 to medium 2. Figure 16.4 shows two examples. Figure 16.4(a) shows a ray moving from medium 1 with a lower optical density into medium 2 with a higher optical density. The light ray is bent toward the dashed normal line. Figure 16.4(b) on the right shows a ray moving from medium 1 with a higher optical density into medium 2 with a lower optical density. This time the light ray bends away from the dashed normal line. In both diagrams, the larger angle is found in the medium that has the lower optical density and the lower index of refraction. The smaller angle is located in the optically denser medium with the higher index of refraction.

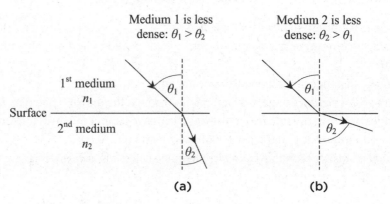

Figure 16.4 Snell's law of refraction

IF YOU SEE
light
changing
mediums

Refraction
and Snell's law

$n_1 \sin \theta_1 =$
$n_2 \sin \theta_2$

The optically denser medium has greater n slower v shorter λ smaller θ

Refraction has some specific characteristics that are worth noting.

1. Refraction occurs only if the two mediums have different optical densities and different indexes of refraction. If the indexes of refraction are equal, no refraction occurs even though the light travels through two different mediums.
2. In order for light to refract, it must strike the boundary between the two mediums at an angle that is not perpendicular to the surface. If light hits the boundary perfectly perpendicular to the surface (the light ray follows the normal), then $\theta_1 = 0°$ and no bending due to refraction occurs. In this case, the light will continue along the normal line without bending, $\theta_2 = 0°$. The speed of the light will change upon entering the new medium. However, the effect will not be visibly noticeable as the angle with reference to the normal line is zero.
3. The optically denser medium will have the larger index of refraction, n, the slower speed of light, v, the shorter wavelength, λ, and the smaller angle, θ.

EXAMPLE 16.1

Index of Refraction and Snell's Law

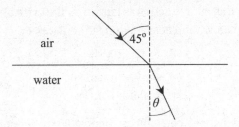

A beam of light moving in air strikes the surface of a lake at an angle of 45°, as shown in the diagram above. Which statement below is true?

(A) $n_{water} < n_{air}$ and $\theta < 45°$

(B) $n_{water} = n_{air}$ and $\theta = 45°$

(C) $n_{water} > n_{air}$ and $\theta > 45°$

(D) $n_{water} < n_{air}$ and $\theta > 45°$

(E) $n_{water} > n_{air}$ and $\theta < 45°$

WHAT'S THE TRICK?

The denser medium has the greater index of refraction, n, the slower speed of light, v, the shorter wavelength λ, and the smaller angle, θ. This problem addresses only the index of refraction and the angle of refraction. Since water is denser than air, it will have a larger n and a smaller θ. The answer is E.

Total Internal Reflection

A special case involves light moving from a medium of higher density into a medium of lower density. Since the second medium has a lower density, the ray of light traveling through it will refract through a larger angle as measured from the normal. If the angle in the first medium is steadily increased, then the larger angle in the second medium will also increase, as shown in Figure 16.5. This figure depicts a light source under water shining upward into air. Several rays with different incident angles are shown for comparison.

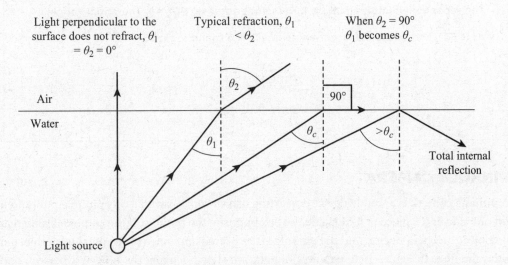

Figure 16.5 Total internal reflection

The light ray labeled A strikes the interface between the mediums perpendicular to the surface and does not refract. Ray B shows traditional refraction. In ray C, the angle in the first medium has become so large that it results in a refracted angle equal to 90°. The angle in the first medium that creates this 90° refraction is known as the **critical angle**, θ_c. Snell's law still applies, but the critical angle, θ_c, replaces θ_1, and 90° replaces θ_2.

$$n_1 \sin \theta_1 = n_2 \sin \theta_2$$
$$n_1 \sin \theta_c = n_2 \sin 90°$$
$$n_1 \sin \theta_c = n_2(1)$$

In the scenario described above, the light ray bent at 90° never enters the second medium. No refraction in the second medium occurs. As mentioned earlier, many things are possible when light strikes a new medium. Not only can light bend due to refraction, but it can also simply reflect. Usually a little of both is happening simultaneously. When the incident angle, θ_1, is small more light refracts into the second medium and less light reflects into the original medium. As the incident angle increases the amount of light refracting decreases and the amount reflecting increases. When the incident angle becomes greater than the critical angle, no light is refracting. Instead all the light is reflected back into the original medium. This phenomenon is known as **total internal reflection**. All of the light (total) stays in initial medium (internal), where it has been reflected (reflection). Total internal reflection occurs when the incident angle is greater than the critical angle.

EXAMPLE 16.2

Critical Angle and Total Internal Reflection

When a light ray is shined from a dense medium into air and the incident angle is slowly increased, total internal reflection is first noticed when the incident angle reaches 53°. Determine the index of refraction of the initial dense medium.

WHAT'S THE TRICK?

Total internal reflection uses an adaptation of Snell's law, where $\theta_1 = \theta_c$ and $\theta_2 = 90°$.

$$n_1 \sin \theta_c = n_2 \sin 90°$$

The second medium is air, $n_{air} = 1$. Remember that $\sin 90° = 1$. The critical angle, θ_c, is 53°, and this is the larger angle in a 3-4-5 triangle, $\sin 53° = \dfrac{4}{5}$.

$$n_1\left(\frac{4}{5}\right) = (1)(1)$$

$$n_1 = \left(\frac{5}{4}\right) = 1.25$$

PINHOLE CAMERA

A **pinhole camera** is a simple device consisting of a closed box with a tiny pinhole punctured on one side and a piece of film inside the box opposite the pinhole. If the camera is aimed at an object such as a person and the pinhole is uncovered, an image of the person is projected onto the film. In Figure 16.6, two rays of light are shown leaving the object, a person, and passing through the pinhole. One ray originates at the head of the person and the other from the feet of the person. The only way a ray of light reflected from the person's head can enter the pinhole is at a downward angle. The only way a ray of light reflected from the person's foot can enter the pinhole is at an upward angle. When these rays pass through the pinhole, they cause the image of the person to be inverted on the film.

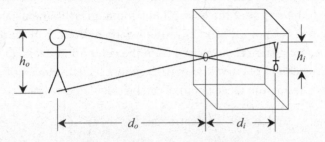

Figure 16.6 Pinhole camera

The **object distance**, d_o, is the distance from the object to the pinhole. The **image distance**, d_i, is the distance from the image to the pinhole. **Object height**, h_o, and **image height**, h_i, are also shown in the diagram above. The **magnification**, M, of the image can be determined using either the ratio of image height to object height or the ratio of image distance to object distance:

$$M = \frac{h_i}{h_o} = -\frac{d_i}{d_o}$$

The signs on the variable are an important aspect when solving problems in geometric optics. In Figure 16.6, the object is upright ($+h_o$), while the image is inverted ($-h_i$). The object distance will always be positive ($+d_o$). However, the sign on the image distance depends on whether the image is **real** or **virtual**. Real images can be projected onto a screen, in this case the film. They are inverted and have a positive image distance ($+d_i$). Virtual images cannot be projected onto a screen. They are upright and have a negative image distance ($-d_i$). The image formed by a pinhole camera is real. It is inverted ($-h_i$) and projected onto a screen ($+d_i$). The signs on image height and image distance are always opposite each other. Therefore, a negative sign has been added to the formula so that the ratios of image height and image distance are equal. The above magnification formula, along with these sign conventions, works for this simple camera and for all lenses and mirrors encountered in the remainder of this chapter.

A pinhole camera works very similarly to the eye. The black pupil at the front of the eye is a small pinhole. Light rays enter the eye and are then projected as an inverted real image on the back of the eye. The brain then inverts the image so it is perceived as right side up.

THIN LENSES

Lenses use refraction to change the way an object appears. When light traveling in air enters the denser lens, it slows and bends due to refraction. When it exits the lens and moves back into air, the light speeds up and experiences a second refraction. The type and degree of curvature of the two lens surfaces dictate where the image will appear and how it will be magnified.

Beginning physics introduces the geometric optics of extremely simplified lenses. The surfaces of these lenses are spherical. As a result, they are not perfect. The images formed by spherical lenses have a slight aberration, causing them to lack sharpness. The spheres forming these lenses are the same size, creating symmetrical lenses. In addition, the lenses

IF YOU SEE
a pinhole
camera

Image is
inverted

$$M = \frac{h_i}{h_o} = -\frac{d_i}{d_o}$$

Inverted images have a negative height ($-h_i$). They are projected on a screen and are real ($+d_i$). The minus sign in the formula implies that h_i and d_i have opposite signs.

are assumed to be extremely thin even though they may not appear very thin in diagrams. These assumptions greatly simplify lens mathematics.

Figure 16.7 demonstrates the symmetrical and spherical nature of a simple **convex lens**. Although the diagram appears to be two circles, the lens is actually formed by the intersection of the two spheres. The optical axis is a horizontal line running through the center of the lens. All vertical measurements are made from the optical axis. A vertical line (usually not drawn) also passes through the center of the lens. All horizontal measurements are made from the center of the lens. Two key points, f and $2f$, are shown on both sides of the lens. The symbols f and $2f$ also represent distances measured from the center of the lens to these key points. The point $2f$ is at the center of curvature of each sphere. The distance $2f$ is, therefore, equal to the radius, R. As a result, the focal length, f, can be calculated as follows.

$$2f = R \qquad \text{rearranges to} \qquad f = \frac{R}{2}$$

This means that the focal point, f, is located midway between the center of curvature and the surface of the lens. The lenses are considered so thin that they have negligible thickness (despite the obvious thickness shown in lens diagrams). As a result, the focal distance is measured from the center of the lens rather than the lens surface. Although only a convex lens is shown in Figure 16.7, the focal-length formula holds true for all spherical lenses and mirrors discussed in this chapter.

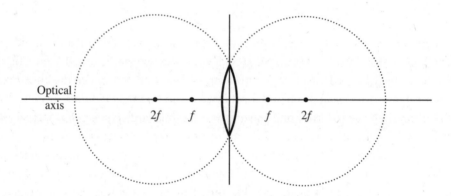

Figure 16.7 A simple convex lens

Converging Lenses, Object Outside f

Converging lenses bring light together. The convex lens is a converging lens. The location and size of the image formed by a lens or mirror can be plotted using ray tracing. In ray tracing, an object is represented by a vertical arrow extending upward from the optical axis, as shown in Figure 16.8. Two different light rays, both starting at the top of the object (tip of the vertical arrow), are drawn moving through the lens. Where they intersect is the location where the image will be focused. Look at the figure and trace the following key rays.

- Light parallel to the optical axis converges on the far focal point.
- Light that passes through the center of the lens moves in a straight line.
- Light passing through the focus exits parallel to the optical axis.

Figure 16.8 shows these three key rays of light originating at the object and intersecting at the image formed by this converging convex lens.

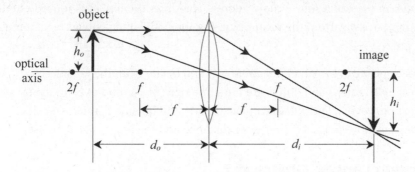

Figure 16.8 Convex lens

The rules for positive and negative signs for d_o, d_i, h_o, and h_i are the same as those for the pinhole camera, discussed earlier. There is one additional sign to consider. Converging optical instruments, such as the convex lens in Figure 16.8, have a positive focal distance ($+f$). When the object is positioned outside of the focal point, f, the image will be inverted and real. For example, movies are projected on a screen. Therefore, they are real images and inverted. In order to show a movie upright in the theater, the film must be fed into the projector inverted.

There is a key geometric relationship between the focal length, f, of the lens, the object distance, d_o, and the image distance, d_i:

$$\frac{1}{f} = \frac{1}{d_o} + \frac{1}{d_i}$$

All of these variables are in the denominator and require a value to be inverted during the solution process. For example, when solving for f, the answer for $1/f$ will be one of the available choices. After solving for $1/f$, remember to invert to find f. The magnification formula and the sign conventions discussed in the "Pinhole Camera" section also apply to lenses and mirrors.

$$M = \frac{h_i}{h_o} = -\frac{d_i}{d_o}$$

Table 16.2 summarizes the signs for the variables in Figure 16.8 and in the equations shown above.

Table 16.2 A Converging Convex Lens with an Object Outside of f

If You See . . .	Key Fact	Result
Converging convex lens	Converging	$+f$
Object arrow points upward	Upright object	$+h_o$
Image arrow points downward	Inverted image	$-h_i$
The sign on magnification matches h_i	Inverted image	$-M$
Distance to the object is always positive	Always	$+d_o$
The sign on image distance is opposite h_i	Inverted image	$+d_i$
If $-h_i$ or $+d_i$, the image is real If $+h_i$ or $-d_i$, the image is virtual	$-h_i$ or $+d_i$	Real
Light transmits through lenses to the far side	Light forms real images	Far side

Problems on the exam may simply address trends as the object is moved either toward the focal point or away from the focal point. For the converging convex lens, remember the following points:

1. Object outside of $2f$: Small image ($M < 1$) and inside $2f$ on the far side.
2. Object at $2f$: Image and object are the same size ($M = 1$) and at $2f$ on the far side.
3. Object between $2f$ and f: Large image ($M > 1$) and outside $2f$ on the far side.
4. As objects move toward f, the image distance and the image size increase.

Converging Lenses, Object at *f*

When the object is placed at the focal point of a converging lens, the ray traces are parallel to each other and never intersect. No image is formed. However, if this scenario is reversed and the object is positioned infinitely far away, then all the light rays arriving at the lens will be parallel to the optical axis. Every ray of light will converge at the focal point, creating an image at *f*. Wavelengths of light are so incredibly small that in comparison, an object 100 meters away might as well be at infinity. You should know that the image of a distant object will be located at the focal point on the opposite side of a converging convex lens.

Converging Lenses, Object Inside *f*

An interesting phenomenon occurs when the object is moved inside the focal point of a converging convex lens. In Figure 16.9, the ray traces used in Figure 16.8 do not intersect on the far side of the lens. However, if they are traced backward (dashed lines), an intersection is found on the near side of the lens. This creates an upright image, which is a virtual image that cannot be projected onto a screen. The formulas used previously remain the same. However, the signs on the upright image height ($+h_i$) and image distance ($-d_i$) reverse when the object moves inside the focal point. Even though the ray traces seem to diverge, this is still a converging convex lens with a positive focal point ($+f$). The actual light rays are still being converged on, and pass through, the focal point. The image will have the greatest magnification when the object is closest to the focal point. As the object is moved from the focal point, *f*, toward the lens, the image decreases in size and moves toward the lens. An example of this is the magnifying lens. An object placed between the focal point and magnifying lens will be upright, virtual, and magnified.

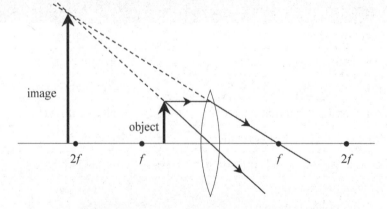

Figure 16.9 Magnifying convex lens

Table 16.3 summarizes the signs for the variables in Figure 16.9 and in the equations shown on page 332.

Table 16.3 A Converging Convex Lens with an Object Inside of *f*

If You See . . .	Key Fact	Result
Converging convex lens	Converging	$+f$
Object arrow points upward	Upright object	$+h_o$
Image arrow points upward	Upright image	$+h_i$
The sign on magnification matches h_i	Upright image	$+M$
Distance to the object is always positive	Always	$+d_o$
The sign on image distance is opposite h_i	Upright image	$-d_i$
If $-h_i$ or $+d_i$, the image is real If $+h_i$ or $-d_i$, the image is virtual	$+h_i$ or $-d_i$	Virtual
The rays trace back to form a virtual image	Opposite side of light	Near side

---

EXAMPLE 16.3

Convex Lenses

The object viewed by a convex lens is positioned outside of the focus, as shown in the diagram above. Which of the following correctly describes the image?

(A) No image is formed
(B) Real and upright
(C) Real and inverted
(D) Virtual and upright
(E) Virtual and inverted

WHAT'S THE TRICK?

Convex lenses produce three possible outcomes. When the object is inside of *f*, the image is virtual and upright. When the object is at *f*, the image cannot be formed. When the object is outside of *f*, the image is real and inverted, which matches the scenario given in the problem. The answer is C.

Diverging Lenses

The **concave lens** is a **diverging lens**. Diverging optical instruments spread rays of light. Instead of the light converging on the far focus, it spreads out from the near focus, as shown in Figure 16.10 and summarized as follows.

- Light parallel to the optic axis diverges in a line from the near focal point.
- Light that passes through the center of the lens moves in a straight line.
- The resulting rays diverge, and their back traces intersect. (Note that the back trace of the straight ray passing through the center of the lens coincides with the ray itself.)

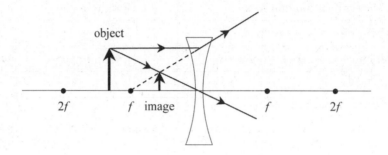

Figure 16.10 Concave lens

The resulting image is upright and virtual. Again, all the equations and rules for variable signs are the same throughout this chapter. They are summarized in Table 16.4. Unlike the converging lens, the diverging concave lens is capable of producing only a small ($M < 1$) image that is upright and virtual. Whether the object is outside f, at f, or inside f has no bearing. As the object moves toward the lens, the image also moves toward the lens and becomes larger.

Table 16.4 A Diverging Concave Lens

If You See . . .	Key Fact	Result
Diverging concave lens	Diverging	$-f$
Object arrow points upward	Upright object	$+h_o$
Image arrow points upward	Upright image	$+h_i$
The sign on magnification matches h_i	Upright image	$+M$
Distance to the object is always positive	Always	$+d_o$
The sign on image distance is opposite h_i	Upright image	$-d_i$
If $-h_i$ or $+d_i$, the image is real If $+h_i$ or $-d_i$, the image is virtual	$+h_i$ or $-d_i$	Virtual
The rays trace back to form a virtual image	Opposite side of light	Near side

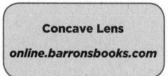

Concave Lens

online.barronsbooks.com

Concave Lens

Which of the following is NOT true for a concave lens?

(A) Concave lens are divergent.

(B) The image is virtual.

(C) The image is upright.

(D) The image is larger than the object.

(E) The image forms on the near side of the lens.

WHAT'S THE TRICK?

In this case, knowing the characteristics of concave lenses is a must. Concave lenses are divergent. They can create only virtual images that are upright and appear on the near side of the lens. These lenses can create only images that are always smaller than the object regardless of the location of the object. Therefore, the correct answer is D.

SPHERICAL MIRRORS

Spherical mirrors are simply small sections of a single sphere that have a reflective surface. For **concave mirrors**, the reflective surface is the inside surface. For **convex mirrors**, it is the outside surface. Mirror-optics problems are nearly identical to lens problems. The equations are the same. The rules for the variable signs are the same. The trends of image location are the same. However, there are a few key differences.

1. Although lenses consist of two intersecting spheres with two focal points, mirrors consist of a single spherical surface with only one focal point.
2. Although converging lenses are convex and diverging lenses are concave, mirrors are the opposite. Converging mirrors are concave, and diverging mirrors are convex.
3. Although light passes through a lens and creates real images on the far side and virtual images on the near side, mirrors reflect light back to the near side. For mirrors, real images form on the near side and virtual images form on the far side.

Converging Mirrors, Object Outside of *f*

The concave mirror converges parallel rays of light through the focal point. Ray tracing for mirrors requires you to use a different strategy than that for a lens.

- Light parallel to the optical axis is reflected through the focus.
- Light through the focus reflects parallel to the optical axis.

The ray trace in Figure 16.11 demonstrates how the image can be found when the object is located outside the focus.

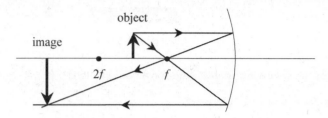

Figure 16.11 Concave mirror

The mathematics, signs, and trends, shown in Table 16.5, are identical to those of the converging lens with an object outside its focus.

Table 16.5 Converging Mirrors with the Object Outside of f

If You See . . .	Key Fact	Result
Converging concave mirrors	Converging	$+f$
Object arrow points upward	Upright object	$+h_o$
Image arrow points downward	Inverted image	$-h_i$
The sign on magnification matches h_i	Inverted image	$-M$
Distance to the object is always positive	Always	$+d_o$
The sign on image distance is opposite h_i	Inverted image	$+d_i$
If $-h_i$ or $+d_i$, the image is real If $+h_i$ or $-d_i$, the image is virtual	$-h_i$ or $+d_i$	Real
Light reflects off of mirrors to the near side	Light forms real images	Near side

Converging Mirrors, Object at f

This is the same as for a converging lens. When the object is positioned at the focus, no image is formed since the light rays are parallel and cannot intersect. However, if the object is positioned far away, the light rays arriving at the mirror will be essentially parallel. Parallel rays striking a converging optical instrument are refracted and focused at the focal point, f. A practical example of this is the collection of light from a distant star through a concave, reflecting telescope.

Converging Mirrors, Object Inside of f

Just as with the converging lens, when the object is moved inside the focus, an upright, virtual image is created. The ray-trace rules are a bit more complicated for this scenario and are depicted in Figure 16.12. The ray parallel to the optical axis reflects through the focus as before. However, a ray drawn from the tip of the object through the focus will not strike the mirror. A new ray must be drawn. The mirror is spherical, and $2f$ is at the center of the sphere. Any ray of light starting at the center of a sphere, $2f$, will be reflected straight back to the center, $2f$. A ray of light starting at $2f$ and passing through the very tip of the object will reflect right back toward $2f$, as shown in Figure 16.12. This ray diverges from the ray passing through f. When back ray traces are drawn they intersect to form an upright image on the far side of the mirror. No light passes through the mirror. Although this image can be seen with the eye, it cannot be projected onto a screen. It is a virtual image.

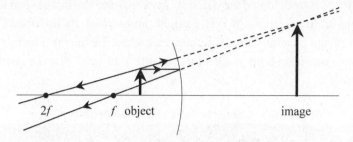

Figure 16.12 Virtual image in a concave mirror

All the equations, variable signs, and image trends are identical to those seen in the converging lens when the object is located inside of *f*. The main difference is that virtual images form on the far side of mirrors. Table 16.6 summarizes this information.

Table 16.6 Converging Mirrors with the Object Inside of *f*

If You See . . .	Key Fact	Result
Converging concave mirrors	Converging	$+f$
Object arrow points upward	Upright object	$+h_o$
Image arrow points upward	Upright image	$+h_i$
The sign on magnification matches h_i	Upright image	$+M$
Distance to the object is always positive	Always	$+d_o$
Sign on image distance is opposite h_i	Upright image	$-d_i$
If $-h_i$ or $+d_i$, the image is real If $+h_i$ or $-d_i$, the image is virtual	$+h_i$ or $-d_i$	Virtual
The rays trace back to form a virtual image	Opposite side of light	Far side

IF YOU SEE

concave mirrors

Convergent

+*f*

When the object is outside the focus, the image is inverted and real.

When the object is inside the focus, the image is upright and virtual.

Diverging Mirrors

The diverging mirror is very similar to the diverging lens. It is convex rather than concave. The ray trace in Figure 16.13 appears very different, but it uses the same logic as seen earlier.

1. Light parallel to the optical axis reflects in a line drawn from the focus.
2. Light aimed at the focus reflects parallel to the optical axis.
3. The reflected rays diverge so the image is formed by their back traces.

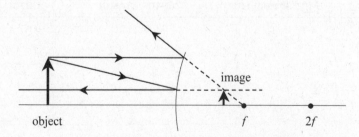

Figure 16.13 Diverging convex mirror

IF YOU SEE

convex

mirrors

Divergent

–f

Always forms
a small,
upright, and
virtual image.

The resulting image is upright and virtual. It is very similar to the image seen in the diverging lens. The image is always small ($M < 1$), upright, and virtual. As the object moves toward the mirror, the image becomes larger and moves toward the mirror. The main difference is that this virtual image is on the far side of the mirror. Table 16.7 lists the characteristics of a diverging mirror.

Table 16.7 Characteristics of a Diverging Mirror

If You See . . .	Key Fact	Result
Diverging convex mirrors	Diverging	$-f$
Object arrow points upward	Upright object	$+h_o$
Image arrow points upward	Upright image	$+h_i$
The sign on magnification matches h_i	Upright image	$+M$
Distance to the object is always positive	Always	$+d_o$
The sign on image distance is opposite h_i	Upright image	$-d_i$
If $-h_i$ or $+d_i$, the image is real If $+h_i$ or $-d_i$, the image is virtual	$+h_i$ or $-d_i$	Virtual
The rays trace back to form a virtual image	Opposite side of light	Far side

Convergent optical instruments, whether they are lenses or mirrors, share many characteristics. This is also true for divergent optical instruments. Table 16.8 summarizes the key characteristics for the lenses and mirrors tested in beginning physics.

Table 16.8 Key Characteristics of Lenses and Mirrors

	Converging $+f$		Diverging $-f$
Lens shape	Convex		Concave
Mirror shape	Concave		Convex
Object location	Outside f	Inside f	Anywhere
Object	$+d_o$ and $+h_o$	$+d_o$ and $+h_o$	$+d_o$ and $+h_o$
Image	Real $+d_i$ Inverted $-h_i$ Lens: far side Mirror: near side	Virtual $-d_i$ Upright $+h_i$ Lens: near side Mirror: far side	Virtual $-d_i$ Upright $+h_i$ Lens: near side Mirror: far side

SUMMARY

1. **LIGHT REFLECTED FROM A SURFACE FOLLOWS THE LAW OF REFLECTION.** The angle at which incident rays of light strike a surface, θ_i, is the same as the angle of the reflected rays of light, θ_r. Expressed as an equation, $\theta_i = \theta_r$, where the angles are measured from a normal drawn perpendicular to the reflective surface.

2. **THE INDEX OF REFRACTION IS A COMPARISON OF SPEEDS OF LIGHT.** The index of refraction is a ratio of the speed of light in a vacuum, c, to the speed of light in a medium, v. It has no units, and it can never be less than 1. The indexes of refraction for a vacuum and for air are both equal to 1. The index of refraction increases in value as the density of the medium increases.

3. **LIGHT ENTERING A NEW MEDIUM WILL REFRACT.** When light enters a medium with a different density and a different index of refraction, it will change speed and wavelength. This causes the ray of light to bend at the surface boundary between the two mediums. The relationship between the indexes of refraction and the angles of the light ray's path is expressed in Snell's law.

4. **A PINHOLE CAMERA CREATES AN INVERTED REAL IMAGE ON THE FILM AT THE BACK OF THE CAMERA.** The ray trace involves two crisscrossing rays and is similar to the function of the pupil of the human eye.

5. **THE LOCATION OF AN OBJECT IN RELATION TO A SPHERICAL LENS OR MIRROR DETERMINES THE LOCATION AND TYPE OF IMAGE CREATED BY THE PARTICULAR INSTRUMENT.** When an object is placed outside of the focal point near a converging lens (convex) or a converging mirror (concave), an inverted, real image is created. If the object is moved inside the focus, then an upright, virtual image is seen instead. However, if a diverging lens (concave) or diverging mirror (convex) is used instead, where the object is placed does not matter. A diverging lens and a diverging mirror can create only a small, upright, virtual image.

If You See	Try	Keep in Mind
A mirror	Law of reflection $$\theta_i = \theta_r$$	All angles in optics are measured from normal lines drawn perpendicularly to the surface of optical devices.
Index of refraction	Compares the speed of light in a vacuum, c, with its speed in a medium, v $$n = \frac{c}{v}$$	The index of refraction is proportional to the density of the medium, and it has no units.
Light-changing mediums	Refraction following Snell's law $$n_1 \sin \theta_1 = n_2 \sin \theta_2$$	A denser medium has a greater n, slower v, shorter λ, and smaller θ.
Critical angle	Total internal reflection, Snell's law where $\theta_1 = \theta_c$ and $\theta_2 = 90°$ $$n_1 \sin \theta_c = n_2 \sin 90°$$	If the incident angle is greater than the critical angle, then all light is reflected back into medium 1, and no light is refracted into medium 2.
Pinhole camera	$$M = \frac{h_i}{h_o} = -\frac{d_i}{d_o}$$ Inverted: $-h_i$ Real: $+d_i$	Image is inverted at the back of the camera. An image projected on a screen is upside down.
Convex lens or a concave mirror	Convergent $+f$ $$f = \frac{R}{2}$$ $$\frac{1}{f} = \frac{1}{d_o} + \frac{1}{d_i}$$ $$M = \frac{h_i}{h_o} = -\frac{d_i}{d_o}$$	If the object is outside f, the image is inverted $(-h_i)$ and real $(+d_i)$. If the object is at f, the image is at infinity. If the object is inside f, the image is upright $(+h_i)$ and virtual $(-d_i)$.
Concave lens or a convex mirror	Convergent $+f$ $$f = \frac{R}{2}$$ $$\frac{1}{f} = \frac{1}{d_o} + \frac{1}{d_i}$$ $$M = \frac{h_i}{h_o} = -\frac{d_i}{d_o}$$	No matter where the object is placed, the image is always small, upright $(+h_i)$, and virtual $(-d_i)$.

PRACTICE EXERCISES

1. An object is 1 meter in front of a plane mirror. The image is

   (A) virtual, inverted, and 1 m behind the mirror
   (B) virtual, inverted, and 1 m in front of the mirror
   (C) virtual, upright, and 1 m in front of the mirror
   (D) real, upright, and 1 m behind the mirror
   (E) none of the above

2. Which diagram below correctly illustrates the path of a light ray moving from point 1 in air to point 2 in water?

   (A)

   (B)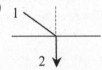

   (C)

   (D)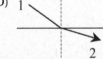

   (E)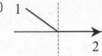

3. Which of the following is true when light enters a denser medium?

   (A) v increases, λ decreases, and n increases.
   (B) v increases, λ decreases, and n decreases.
   (C) v decreases, λ increases, and n increases.
   (D) v decreases, λ decreases, and n increases.
   (E) v decreases, λ decreases, and n decreases.

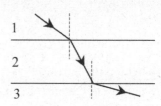

4. In the diagram above, a light ray refracts as it travels through three mediums: 1, 2, and 3. Rank the indexes of refraction from greatest to least.

(A) $n_1 > n_2 > n_3$
(B) $n_2 > n_1 > n_3$
(C) $n_3 > n_1 > n_2$
(D) $n_1 > n_3 > n_2$
(E) $n_2 > n_3 > n_1$

5. A light ray is incident on a boundary between two mediums where the optical density is greater in the first medium. If the angle of the incident ray is increased beyond the critical angle, then light striking the boundary between two substances is

(A) refracted
(B) reflected
(C) absorbed
(D) dispersed
(E) transmitted

6. The image of a distant object formed by a pinhole camera is

(A) upright, real, and larger than the object
(B) upright, virtual, and larger than the object
(C) inverted, real, and larger than the object
(D) inverted, real, and smaller than the object
(E) inverted, virtual, and smaller than the object

7. The type of lens that refracts parallel light rays to the far focal point is a

(A) converging, concave lens
(B) converging, convex lens
(C) diverging, concave lens
(D) diverging, convex lens
(E) All spherical lenses refract parallel rays to the far focus.

8. An image appearing on a screen is

   (A) real and upright
   (B) real and inverted
   (C) virtual and upright
   (D) virtual and inverted
   (E) none of these

9. An image formed by a convex mirror is

   (A) real and upright
   (B) real and inverted
   (C) virtual and upright
   (D) virtual and inverted
   (E) No image is formed by this mirror.

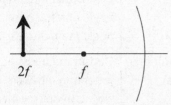

10. In the diagram above of a concave mirror, an object is initially located at $2f$.
    How will the image change as the object moves toward the focal point f?

    (A) The image will increase in size and move away from the mirror.
    (B) The image will increase in size and move toward the mirror.
    (C) The image will decrease in size and move away from the mirror.
    (D) The image will decrease in size and move toward the mirror.
    (E) The image size remains constant but moves away from the mirror.

ANSWERS EXPLAINED

		Key Words	Needed for Solution	Now Solve It
1.	**(E)**	Plane mirror . . . image	Knowledge/definitions Reflection of a plane mirror	The image created by a plane mirror is nearly identical to the object. The image is upright, which means that it is a virtual image. The image is also the same size ($M = 1$) as the object. The image forms behind the mirror at the same distance as the object ($d_i = d_o$). Answers A through D are all incorrect.
2.	**(C)**	Light ray . . . from point 1 in air to point 2 in water	Knowledge/definitions Refraction	Light moving in a denser medium has a lower speed and a shorter wavelength. This causes the angle of its path, measured from the normal, to be smaller than the angle measured in a less dense medium. Medium 2 is denser. So, the angle of refraction is smaller than the incident angle in medium 1.
3.	**(D)**	Light enters a denser medium	Knowledge/definitions Refraction	Denser mediums have higher indexes of refraction (answers A, C, and D). Light moves slower in denser mediums (answers C, D, and E). When light changes mediums, its frequency remains constant. As a result, wavelength is directly proportional to speed. When light slows, its wavelength decreases (answers A, B, D, and E). Only answer D satisfies all these conditions.
4.	**(B)**	Light ray refracts . . . three mediums . . . rank the indexes of refraction; greatest to least	Knowledge/definitions Snell's law $n_1 \sin \theta_1 = n_2 \sin \theta_2$	Snell's law indicates that the index of refraction, n, and the angle of refraction, θ, are inversely proportional. Although denser mediums have higher indexes of refraction, they have smaller angles of refraction. The smallest angle is seen in medium 2, which has the greatest index of refraction. Medium 1 has the next smallest angle, followed by medium 3.
5.	**(B)**	Critical angle	Knowledge/definitions Critical angles Total internal reflection	If the angle of the incident ray of light is increased, the angle of refraction increases. When the incident angle becomes critical, the refracted angle is 90°. At the critical angle, light enters the second medium and all light is reflected back into the first medium.
6.	**(D)**	Pinhole camera	Knowledge/definitions Pinhole cameras	Pinhole cameras focus a small, inverted image on the film at the rear of the camera. Inverted images are real images that are projected onto a screen (the film).

	Key Words	Needed for Solution	Now Solve It
7. **(B)**	Type of lens . . . refracts light . . . to the far focal point	Knowledge/definitions Lenses	Converging instruments refract parallel light to the focal point. The converging lens is convex.
8. **(B)**	Screen	Knowledge/definitions	Images projected onto a screen are real. Real images are always inverted.
9. **(C)**	Convex mirror	Knowledge/definitions Convex mirror	Convex mirrors are divergent instruments. Divergent optical instruments can create only small, virtual, upright images.
10. **(A)**	Diagram of a concave mirror . . . object is initially at $2f$ . . . how will the image change as the object moves toward . . . f	Knowledge/definitions Concave mirror $$\frac{1}{f} = \frac{1}{d_o} + \frac{1}{d_i}$$	Concave mirrors are convergent instruments. When the object is outside the focal point, the image is real, is inverted, and appears on the near side of the mirror. When the object is at $2f$, the image will be the same size and will also be at $2f$. According to $$\frac{1}{f} = \frac{1}{d_o} + \frac{1}{d_i}$$ object distance, d_o, and image distance, d_i, vary inversely. If the object moves toward the focus, the image moves away from the focus and away from the mirror. If image distance is increasing while object distance is decreasing, then according to $$M = \frac{h_i}{h_o} = -\frac{d_i}{d_o}$$ both magnification and image height increase.

Physical Optics

17

→ **DIFFRACTION**

→ **INTERFERENCE OF LIGHT**

→ **POLARIZATION OF LIGHT**

→ **COLOR**

Physical optics, or wave optics, involves the effects of light waves that are not related to the geometric ray optics covered in the previous chapter. Physical optics involves such topics as diffraction, interference, and polarization. These effects are usually analyzed using either the wave-front model of light or the sinusoidal form. In addition, this chapter includes a section involving the color of visible light, which contains some elements of geometric optics. This chapter will do the following:

- Explain the nature of diffraction and identify its characteristics.
- Analyze the double- and single-slit interference patterns.
- Examine the polarization of light.
- Explain why objects appear as a specific color and how light can be dispersed or scattered.

Table 17.1 lists the variables used in this chapter.

Table 17.1 Variables for Physical Optics

New Variables	Units
m = a specific maximum or minimum	No units
x_m = distance measured to m above	m (meters)
θ_m = angle measured to m above	Degrees
d = slit spacing	m (meters)
L = distance from slits to screen	m (meters)

DIFFRACTION

When waves pass near a barrier or through an opening (slit), they bend and spread out to fill the space behind the barrier or the slit. The bending of a wave due to a barrier or opening is known as **diffraction**. The amount of bending has to do with the size of the obstacle or opening compared with the wavelength of the wave. In the diagrams in Figure 17.1, parallel wave fronts are shown approaching various openings. The bending due to diffraction increases as the openings become smaller.

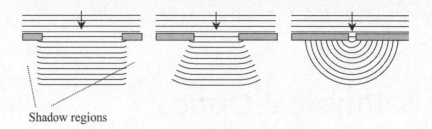

Figure 17.1 Diffraction

When the opening is large as compared with the wavelength of the waves, the waves move through the opening with little diffraction. This creates large shadow regions with no wave activity to the left and right of the opening. When light waves are diffracted, the absence of light in the shadow regions leaves these areas dark. However, when the size of the opening is similar to the size of the wavelength of the waves, the diffraction is so pronounced that the spreading wave fronts form a circular pattern with no shadow regions. Since light is composed of very small wavelengths, openings that cause significant diffraction must be incredibly narrow.

A geometric explanation for the circular nature of the resulting diffraction pattern was proposed by Christian Huygens and is known as the **Huygens' principle**. There are two main aspects of his principle.

1. Every oscillator in a wave creates spherical wavelets that propagate outward.
2. The wave front created by these oscillators is due to the combined interference of the wavelets.

Figure 17.2 shows a diagram of a portion of a linear wave. The dots lying along the crest of the wave represent the individual oscillators, each creating circular wave fronts.

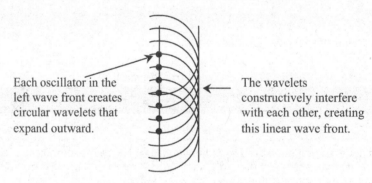

Each oscillator in the left wave front creates circular wavelets that expand outward.

The wavelets constructively interfere with each other, creating this linear wave front.

Figure 17.2 Wave fronts

This provides an explanation for the circular wave fronts seen when waves move through narrow openings. If the opening is very small, only one, or very few, oscillator(s) propagate(s) the wave through the opening. As a result, circular wave fronts are generated on the other side of the opening.

INTERFERENCE OF LIGHT

When light waves interact, they can interfere constructively or destructively. If identical wave crests (represented by wave fronts) meet, **constructive interference** adds them together to create a single larger wave. Constructive interference causes water waves to increase in height. It causes sound to become louder and light to become brighter. If, on the other hand, a wave crest meets a wave trough of identical size, **destructive interference** will cancel these waves entirely. This creates an absence of wave activity. It causes water to be flat, quiet instead of loudness, and darkness instead of brightness.

Young's Double-Slit Experiment

Thomas Young constructed an experiment, known as **Young's double-slit experiment**, involving the interference of light. He shined **monochromatic light**, which is light composed of only one wavelength, on two incredibly narrow openings (slits) that were near each other. The light diffracted through each slit, causing circular wave fronts to spread outward. These patterns overlapped each other and created both constructive and destructive interference. The resulting pattern was projected onto a screen, where constructive interference created bright regions (bright fringes or maximums) and destructive interference created dark regions (dark fringes or minimums). Figure 17.3 shows the interference pattern created by the circular wave fronts spreading from each slit.

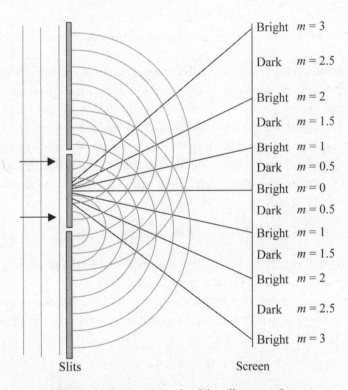

Figure 17.3 Young's double-slit experiment

The expanding circular wave fronts represent the wave crests. Where wave crests intersect, they interfere constructively. In Figure 17.3, lines through successive wave-crest intersections have been added. They extend from a point midway between the slits toward the screen. Where these lines of constructive interference hit the screen is where bright bands (maximums, m) of light will be seen. In the regions in between, destructive interference creates bands that are dark (minimums, m). The maximums and minimums are numbered starting with the central maximum, $m = 0$. The bright maximums are numbered as whole numbers ($m = 1, 2,$ or $3 \ldots$), while the minimums are numbered with halves ($m = 0.5, 1.5,$ or

2.5 . . .). Be very careful with the values assigned to dark minimums. The value for the third dark minimum (third dark fringe) is $m = 2.5$ and not 3.5 as students often incorrectly believe. The value representing the maximums and minimums has no units, and the reason behind the numbering system will be discussed below.

A mathematical relationship describes the resulting bright and dark fringes. This will be easier to explain if the diagram is simplified to solve for one specific maximum. Figure 17.4 shows the mathematical relationships for the first bright maximum, $m = 1$. Only the lines extending to the central (reference) maximum and the first maximum are shown.

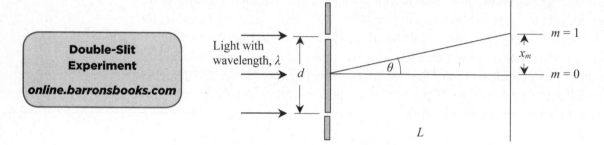

Figure 17.4 First bright fringe position for the double-slit experiment

Various key variables have been labeled in Figure 17.4. These include the number for the maximum to be analyzed, m, the distance from the central maximum to the maximum being investigated, x_m, the spacing between the slits, d, the distance from the slits to the screen, L, and the angle, θ. The actual widths of the slits themselves are not needed. The variable m has no units, and the angle is measured in degrees. All other variables are lengths measured in meters. Students often confuse m and x_m. Remember that m is the number for the maximum and x_m is the distance to that maximum. There are two mathematical relationships for Young's double-slit experiment:

$$x_m \approx \frac{m\lambda L}{d} \qquad \text{and} \qquad d \sin \theta = m\lambda$$

The formula on the left is actually an approximation that works well as long as the screen distance, L, is very large, which is typically the case on exams.

When the diagram is viewed differently, an explanation for the numerical values of m is apparent. In Figure 17.5, two rays of light are shown moving from each slit toward the maximum, m.

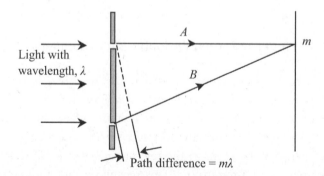

Figure 17.5 Path difference

Light ray *B* must travel a longer distance than light ray *A*. The difference between the distance the two rays travel is known as the **path difference**. Light reaching the first maximum, $m = 1$, has a path difference equal to 1 wavelength. Light reaching the second maximum, $m = 2$, has a path difference equal to 2 wavelengths. In other words, the maximum numbers, m, represent the number of wavelengths by which the paths differ. Bright maximums occur when crests meet and interfere constructively. At maximum $m = 1$, a wave following path *A* must meet up with a wave following path *B* that is exactly 1 wavelength off. Likewise, all the other maximums must occur when the paths of light differ by whole numbers of wavelengths. Otherwise, their crests will not match. The dark minimums appear when the waves are off by half of a wavelength. This is when crests meet troughs and destructively interfere.

Young's double-slit experiment is considered experimental evidence that light possesses a wave characteristic. The double-slit effects can be seen with water waves, and it can be heard with sound waves. Sometimes on an exam, the two closely spaced slits may be replaced with two closely spaced loudspeakers. The effect is the same with sound as it is for light.

A key element in a double-slit experiment is using waves that have a single wavelength. If more than one wavelength is present, a more complex and less symmetrical interference pattern would occur. For sound waves, the experiment can be conducted using two speakers both playing a tone with the same frequency. For light waves, the experiment must be conducted using monochromatic light, which is light with a single constant wavelength and a specific color.

IF YOU SEE a double slit

Young's double-slit experiment

The pattern seen is a result of both diffraction and interference.

This is experimental evidence that light has wave characteristics.

EXAMPLE 17.1

Young's Double-Slit Experiment

Monochromatic light passes through two narrow slits and is projected onto a screen, creating a double-slit interference pattern. The first bright maximum occurs at a distance of 0.004 meters from the central maximum. The slit spacing is 3.0×10^{-6} meters, and the distance to the screen is 2.0 meters. Determine the wavelength of light used in this experiment.

WHAT'S THE TRICK?

If no angle is given, then the formula solving for double-slit interference is:

$$x_m = \frac{m\lambda L}{d}$$

$$\lambda = \frac{x_m d}{mL}$$

$$= \frac{(0.004 \text{ m})\left(3.0 \times 10^{-6} \text{ m}\right)}{(1)(2.0 \text{ m})}$$

$$= 6.0 \times 10^{-9} \text{ m}$$

$$= 6 \text{ nm}$$

EXAMPLE 17.2

Trends in Double-Slit Interference Patterns

Light incident on two narrow slits is used to project an interference pattern onto a screen. How will the interference pattern change if the distance between the slits is doubled?

WHAT'S THE TRICK?

The appearance of the interference pattern is tied to the distance between the maximums, x_m. If the distance between the maximums increases, the maximums are moving farther apart. The bright and dark regions will spread out. If the distance between maximums decreases, the maximums will move closer together, compressing the appearance of the interference pattern. Analyze how the change in slit spacing, d, affects the interference-pattern spacing, x_m.

$$x_m = \frac{m\lambda L}{d}$$

$$\left(\frac{1}{2}\right)x_m = \frac{m\lambda L}{(2d)}$$

The interference-pattern spacing, x_m, is halved. Therefore, the pattern compresses.

Single-Slit Interference

A pattern similar to Young's double-slit experiment can be seen even when only one slit is present. Huygens' principle offers an explanation. Two points at opposite ends of the slit are sources of circular wavelets that will interfere with each other in the same manner as two points in separate slits. However, when a single slit is involved, a large amount of wave activity moves straight through the center of the slit with minimal interference. The result is a very large central maximum. This maximum is not only wider but is also more intense than the central maximum produced with a double slit. The intensity of light is associated with its brightness. The wider and brighter central maximum is surrounded by much smaller maximums that are dimmer.

The intensity of light is often indicated as a curve superimposed on the double- and single-slit diagrams, as shown in Figure 17.6.

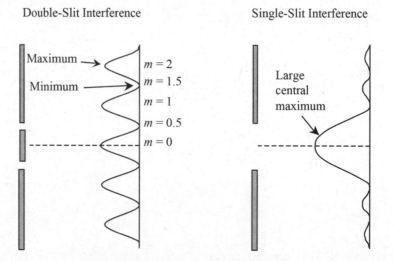

Figure 17.6 Double- and single-slit interference patterns

Intensity graphs have been superimposed on the diagrams in Figure 17.6. They indicate where bright maximums and dark minimums occur. For both patterns, the central maximum is the brightest and largest. However, the maximums for a double-slit interference pattern are more evenly spaced and are more similar in size. The single-slit pattern has a distinctly oversized central maximum.

Another important difference is that the double-slit pattern depends on the space between the slits, d, and not on the actual width of either slit. However, the single-slit pattern depends on the actual width of the slit opening. Increasing the slit width in the single-slit pattern has the same effect as increasing the slit spacing for the double-slit pattern. In both cases, the interference pattern compresses so that the space between maximums is decreased. The reverse is true if the double slits are moved closer together or if the single-slit width is reduced. This will cause both patterns to spread out, moving the maximums away from each other.

IF YOU SEE
a single slit

Large central maximum

Small side maximums are present.

POLARIZATION OF LIGHT

Light waves are electromagnetic waves consisting of both an oscillating electric field and an oscillating magnetic field. These two oscillations sustain each other and allow light to propagate independently, even through a vacuum. The electric-field and magnetic-field waves are both transverse waves where the oscillating fields are perpendicular to the direction of wave travel. In addition, the electric-field oscillation is perpendicular to the magnetic-field oscillation. Working with both of these perpendicular oscillations is complex. As a result, light waves are often simplified as a single transverse wave involving only the electric-field oscillation.

The electric field is capable of doing work on charges. As a result, the polarization of electromagnetic waves is viewed from the perspective of the electric field's transverse wave form. The poles of the oscillation are the crests and troughs of the electric-field wave, which maintain their orientation as light waves propagate. The plane in which the electric field oscillates is known as the **plane of polarization**. In Figure 17.7, the electric field is oscillating in the x-y plane, and this electromagnetic wave is polarized along the y-axis.

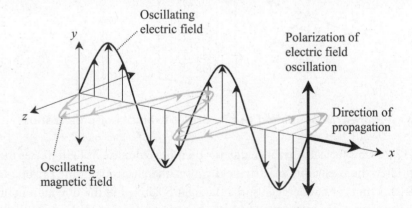

Figure 17.7 Electromagnetic wave

Light waves can be polarized in any direction. Light arriving from the Sun consists of countless waves, each polarized in random directions. Light from the Sun, as well as light from other conventional light sources, is unpolarized. The diagrams in Figure 17.8 are simplified by showing the polarization of light for two light sources coming directly out of the page. Figure 17.8(a) represents a vertically polarized electric-field oscillation. Figure 17.8(b) consists of electric-field waves polarized in a variety of directions and is an example of unpolarized light.

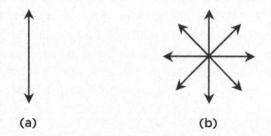

(a)   (b)

Figure 17.8 Polarized (a) and unpolarized (b) light

If unpolarized light passes through a **polarizing filter**, it will become polarized. A polarizing filter contains a transparent sheet imbedded with long, organic molecules oriented in only one direction. It is similar to the bars of a jail cell, where some things can pass through and others cannot. Only the light rays oscillating in one direction will pass through the polarizing filter. Light oscillating in all other directions is blocked.

Unfortunately, polarization is often shown incorrectly in diagrams in order to simplify the concept for beginning students. Figure 17.9 is the simplified, incorrect example. It shows unpolarized light passing through a polarizing filter where the organic molecules are aligned in the y-direction.

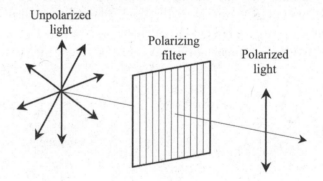

Figure 17.9 Simplified but incorrect example of polarization

Figure 17.9 seems logical, but the reality is a bit more complex. As light passes through the polarizing filter, the oscillating electric field causes the electrons in the organic molecules to vibrate. This transfer of energy absorbs the light polarized in the direction matching the alignment of the organic molecules. The light that actually transmits through the filter is the light perpendicular to the strands of organic molecules, as shown in Figure 17.10.

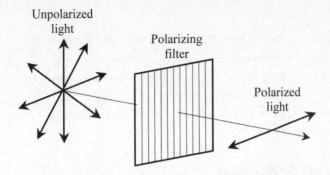

Figure 17.10 Correct example of polarizing light

Exams are more likely to focus on the fact that a single polarizing filter allows light waves oscillating in only one direction to pass through. You should also note that this phenomenon experimentally demonstrates that light is a transverse wave. In addition, exams will often test the effect of two polarizing filters used together, as shown in Figure 17.11.

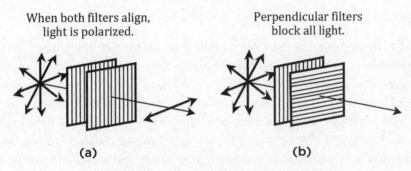

Figure 17.11 Effect of two polarizing filters

When two polarizing filters have the same orientation, light is polarized in the same manner as if only one filter is present. When two polarizing filters are perpendicular, they block all waves in any orientation. No light passes through. If the filters start in the position shown in Figure 17.11(a) and one filter is turned about the axis of propagation, the light passing through will become dimmer and dimmer. When the filter has turned 90°, all light will be blocked. Polarizing filters are used in 3-D glasses to view movies. A 3-D movie is actually two polarized movies superimposed on the screen. One lens of the 3-D glass allows vertically polarized light to pass through, and the other lens allows horizontally polarized light to pass through. Each eye is watching a different movie. The brain interprets the resulting images as three dimensional.

When light strikes surfaces at certain angles it can result in some of the reflected light's becoming polarized. The direction of the polarized waves is parallel to the reflecting surface. This intensifies the brightness of the light entering the eye and is often referred to as "glare." This can be seen when the sun is low in the sky and the sunlight's reflecting off the horizontal surface of water or the hood of a car makes it difficult to look in the direction of the setting or rising sun. Polarizing sunglasses are actually polarizing filters that block the intensified horizontally polarized light seen under these conditions.

IF YOU SEE
a polarizing
filter

Only light waves with the correct direction of oscillation pass through.

Polarization demonstrates that light is a transverse wave.

IF YOU SEE
an object
with a
specific color

**The object
is reflecting
that color and
absorbing all
other colors.**

Leaves appear
green because
they reflect
green light
while absorbing
red and blue
light, which
are used in
photosynthesis.

COLOR

Visible light consists of the colors extending from red to violet in the electromagnetic spectrum. To see specific colors, light waves must be aimed at and enter the eye in order to stimulate the photoreceptors in the retina. There are two ways to see light. Either a beam of light is shined directly into the eye from a light source, or light can reflect off of a surface and then enter the eye. If a light source is red, an observer will see red. However, the reflection of light is more complicated.

Absorption and Reflection

When light strikes a surface, some wavelengths will be absorbed and others will reflect. If light is absorbed, it cannot be seen. Only the reflected rays bouncing off of an object are capable of entering the human eye. When objects are observed with the eye, the reflected colors are seen. For example, think of green leaves. Leaves reflect the wavelengths of light that appear green. This means that other wavelengths of light are being absorbed. Leaves absorb the wavelengths consistent with red and blue light, which are the wavelengths needed for photosynthesis.

Dispersion

White light arriving from the Sun is composed of all the wavelengths of visible light. These wavelengths can be separated into distinct colors of light through a process known as **dispersion**. This process is commonly demonstrated using a prism. Dispersion takes advantage of both refraction and geometric optics. When white light strikes a prism at an angle, the light separates (disperses) into individual colors. Each color of light has a slightly different wavelength and index of refraction when it moves through the prism. As a result, each color bends at a slightly different angle as it enters and leaves the prism, as shown in Figure 17.12. The short wavelengths of light (violet and blue) have the highest index of refraction, so they bend the most. All the colors of the spectrum are separated. A rainbow effect is seen that extends from red to violet.

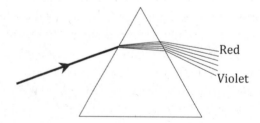

Figure 17.12 Dispersion of light through a triangular prism

Dispersion in water droplets together with refraction and reflection are responsible for the appearance of rainbows.

Scattering

When light rays in the atmosphere strike particles in the air, the light rays are reflected in various directions. This process is known as **scattering**. Shorter wavelengths of light are scattered the most. If atmospheric particles are large compared to the wavelength of light, such as water droplets in clouds, then scattering is simply due to the reflection of sunlight. Since sunlight includes all colors in the visible spectrum, clouds appear white on sunny days. If atmospheric particles are small compared to the wavelength of light, such as the transparent molecules that make up air, then scattering is due to the interaction between the light waves and electrons in the atoms of these molecules. This type of scattering is dependent on the wavelength of light. All wavelengths of light in the visible spectrum are scattered. However, shorter-wavelength blue light is scattered the most. As a result, more blue light reaches observers on Earth, giving the sky an overall blue appearance. Due to the curvature of Earth, the distance light must travel through the atmosphere is greater at the horizons than it is overhead. When the sun is low in the sky, such as at sunset, most of the blue light is scattered away from us as it travels the longer distance through the atmosphere. Since red light is scattered the least, it becomes the more prominent color of the sky at sunset.

SUMMARY

1. **DIFFRACTION IS THE BENDING OF LIGHT WHEN IT ENCOUNTERS AN OBSTACLE OR AN OPENING.** When light passes through an opening or slit, the diffraction pattern is circular if the slit width is nearly the same as the wavelength of the light.

2. **YOUNG'S DOUBLE-SLIT EXPERIMENT SHOWED THAT LIGHT HAS A WAVE CHARACTERISTIC.** When light diffracts through two narrow and adjacent slits, each opening creates circular wave fronts that interfere with one another. When displayed on a screen, an alternating pattern of bright regions (constructive interference) and dark regions (destructive interference) is seen. These effects are consistent with wave behavior. They can be seen in water waves and sound waves as well as light waves.

3. **LIGHT IS AN ELECTROMAGNETIC WAVE WITH POLARIZED OSCILLATION.** Light is made up of an oscillating electric and magnetic field. The direction in which the electric field vibrates is the direction of polarization. Light from a source such as the Sun consists of many waves all oscillating in various directions. A polarizing filter is embedded with long, linear, organic molecules that allow only light with a matching polarization to pass through it.

4. **THE COLOR OF AN OBJECT IS THE COLOR OF LIGHT THAT IT REFLECTS.** In order to see light, the light must enter the eye in a direct path to stimulate the retina. When an object is observed, the reflected colors enter the eye. The colors that are not seen have been absorbed by the object and cannot enter the eye.

If You See	Try	Keep in Mind
An obstacle or a slit	Diffraction Huygens' principle	As the width of the slit approaches the wavelength of light, the diffraction pattern becomes more circular and the shadow region decreases.
Young's double slit	$$x_m = \frac{m\lambda L}{d}$$ $$d\sin\theta = m\lambda$$	This experiment is evidence that light is a wave phenomenon.
Single-slit interference	A single slit creates a pattern similar to a double slit. However, the central maximum is wide and has a large intensity.	All other maximums are very small.
Polarizing filter	Only light with the correct direction of vibration will pass through the filter	This is evidence that light is a transverse wave.
An object with a specific color	Light must reflect off of an object in order for eyes to see it. When an object appears to have a specific color, it is reflecting the wavelengths of light that make up that color.	The object has absorbed the wavelengths corresponding to colors that are not seen.

PRACTICE EXERCISES

1. The bending of light around obstacles is called

   (A) refraction
   (B) reflection
   (C) diffraction
   (D) interference
   (E) polarization

2. If linear wave fronts are incident on a barrier that has a very small opening, the waves moving through the opening will

   (A) become polarized
   (B) converge on a single point
   (C) continue moving as linear wave fronts
   (D) form circular wave fronts
   (E) destructively interfere and cancel each other completely

3. The shadow region associated with diffraction is more pronounced when

   (A) a slit is narrower than the wavelength
   (B) a slit is equal to the wavelength
   (C) a slit is slightly larger than the wavelength
   (D) a slit is many times larger than the wavelength
   (E) The shadow region does not depend on the width of the slit.

4. Young's double-slit experiment provided evidence that light

   (A) refracts
   (B) reflects
   (C) transmits
   (D) acts like a particle
   (E) acts like a wave

5. Light incident on two slits is used to project an interference pattern onto a screen. The distance between the bright maximums observed on the screen can be increased by

   (A) moving the slits closer together
   (B) moving the slits farther apart
   (C) making the slit openings narrower
   (D) making the slit openings wider
   (E) increasing the intensity of the light source

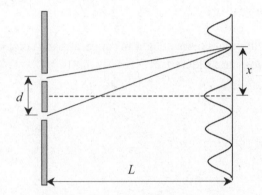

6. Light with wavelength λ is incident on two narrow slits spaced d meters apart. An interference pattern is visible on a screen located L meters from the slits. The second dark fringe is observed to be x meters from the prominent central maximum. The diagram above shows the path of light rays extending from each slit to the second dark fringe. What it the path difference?

   (A) $1.5x$
   (B) $1.5\,\lambda$
   (C) $2.5x$
   (D) $2.5\,\lambda$
   (E) $2.5d$

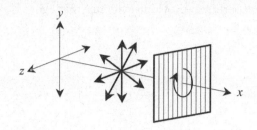

7. A polarizing filter is positioned as shown in the diagram above. Unpolarized light waves are passing through the filter with maximum intensity. How will the light passing through the filter be affected if the polarizing filter is rotated about the *x*-axis while the filter remains in the *y-z* plane?

(A) The amount of light passing through the filter will gradually diminish.
(B) When the filter has rotated 90°, no light will pass through the filter.
(C) Rotating the filter will have no effect on the amount of light passing through it.
(D) The direction of polarization will change to match the orientation of the filter.
(E) Both C and D are correct.

8. A prism disperses different colors of light because, as each color of light moves through the prism, each color has a different

(A) amplitude
(B) initial incidence
(C) wavelength
(D) critical angle
(E) oscillation

9. Light striking the surface of an object can be

(A) reflected
(B) scattered
(C) polarized
(D) absorbed
(E) All of these are possible to some degree.

ANSWERS EXPLAINED

	Key Words	Needed for Solution	Now Solve It
1. **(C)**	Bending of light around obstacles	Knowledge/definitions Diffraction	Diffraction is the bending of light around obstacles or through openings. Refraction is the bending of light due to the change in speed when light enters a new medium.
2. **(D)**	Very small opening	Knowledge/definitions Diffraction	The smaller an opening becomes, the more circular the wave fronts leaving the opening become.
3. **(D)**	Shadow region . . . more pronounced	Knowledge/definitions Diffraction and their shadow regions	The shadow region occurs where there is little or no diffraction. The question is the complete opposite of the previous question. Wider slits cause less diffraction, resulting in more distinct shadow regions.
4. **(E)**	Double-slit experiment . . . evidence that light	Knowledge/definitions Double-slit experiment	The double-slit interference pattern is a characteristic of a wave phenomenon. Young experimentally showed that light acts like a wave.
5. **(A)**	Interference pattern . . . distance between the bright . . . maximums . . . can be increased	$$x_m = \frac{m\lambda L}{d}$$	In the equation, the spacing between the maximums, x_m, is inversely proportional to the slit separation, d. The slit spacing must be decreased in order to increase the space between maximums. Be careful to distinguish between the space between slits, d, and the width of the slits. Answers C and D are distracters.
6. **(B)**	Diagram above shows the paths of light . . . second dark fringe . . . path difference	Path difference = $m\lambda$	Students should know that the dark fringes are assigned numbers ending in one-half. There is a tendency to number the second dark fringe incorrectly as $m = 2.5$. The second dark fringe is actually numbered as $m = 1.5$. Therefore, the path difference is 1.5 λ.
7. **(E)**	Polarizing filter . . . light . . . affected . . . ; filter rotated about the x-axis	Knowledge/definitions Polarization of light Polarizing filters	Answers A and B are true when two filters are used and only one of them is rotated. However, a single polarizing filter rotated as shown in the diagram cannot change the amount of light passing through it. Rotating the filter will change which light waves pass through the filter. At every instant, only the correctly aligned light waves will pass through. If the filter's orientation is changing, then so is the orientation of the light moving through it. C and D are both correct.

	Key Words	Needed for Solution	Now Solve It
8. **(C)**	Prism disperses . . . as each color of light moves through the prism . . . different	Knowledge/definitions Dispersion	Prisms disperse light because each color of light has a different wavelength and, as a result, a slightly different index of refraction in the prism. A different index of refraction leads to a different angle of refraction for each wavelength of light.
9. **(E)**	Light striking . . . surface	Knowledge/definitions Reflection, scattering, polarization, and absorption	Choices A to D can all occur when light strikes the surface of an object. Some may be more prominent in certain cases, but they are all a possibility.

Thermal Properties

18

- → **THERMAL SYSTEMS**
- → **THERMAL ENERGY**
- → **TEMPERATURE**
- → **THERMAL EXPANSION**
- → **IDEAL GASES**
- → **HEAT AND HEAT TRANSFER**
- → **HEATING AND COOLING**

Thermal systems consist of all the atoms that make up solids, liquids, and gases. These atoms are in constant random motion. That motion affects the temperature and thermal energy of all of the atoms grouped together as a larger system. Temperature differences between systems will cause energy to flow as heat from the hotter system to the colder system. Temperature, thermal energy, and heat are affected by the physical properties of the atoms making up each system. Table 18.1 lists the variables discussed in this chapter. By the end of the chapter you will be able to do the following:

- Define and explain thermal systems and the difference between thermal energy and temperature.
- Explain the mechanisms of thermal expansion.
- Identify the properties of ideal gases.
- Define heat and describe heat transfer.
- Analyze the heating and cooling of solids, liquids, and gases.

THERMAL SYSTEMS

In mechanics, the focus was often on the motion of a single object, such as a block sliding down an incline. However, when several objects experience the same motion, such as two blocks tied together by a string, then they can be treated as a single system. This allows the problem to be solved as though the blocks and the string are one larger object. Treating several objects experiencing the same conditions as one aggregate object is extremely useful in thermal physics.

The objects involved in thermal-physics problems are the countless atoms and molecules that make up substances. Working with single atoms is not possible. As a result, quantities and equations that describe the particles acting together as a single system have been established. The system might be an ice cube that melts into a liquid and vaporizes into a gas. It could be an iron rod that is heated and expands. It could even be a gas trapped in a cylinder with a movable piston. Some aspects of thermal physics involve analyzing the individual objects (atoms) within the larger system. Other problems will be concerned with the properties, trends, and quantities that describe the system as a whole.

Table 18.1 Variables That Describe Thermal Properties

New Variables	Units
T = temperature	kelvin
E_{th} = thermal energy	J (joules)
M = molar mass	kg/mol (kilograms per mole)
α = coefficient of linear expansion	1/K or K^{-1} or 1/°C or $°C^{-1}$
L = length	m (meters)
P = pressure	Pa (pascals)
V = volume	m^3 (meters cubed)
n = number of moles	mol (moles)
R = universal gas constant	8.31 J/mol • K
Q = heat	J (joules)
k = thermal conductivity	W/m • K (watts per meter • kelvin)
c = specific heat capacity	J/kg • K (joules per kilogram • kelvin)
L = latent heat	J/kg (joules per kilogram)

THERMAL ENERGY

The atoms composing solids, liquids, and gases are in constant motion. Thus, each atom has a tiny amount of kinetic energy. The chemical bonds and intermolecular forces holding substances together store tiny amounts of potential energy. **Thermal energy** is the sum of the microscopic kinetic and potential energies of all the atoms making up the system under investigation.

Thermal energy increases when substances become hotter. When solids become hotter, their molecules vibrate faster and increase in kinetic energy. When liquids and gases become hotter, the randomly moving molecules move faster and increase in kinetic energy.

Converting Mechanical Energy into Thermal Energy

Thermal energy and mechanical energy are both measured in joules. According to the law of conservation of energy, these two forms of energy are capable of transforming from one type into the other as long as the total amount of energy in the system is conserved. This chapter will explore instances where mechanical energy is transformed into thermal energy. This transfer occurs in two common ways: friction and inelastic collisions.

When a block slides down a rough incline, the rubbing of the block against the incline makes the molecules in both the block and the incline vibrate faster. Their thermal energies increase. Conservation of energy dictates that the thermal energy gained must equal the kinetic energy lost by the slowing block. The kinetic energy lost by the block is an energy change, and an energy change is known as work. In this case, the work of friction transfers kinetic energy from the block into thermal energy. The work of friction is equal to the kinetic energy lost by the block and is also equal to the thermal energy generated.

In inelastic collisions, the colliding objects impact one another. The impact causes the atoms within the objects to vibrate faster. In an inelastic collision, kinetic energy is lost. As with friction, the thermal energy generated in the inelastic collision is equal to the kinetic energy lost during the collision.

TEMPERATURE

Temperature is a relative measure of hot and cold. The **Celsius** and Fahrenheit temperature scales were established to encompass common extremes of hot and cold experienced by humans. The extremes of freezing and boiling water were used to set the 0° and 100° marks on the Celsius scale. Once established and accepted, the Celsius temperature scale became a means to quantify temperatures so that hot and cold objects could be numerically compared with one another.

Temperature reflects the speed and microscopic kinetic energy of the molecules of a substance. This means that the temperature of an object reflects the thermal energy of the object. Raising the temperature of an object increases its hotness. This causes the atoms of the substance to move faster, increasing the thermal energy. In order to relate temperature directly to thermal energy mathematically, a less arbitrary temperature scale is needed. Experimentation with gases resulted in projecting a temperature at which all molecular motion should cease. At this temperature, gas molecules would have no microscopic kinetic energy and therefore no thermal energy. The resulting temperature scale is the **Kelvin** temperature scale. Absolute zero, 0 kelvin, on this scale is equal to −273°C. It is the point where both temperature and thermal energy equal zero. As a result, formulas could be developed that use the Kelvin temperature to calculate thermal energy.

Which scale should be used for calculations? If a formula contains a T for temperature, then the Kelvin temperature must be used. However, if a formula contains a ΔT, then either the Celsius or Kelvin scale can be used. Why? Although the Celsius and Kelvin temperature scales have different zero points, 1 degree of change on the Celsius temperature scale is equal to 1 degree of change on the Kelvin temperature scale. Therefore, a change in temperature, ΔT, will be the same using either scale. When in doubt, it is always safer to use the Kelvin scale. To convert from degrees Celsius to kelvin, add 273.

THERMAL EXPANSION

When a solid or a liquid is heated, the atoms and molecules vibrate faster, causing a substance that is heated to expand. This process is known as **thermal expansion**. Conversely, a substance that is cooled will contract. Heating and cooling will change the length of linear objects, the surface area of two-dimensional objects, and the volume of three-dimensional objects.

In conceptual problems any type of object can be given and the effects of heating or cooling it may be asked. Essentially the entire object expands or contracts proportionally.

Linear Expansion

When a linear object, such as a metal rod, is heated, its length will increase, as shown in Figure 18.1.

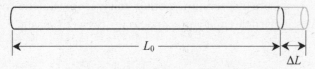

Figure 18.1 Linear expansion

IF YOU SEE
a solid
object being
heated or
cooled

Thermal expansion

$\Delta L = \alpha L_0 \Delta T$

α = coefficient of linear expansion

Heated objects increase in length by ΔL. Cooled objects decrease in length by ΔL.

The amount that the rod increases in length, ΔL, is calculated as follows:

$$\Delta L = \alpha L_0 \Delta T$$
$$\Delta L = \alpha L_0 (T_f - T_i)$$

The change in length, ΔL, is equal to the product of the coefficient of linear expansion, α, the original length, L_0, and the change in temperature, ΔT. The coefficient of linear expansion is a physical property of the substance that is expanding. Its value depends on the composition of the object. When a substance is heated, its final temperature, T_f, will be larger than its initial temperature, T_i. This heating will cause ΔL to become positive, indicating that length is increasing. However, when a substance is cooled, T_i will be larger than T_f. This cooling will result in a negative ΔL, indicating that length is decreasing. To find the new length, L, simply add the change in length, ΔL, to the starting length, L_0.

The linear-expansion formula can be used to solve for the expansion of a rectangular area. A rectangle has a length and a width. Simply solve for linear expansion twice, once for each dimension. Then, multiply the new length by the new width to find the new area.

EXAMPLE 18.1

Linear Expansion

A 5.0-meter-long iron rod is heated from 0°C to 100°C. The linear coefficient of expansion for iron is 12×10^{-6} K^{-1}. Determine the change in length of the rod.

WHAT'S THE TRICK?

Use the linear-expansion formula, but read the question carefully. Is the question looking for the change in length or the new length after expanding or contracting? This question simply requests the change in length.

$$\Delta L = \alpha L_0 \Delta T$$
$$\Delta L = \alpha L_0 (T_f - T_i)$$
$$\Delta L = (12 \times 10^{-6}°C^{-1})(5.0 \text{ m})(100°C - 0°C)$$
$$\Delta L = 6.0 \times 10^{-3} \text{ m}$$

IDEAL GASES

Gases are extremely chaotic. They consist of countless particles moving at different speeds in every possible direction. These particles collide with each other and with surfaces. Analyzing the behavior of a gas by examining individual gas molecules is impossible. Instead, a gas is treated as a single aggregate system. Due to its complexity, a simplified model of a gas system was developed. This model is known as the **ideal gas** model. Certain assumptions are made about ideal gases:

- The particles are so small that the volume of the particles is negligible.
- The attraction between particles is zero (no microscopic potential energy).
- The particles are in constant random motion (microscopic kinetic energy).
- Collisions between particles and with surfaces are perfectly elastic.

These assumptions make it possible to analyze ideal gases using the mathematical relationships discussed in the sections that follow. The behavior of a real gas at normal temperatures is nearly identical to that of an ideal gas.

Since the particles of a gas move at varying speeds in all directions, calculations involving specific directional velocities are impossible. As a result, only the average speed of the gas particles can be determined. As temperature increases, average speed increases. The average speed, v_{rms}, is related to the Kelvin temperature, T, of the gas according to the following formula:

$$v_{rms} = \sqrt{\frac{3RT}{M}}$$

This formula is unlikely to be used on the exam because it requires the use of a calculator. However, you should know that the speed of a gas is changed by the square root of the factor applied to the Kelvin temperature. If the Kelvin temperature is doubled, the average speed of the gas particles increases by $\sqrt{2}$. The variable R is the universal gas constant. In physics, it has the value of 8.31 joules per mole • kelvin. The variable M is the molar mass of the gas particles in kilograms per mole.

The energy of a gas is related to the average speed and kinetic energy of the gas particles. Since the attraction between the particles of an ideal gas is assumed to be zero, the only type of energy that gas particles possess is the microscopic kinetic energy due mainly to their speed. Since all the particles have different speeds, the kinetic energy, K, of the gas particles is also an average. It is calculated as follows:

$$K_{avg} = \frac{3}{2} k_B T$$

The variable k_B is Boltzmann's constant, $k_B = 1.38 \times 10^{-23}$ joules/kelvin. The nature of this value makes this formula unlikely to appear on an exam that does not allow you to use a calculator. However, the formula is valuable in demonstrating that the average kinetic energy of a gas is directly proportional to its Kelvin temperature. If the Kelvin temperature of a gas is doubled, the average kinetic energy of the gas particles also doubles. **Be careful.** This is not true for temperatures involving the Celsius scale. Doubling the Celsius temperature does not double the Kelvin temperature. Again, when in doubt, use the Kelvin temperature scale.

The trends relating temperature, gas-particle speed, and the energy of the gas may be the most important information to retain. As temperature increases, molecules move faster and their kinetic energy increases. If the attraction between the gas particles is assumed to be zero, the microscopic potential energy between gas particles is also zero. This means that the thermal energy of a gas is essentially the sum of the kinetic energies of all the gas particles.

Pressure

When a particle of gas collides with a surface, such as the walls of a container, the momentum change (impulse) experienced by the gas particle causes a force to be applied to the surface. Calculating the force of each particle of gas and adding them up would be an impossible task. However, the aggregate force of all gas particles treated as a single system striking a specific area can be measured and is a valuable quantity. The force, F, due to all the collisions of gas particles striking 1 square meter of surface area, A, is known as **pressure**, P.

$$P = \frac{F}{A}$$

Pressure is measured in pascals (Pa). A pascal is equal to a newton per meter squared (N/m^2).

Once again, the overall trends are extremely important. As the temperature of a gas increases, the particles of the gas move faster and increase their speed. The kinetic energy of the gas particles increases as does the thermal energy of the gas as a whole. In addition, the faster and more energetic gas particles experience greater momentum changes when they strike surfaces, resulting in an increase in the pressure of the gas.

Ideal Gas Law

The ideal gas law is a mathematical relationship relating the pressure, P, volume, V, number of moles, n, and temperature, T, of an ideal gas.

$$PV = nRT$$

IF YOU SEE
a gas

Ideal gas law

PV = nRT

Pressure and volume vary inversely. Pressure and temperature vary directly. Volume and temperature vary directly.

Volume in physics is measured in meters cubed, m^3. The variable n represents the number of moles of gas particles. For students who have not yet completed chemistry, a mole is similar to terms such as "pair" or "dozen." These terms bring specific numbers to mind (pair = 2 and dozen = 12). A mole is a specific number of particles. One mole is 6.02×10^{23} particles, and this value is known as Avogadro's number. Avogadro's number is not needed for the ideal gas law; just the number of moles is needed. If a problem specifies one mole (1 mol) of gas, then $n = 1$. For two moles of gas, $n = 2$, etc. The ideal-gas constant, R, is equal to 8.31 joules per mol • kelvin.

Conceptual problems will focus on relationships between key variables in the ideal gas law: $PV = nRT$.

1. Pressure and volume are inversely proportional to each other. For example, if the temperature and the moles of gas are held constant, then a decrease in volume is compensated for by an increase in pressure.
2. Pressure is directly proportional to the Kelvin temperature of a gas. For example, increasing the temperature while holding the volume and the number of moles constant will increase the pressure of a gas.
3. Volume is directly proportional to the Kelvin temperature of a gas. For example, increasing the temperature while holding the pressure and the number of moles constant will increase the volume of a gas.

EXAMPLE 18.2

Ideal Gas Law

A gas is trapped in a cylinder with a movable piston. How is the pressure of the gas affected if the temperature of the gas doubles while the piston moves inward, reducing the volume by half?

WHAT'S THE TRICK?

The ideal gas law is the key. The problem states that the gas is trapped, implying that the number of moles remains constant. Any value remaining constant, such as the number of moles and the gas constant, cannot cause change. Ignore them.

$$PV = nRT$$

$$(4P)\left(\frac{1}{2}V\right) = nR(2T)$$

The pressure must quadruple in order to maintain the equality.

HEAT AND HEAT TRANSFER

Heat is often misinterpreted by beginning physics students since the word *heat* is used incorrectly in everyday life. When students touch a warm object, they think it contains a large quantity of heat energy. This is not necessarily correct. The temperature felt by touching a warm object is a result of the thermal energy of the vibrating particles in that object. Try not to confuse the word *heat* with the word *hot*. **Hot** refers to the temperature of an object alone. Temperature is a measure of the average kinetic energy of the particles in that object. **Heat** refers not only to temperature but also to the number of particles (mass) that are involved. Consider a hot cup of coffee and the ocean. The coffee is hotter in temperature. However, the ocean contains more heat since it contains significantly more particles than the coffee. Then what is heat? Heat is similar to work. Work is a mechanical change in energy that can be seen by the eye (macroscopic). When an object is pushed by a force through a distance, its kinetic energy changes. Work equals the change in macroscopic kinetic energy. **Heat**, Q, is a change in thermal energy that cannot be seen by the eye (microscopic). When a flame is applied to an object, the object's thermal energy changes. Heat equals the change in thermal energy. While work is the quantity of mechanical energy transferred from one system to another, heat is the quantity of thermal energy transferred from one system to another system.

Heat Transfer

Heat transfer is the process of transferring thermal energy from one system to another. In order for heat transfer to take place between two systems, the systems must be at different temperatures. The natural direction of heat transfer is from the high-temperature system to the low-temperature system. The particles in the high-temperature system are vibrating or moving faster. When they come into contact with the slower particles in the low-temperature system, collisions transfer energy (heat) from the high-temperature system to the low-temperature system. The particles in the high-temperature system lose energy, slow down, and become cooler. The particles in the low-temperature system gain energy, speed up, and become hotter. Energy transfer continues until both systems reach the same temperature. **Thermal equilibrium** describes the condition when two objects have the same temperature, and no net heat transfer will take place between them on their own. Consider the coffee and ocean example. Pouring the coffee into the ocean transfers heat from the coffee, which is hotter, to the ocean, which is cooler. Both will ultimately achieve the same temperature, or thermal equilibrium. However, the direction of heat transfer always goes from the object with the higher temperature to that with the lower temperature.

Heat transfer occurs by three methods. In **conduction**, heat is transferred when two objects at different temperatures physically touch each other. **Convection** is heat transfer by fluids (liquids and gases). **Radiation** is heat transfer due to the absorption of light energy.

Conduction

Transfer of heat by **conduction** requires objects to touch each other physically. This normally involves a hotter solid object's touching a colder solid object. This is similar to the process of conduction in electricity, where charges transfer energy from one conductor to another conductor when they touch each other. In fact, substances that are good conductors of electricity, such as metals, are also good conductors of heat.

IF YOU SEE
heat transfer

Conduction:
solids
touching

Convection:
fluids

Radiation:
light

The natural direction of heat flow is from high temperature to low temperature.

Rate of Heat Transfer and Thermal Conductivity

The **rate of heat transfer**, $Q/\Delta t$, through an object is dependent on the length, L, the heat must travel through the object; the cross-sectional area of the object, A; the temperature difference between the ends of length L; and the thermal conductivity, k, of the object. The following formula solves for the rate of heat transfer:

$$\frac{Q}{\Delta t} = \frac{kA\Delta T}{L}$$

IF YOU SEE
heat
transfer by
conduction

Rate of heat transfer

$$\frac{Q}{\Delta t} = \frac{kA\Delta T}{L}$$

Substances with higher thermal conductivities, k, transfer heat at a faster rate.

Thermal conductivity is a physical property of an object. It indicates how well heat is conducted through an object. Every substance has a unique thermal conductivity. The higher the thermal conductivity of the substance, the faster heat transfers through it. Copper has a thermal conductivity of 400 watts/meter • kelvin, while stainless steel has a thermal conductivity of 14 watts/meter • kelvin. This means that copper pots and pans transfer heat through them at a much faster rate, allowing for faster meal preparation and quicker adjustments in temperature when cooking.

Insulators can be used to reduce the rate of heat transfer. This is similar to the process of insulation in electricity. In fact, substances that are good insulators of electricity, such as nonmetals, are also good insulators of heat. Insulation is used in homes to prevent summer heat from entering and winter heat from escaping. Insulating materials need to prevent the rapid transfer of heat, and these substances will have very low thermal conductivities.

Thermal conductivity also explains why objects at the same temperature can feel like they have different temperatures when touched. Metals have high thermal conductivities, and wood has a very low thermal conductivity. Even if a piece of metal and a piece of wood are both at room temperature, the metal will feel cooler to the touch. The human body is warmer than room temperature. When these objects are touched, heat transfers from the fingers into both the metal and the wood. However, heat transfers into the metal at a faster rate. The rapid loss of body heat as it moves into the piece of metal makes the fingers feel colder.

Convection

Heat transfer through **convection** involves fluids. Fluids can flow. This means both liquid and gases are fluids. When a hot fluid mixes with a cold fluid, thermal energy is transferred as the faster-moving particles collide with the slower-moving particles. Heating a home is an example of convection. Hot air flows through ducts in the home, eventually pouring into a room, much like a river pours into a lake. A moving fluid with a different temperature than its surroundings is known as a **convection current**.

Radiation

Heat can also be transferred by electromagnetic **radiation**. Light waves carry energy that can be absorbed when the waves strike objects. The absorption of light will increase the thermal energy of an object, causing its temperature to increase.

HEATING AND COOLING
Specific Heat

If a substance is heated, it will absorb energy, and its temperature will increase. If a substance is cooled, it will lose energy, and its temperature will decrease. Every substance absorbs or loses energy at a set rate, which can be quantified. **Specific heat**, c, is the amount of heat needed to raise the temperature of 1 kilogram of a substance by 1 kelvin. For example, the specific heat of liquid water is 4,190 joules per kilogram • kelvin. In order to raise the temperature of 1 kilogram of water by 1 kelvin, 4,190 joules of heat must be transferred to the water. The following equation is used to solve for the heat needed to change the temperature of a substance with a mass of m and a temperature difference of ΔT.

$$Q = mc\,\Delta T$$

The specific heat, c, is a physical property that is unique for every substance. In addition, each state of matter has a unique specific heat. For example, the specific heat of solid water, c_s, is 2,090 joules per kilogram • kelvin. However, the specific heat of liquid water, c_L, is 4,190 joules per kilogram • kelvin.

Liquid water has the highest specific heat of substances commonly found on Earth. As a result, a tremendous amount of energy is needed to increase the temperature of water. Similarly, water can retain a tremendous amount of energy. This allows for the temperature of water to change quite slowly in either direction. It also allows water to be a **heat sink**, whereby it can store large quantities of energy in places such as rivers, lakes, and particularly oceans.

IF YOU SEE
the heat needed to change the temperature of a system

$Q = mc\,\Delta T$

c = specific heat

Specific heat is the amount of heat in joules needed to raise 1 kilogram of a substance by 1 kelvin.

EXAMPLE 18.3

Specific Heat

The specific heat of aluminum is 900 joules/kilogram • kelvin. How much heat is required to raise 2.0 kilograms of aluminum 10°C?

(WHAT'S THE TRICK?)

For quantities other than 1 kilogram and 1 kelvin, use the equation:

$$Q = mc\,\Delta T$$
$$Q = (2.0\ \text{kg})(900\ \text{J/kg} \cdot \text{K})(10\ \text{K})$$

Note that a change of 10°C is the same as a change of 10 K.

$$Q = 18{,}000\ \text{J}$$

Phase Changes

Matter on Earth is commonly found in three phases: solid, liquid, or gas. A **phase change** is when a substance physically changes from one phase to another. Phase changes occur when substances reach critical temperatures. Solids change into liquids at their melting point, and liquids change into gases at their boiling point. The melting and boiling points are two important physical properties of a substance. Different substances will have different melting and boiling points. However, every substance has a set melting and boiling point under specific environmental conditions.

Heat of Transformation (Latent Heat)

When a substance is heated, its temperature will increase until it reaches the critical temperature (melting point or boiling point) at which a phase change can occur. At this critical temperature, all the thermal energy added to the substance is used to conduct the phase change. The **heat of transformation**, also known as **latent heat**, is the energy added

$Q = mL$

L = latent heat

Latent heat is the amount of heat in joules needed to change the phase of 1 kilogram of a substance.

during the phase change. This energy is used to weaken the intermolecular forces that hold molecules together as solids and liquids. Since all the energy added is involved in the transformation, the temperature of the substance does not change during the phase change. When all of the substance has completed the phase change, the temperature of the substance can then resume its rise.

The heat of transformation, or latent heat, L, is a physical property of a substance. Each substance has a unique value for this quantity. The **heat of fusion** or **latent heat of fusion**, L_f, is the heat energy needed to convert 1 kilogram of a substance from its solid form to its liquid form. The **heat of vaporiztion** or **latent heat of vaporization**, L_v, is the heat energy needed to convert 1 kilogram of a substance from its liquid form to its gaseous form. The heat of vaporization is always significantly larger than the heat of fusion. More energy is required for the phase change from liquid to gas. The values given for latent heat will be the amount of heat needed for exactly 1 kilogram of a substance. To solve for the heat needed in phase changes involving a substance with mass m, use the following formula:

$$Q = mL$$

EXAMPLE 18.4

Heat of Transformation

Water has a heat of fusion of 3.33×10^5 joules per kilogram and a heat of vaporization of 22.6×10^5 joules per kilogram. How much heat energy is needed to melt 2.0 kilograms of ice at 0°C?

WHAT'S THE TRICK?

Melting involves the heat of fusion. The given heat of vaporization is a distracter.

$$Q = mL$$
$$Q = (2.0 \text{ kg})(3.33 \times 10^5 \text{ J/kg}) = 6.66 \times 10^5 \text{ J}$$

Heating and Cooling Curve

Heating a substance from a solid to a liquid can be summarized in a graph known as a **heating and cooling curve**. In Figure 18.2, a substance starting as a solid is heated by adding heat at a constant rate. The temperature of the substance is graphed versus time.

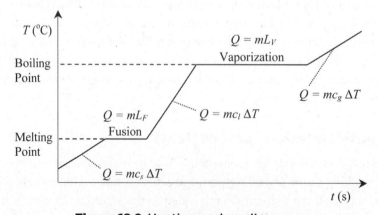

Figure 18.2 Heating and cooling curve

The temperature of the solid rises as heat is added, $Q = mc_s\Delta T$. When the substance reaches its melting point, the temperature becomes constant, while the heat of fusion, $Q = mL_F$, is added to convert the solid into a liquid. When the entire substance has become a liquid, the heat added, $Q = mc_l\Delta T$, raises the temperature until the substance reaches its boiling point. At the boiling point, the temperature is again constant while the heat of vaporization, $Q = mL_v$, converts the substance into a gas. Once the phase change is completed, the gas changes temperature with the addition of more heat, $Q = mc_g\Delta T$. When boiling water on a stove, the last line of the heating curve would not be possible as the steam attains the highest temperature at the boiling point. However, it is possible to capture the steam in a pipe and add more thermal energy by applying a flame to increase the water's temperature beyond the boiling point.

Note how the sloped sections differ. This is because the specific heat differs for a solid, c_s, a liquid, c_l, and a gas, c_g. During the phase changes, the temperature does not rise, and the graph remains horizontal. The energy added is used to conduct the phase change. The temperature will not rise until the entire substance has completed the change. In addition, the horizontal sections have different lengths. More energy is required to change a substance into a gas than to change it into a liquid. Since heat is added at a continuous rate, more time is needed to change the substance into a gas than to change it into a liquid.

The graph is identical for a substance that is cooled, but is reversed in terms of time. When cooled, a substance could start as a gas and cool until it condenses into a liquid and then freezes into a solid. The mathematics for heating and cooling are the same. It is just a question of whether heat is added or removed and whether temperature rises or decreases. Either way, the formulas and calculated values are the same.

SUMMARY

1. **A THERMAL SYSTEM CONSISTS OF THE ATOMS MAKING UP SUBSTANCES.** The atoms are the objects, but together they form a single complex system.

2. **THERMAL ENERGY IS THE SUM OF THE MICROSCOPIC KINETIC AND POTENTIAL ENERGIES OF ALL THE ATOMS MAKING UP A THERMAL SYSTEM.** The microscopic kinetic energy results from the vibration of atoms in solids and the motion of atoms in liquids and gases. The microscopic potential energy is associated with the bonds and intermolecular forces holding atoms together. When atoms vibrate or move faster, the thermal energy of the entire system increases.

3. **TEMPERATURE IS RELATED TO THE SPEED AND KINETIC ENERGY OF THE MOLECULES MAKING UP A SUBSTANCE.** The Kelvin scale, where absolute zero coincides with the lack of motion at the atomic level, can be used to calculate thermal energy and other thermal quantities.

4. **THERMAL EXPANSION OCCURS WHEN TEMPERATURE INCREASES.** At higher temperatures, atoms vibrate and move faster. This results in an expansion of the entire system. Linear objects become longer, and areas and volumes increase proportionally. Systems that are cooled contract.

5. **AN IDEAL GAS IS A MODEL ALLOWING ANALYSIS OF A COMPLEX SYSTEM.** Gas particles are in constant random motion, moving in different directions at different speeds. When they strike surfaces, the change in momentum creates pressure. Increasing temperature increases the pressure of a gas. At normal temperatures, real gases behave like an ideal gas.

6. **HEAT IS THE TRANSFER OF MICROSCOPIC ENERGY INTO OR OUT OF A SYSTEM.** Heat is similar to work. However, while work is concerned with large objects where the acting forces and distances can be seen, heat is the transfer of energy due to collisions between atoms that spread from one atom to the next. Heat can be transferred by conduction, convection, and radiation. The rate at which heat is transferred by conduction depends on the thermal conductivity of a substance.

7. **THE HEATING AND COOLING OF SUBSTANCES DEPENDS ON THEIR PHYSICAL PROPERTIES AND CAN RESULT IN PHASE CHANGES.** The amount of heat needed to raise the temperature of 1 kilogram of a substance 1 kelvin is known as the specific heat. The amount of heat needed to change 1 kilogram of a solid into a liquid is known as the heat of fusion (latent heat of fusion). The amount of heat needed to change 1 kilogram of a liquid into a gas is known as the heat of vaporization (latent heat of vaporization).

If You See	Try	Keep in Mind
A solid object being heated or cooled	Thermal expansion $$L = \alpha L_0 \Delta T$$ $$\Delta L = \alpha L_0 (T_f - T_i)$$ α = coefficient of linear expansion	Heated objects increase in length by ΔL, whereas cooled objects decrease by this length. Areas and volumes increase and decrease proportionally.
A gas	Ideal gas law $$PV = nRT$$ Temperature must be in Kelvin scale	Pressure and volume vary inversely. Pressure varies directly with temperature. Volume varies directly with temperature.
Heat transfer in general	Conduction: solids Convection: fluids Radiation: light	The natural direction of heat flow is from systems with high temperatures to systems with low temperatures.
Heat transfer by conduction	Rate of heat transfer (conduction) $$\frac{Q}{\Delta t} = \frac{kA\Delta T}{L}$$ k = thermal conductivity A = cross-sectional area L = length the heat transfers through	Substances with higher thermal conductivities transfer heat at a faster rate. Thermal insulators have low thermal conductivity.
Heat needed to change the temperature of a system	$$Q = mc\,\Delta T$$ c = specific heat	The specific heat, c, is the amount of heat in joules needed to raise 1 kilogram of a substance 1 kelvin.
A phase change from solid to liquid or from liquid to gas	$$Q = mL$$ L_f = latent heat of fusion L_v = latent heat of vaporization	Latent heat, L, is the amount of heat in joules needed to change the phase of 1 kilogram of a substance.

PRACTICE EXERCISES

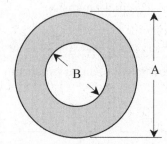

1. A metal washer is a flat, circular piece of metal with a hole through its center, as pictured above. What will be the effect of heating this washer?

   (A) No change occurs as the effects on the diameters cancel.
   (B) Diameter A will decrease, and diameter B will decrease.
   (C) Diameter A will decrease, and diameter B will increase.
   (D) Diameter A will increase, and diameter B will decrease.
   (E) Diameter A will increase, and diameter B will increase.

2. Which physical property affects the rate of heat transfer?

   (A) Coefficient of linear expansion
   (B) Specific heat
   (C) Heat of fusion
   (D) Heat of vaporization
   (E) Thermal conductivity

3. How is the average kinetic energy of gas particles affected by changing the temperature of the gas from 50°C to 100°C?

   (A) The average kinetic energy is cut in half.
   (B) The average kinetic energy remains constant.
   (C) The average kinetic energy increases by less than $\sqrt{2}$.
   (D) The average kinetic energy increases by $\sqrt{2}$.
   (E) The average kinetic energy is doubled.

4. A gas with pressure P, volume V, and temperature T is trapped inside a cylinder with a movable piston. If the temperature of a trapped gas is tripled while the volume of the gas doubles, what will be the new pressure?

   (A) $\frac{2}{3}P$

   (B) $\frac{3}{2}P$

   (C) $2P$

   (D) $3P$

   (E) $6P$

5. A container is filled with a liquid and is sealed. The container is then suspended in a chamber so that it does not contact any surfaces. Air is pumped completely out of the chamber, creating a vacuum around the container and its contents. The temperature of the liquid inside the container is observed to rise. What method of heat transfer is responsible for increasing the thermal energy of the liquid?

(A) Conduction
(B) Convection
(C) Radiation
(D) Both A and C
(E) Both B and C

6. The specific heat of a liquid is 3,000 joules/kilogram • kelvin. How much heat is required to raise 5.0 kilograms of this liquid 20°C?

(A)   300 J
(B)   500 J
(C)   12,000 J
(D)   300,000 J
(E) 4,395,000 J

7. A substance has a specific heat of 400 joules/kilogram • kelvin, a latent heat of fusion of 2.5×10^5 joules per kilogram, and a latent heat of vaporization of 10×10^5 joules per kilogram. In order to melt a 2.0-kilogram solid sample of this substance, 5.0×10^5 joules of heat must be added. Determine the change in temperature as this 2.0-kilogram sample melts.

(A)   0 K
(B)   312.5 K
(C)   625 K
(D) 1,250 K
(E) 2,500 K

8. When two systems are in thermal equilibrium with each other

(A) heat flows from the system of higher temperature to the one of lower temperature
(B) heat can be made to flow from the system of lower temperature to the one of higher temperature
(C) both A and B
(D) no net heat flow occurs
(E) None of these is correct.

9. A gas is sealed in a container that has fixed walls, and heat is added to the system at a constant rate. Which statement below is NOT true?

(A) The temperature of the gas increases at a constant rate.
(B) The thermal conductivity of the gas particles increases.
(C) The average speed of the gas particles increases.
(D) The average kinetic energy of the gas particles increases.
(E) The pressure of the gas increases at a constant rate.

10. Several materials along with their thermal conductivities are listed below. Which material should be used to insulate a home?

(A) Concrete, $k = 0.800$ W/m • K
(B) Glass windows, $k = 0.750$ W/m • K
(C) Fat, $k = 0.210$ W/m • K
(D) Feathers, $k = 0.040$ W/m • K
(E) Air, $k = 0.026$ W/m • K

ANSWERS EXPLAINED

	Key Words	Needed for Solution	Now Solve It
1. **(E)**	Effect of heating	Knowledge/definitions Thermal expansion	When an object is heated, the entire object expands proportionally. Both diameters A and B will increase. Students are often fooled by answer D. An analogy may help. What would happen if a picture of a person were resized? Would the head expand while the mouth became smaller? No. Both would expand.
2. **(E)**	Rate of heat transfer	Knowledge/definitions Thermal conductivity	Thermal conductivity is a physical property indicating how well heat transfers. Note that this question can be easily reworded into four additional questions testing all of the given terms. It is important to know the definitions and the effects of each of the physical properties listed as choices in this question.
3. **(C)**	Average kinetic energy . . . from 50°C to 100°C	$$K_{avg} = \frac{3}{2}k_B T$$	Although the temperature appears to be doubling, this is actually a trick question. Formulas containing T require the Kelvin temperature scale (ΔT uses either scale). $$50°C = 323 \text{ K} \quad \text{and} \quad 100°C = 373 \text{ K}$$ Increasing temperature from 323 K to 373 K is a 15.5% increase. Average kinetic energy is directly proportional to temperature, an increase by a factor of 0.155. Answer C is the only possible answer. When in doubt, or to be safe, use Kelvin temperatures.
4. **(B)**	Temperature of a trapped gas is tripled . . . volume of the gas doubles . . . new pressure	Ideal gas law $$PV = nRT$$	Modify the variables that change and determine the coefficient for P that will maintain the equality. $$\left(\frac{3}{2}P\right)(2V) = nR(3T)$$ Pressure is $\frac{3}{2}$ times its original value: $\frac{3}{2}P$.
5. **(C)**	Does not contact any surfaces . . . vacuum . . . method of heat transfer	Knowledge/definitions Conduction Convection Radiation	If the container does not touch any surfaces, then heat transfer by conduction cannot take place. Heat transfer by convection requires fluids. Although a fluid is inside the container, the vacuum surrounding the container has no fluids to transfer heat into the container. This leaves radiation. There are many forms of electromagnetic radiation. All can move through a vacuum, and some can penetrate walls and containers.

	Key Words	Needed for Solution	Now Solve It
6. **(D)**	Specific heat . . . how much heat	$Q = mc\,\Delta T$	$Q = (5.0\text{ kg})(3{,}000\text{ J/kg} \cdot \text{K})(20°\text{C})$ $Q = 300{,}000\text{ J}$
7. **(A)**	Determine the change in temperature . . . melts	Knowledge/definitions Heating and cooling	The temperature does not change during a phase change. The values for the specific heat, heat of fusion, and heat of vaporization are all distracters inviting unnecessary calculations.
8. **(D)**	Thermal equilibrium	Knowledge/definitions	When two systems are in thermal equilibrium, they are at the same temperature and no net flow occurs between them.
9. **(B)**	Gas . . . sealed in a container . . . heated . . . temperature to increase at a constant rate	Knowledge/definitions Heat Temperature Thermal conductivity Average speed Average kinetic energy Pressure	When heat is added to a gas, the temperature of the system increases. As a result, the gas particles move faster and have a higher average kinetic energy. However, thermal conductivity is a property of a substance and remains constant for that substance. Thermal conductivity influences the rate at which heat transfers.
10. **(E)**	Thermal conductivities . . . insulate	Knowledge/definitions Heat transfer	Air has a very low thermal conductivity. As a result, it transfers heat very slowly. This makes air an ideal insulator. Insulation materials are often designed to trap air pockets in order to slow heat transfer.

Thermodynamics

19

- → **INTERNAL ENERGY**
- → **ENERGY TRANSFER IN THERMODYNAMICS**
- → **ENERGY MODEL SUMMARIZED**
- → **FIRST LAW OF THERMODYNAMICS**
- → **HEAT ENGINES**
- → **ENTROPY**
- → **SECOND LAW OF THERMODYNAMICS**

Thermodynamics is the branch of physics dealing with thermal energy and heat. In beginning physics classes, thermodynamics focuses on a system consisting of gas particles that are contained in a cylinder with a movable piston. As discussed in the previous chapter, working with countless particles in random motion is an impossible task. Therefore, thermodynamics examines gases as a single system. Table 19.1 lists the variables used in thermodynamics. In this chapter, we will do the following:

- Define internal energy and examine its role in thermodynamics.
- Summarize how work and heat affect internal energy.
- Apply the first law of thermodynamics to solve problems.
- Calculate the efficiencies of heat engines.
- Explain the nature of entropy.
- Analyze the consequences of entropy and the second law of thermodynamics.

Table 19.1 Variables Used In Thermodynamics

New Variables	Units
U = internal energy	J (joules)
e = efficiency	%

INTERNAL ENERGY

Internal energy, U, is the total energy of a system. Adding up all possible energies that compose the total energy of a system would be extremely complex. However, doing this is not necessary. Thermodynamics is not concerned with the actual value of internal energy, U, but rather with the change in internal energy, ΔU. It is the change in internal energy that makes engines operate. In thermodynamics, all but one of the energies making up internal energy are held constant. Thermal energy, E_{th} or $E_{thermal}$, is the only energy that is changing. The change in all other energies will be equal to zero.

$$\Delta U = \Delta E_{thermal}$$

The thermal energy of a gas is the sum of all the microscopic kinetic energies of the randomly moving gas particles. This energy can be calculated and results in the following equation for the change in internal energy of a gas:

$$\Delta U = \frac{3}{2}\, nR\,\Delta T$$

It is unlikely that this equation will appear on the exam. However, it has been included here to show the important direct relationship between temperature change and the change in internal energy. In addition, the sign on the change in internal energy in this equation is the key to setting the correct sign on other thermodynamic variables.

During a thermodynamic process there are three possibilities for the value of internal energy.

1. **INTERNAL ENERGY INCREASES.** If energy is added to the system, the internal energy increases, $+\Delta U$. This is accompanied by an increase in the temperature of the system, $+\Delta T$.

2. **INTERNAL ENERGY DECREASES.** If energy is removed from the system, the internal energy decreases, $-\Delta U$. This is accompanied by a decrease in the temperature of the system, $-\Delta T$.

3. **ISOTHERMAL PROCESS.** This is a thermodynamic process where the temperature remains constant, $\Delta T = 0$. If the temperature is not changing, then the internal energy is not changing, $\Delta U = 0$.

ENERGY TRANSFER IN THERMODYNAMICS

To visualize a thermodynamic system, think of a gas trapped in a cylinder with a movable piston. Only the gas particles constitute the system. The cylinder and piston are part of the environment. They simply contain the gas in an adjustable and measurable volume. This system is shown in Figure 19.1.

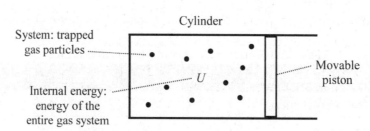

Figure 19.1 A thermodynamic system

Work and heat are the two processes that can transfer energy into and out of the gas system shown in Figure 19.1. Work and heat are in units of joules. Work and heat are the changes to state functions, such as internal energy U, kinetic energy K, and the energy of a spring U_s.

Work

Work, W, is a macroscopic (visible) transfer of energy involving a net force, ΣF, acting through a distance, d. Work is a mechanical way to add or remove energy from the gas system by physically moving the piston. The force that moves the piston is created by pressure acting on the inside and outside areas of the piston. The pressure of the gas inside the cylinder pushes

IF YOU SEE
internal
energy

Change in internal energy, ΔU, is the change in the total energy of the system.

ΔU is linked to changes in temperature.

Increase: $+\Delta U$

Decrease: $-\Delta U$

Isothermal: $\Delta U = 0$

the piston outward. Forces from the environment, such as the pressure of the atmosphere surrounding the cylinder, push the piston inward.

Moving the piston definitely involves a volume change, ΔV, but what about pressure? Under some conditions, the pressure of the gas changes when work moves the piston. Under special conditions, the pressure remains constant as the piston moves. This is known as an **isobaric** process. Under these special conditions, work is a product of the constant pressure and changing volume.

$$W = -P\Delta V$$

Whether the pressure of the gas is changing or is constant, there are three possible outcomes. Which particular outcome occurs depends on how the magnitudes of the pressures inside and outside the cylinder compare to one another.

1. **WORK DONE ON THE GAS.** If the pressure outside the cylinder is greater than the pressure inside the cylinder, then work is done *on the gas* while moving the piston inward. The volume of the gas is compressed, $-\Delta V$. The environment adds energy to the system, increasing the internal energy, $+\Delta U$. Work is a change in energy. The work done *on the gas* is positive, $+W$, increasing the internal energy of the system.

2. **WORK DONE BY THE GAS.** If the pressure inside the cylinder is greater than the pressure outside the cylinder, then work is done *by the gas* while moving the piston outward. The volume of the gas expands, $+\Delta V$. In the process, the gas uses its energy to push the piston outward, which causes internal energy to decrease, $-\Delta U$. Therefore, the work done *by the gas* is negative, $-W$.

3. **ISOMETRIC (ISOCHORIC) PROCESS.** This is a thermodynamic process where the volume of the gas remains constant. This occurs when the pressure of the gas system inside the cylinder is the same as the pressure of the environment outside the cylinder. The net force is zero. The piston cannot move, $\Delta V = 0$, and *no work is done*, $W = 0$.

Heat

As discussed in the previous chapter, **heat** (Q) is a microscopic (invisible) transfer of thermal energy between objects having different temperatures. Heat can be added or removed by touching a **heat reservoir** to the cylinder. An example of a heat reservoir could be air or water surrounding the cylinder. The energy transferred into or out of the reservoir is negligible compared with the huge size of the reservoir. As a result, heat reservoirs have constant temperature and transfer heat at a constant rate. A hot reservoir has a temperature greater than the gas in the cylinder. A cold reservoir has a temperature lower than the gas in the cylinder.

Heat flows from high temperature to low temperature. There are three possible outcomes for heat transfer depending on the temperature of the heat reservoir as compared with the temperature of the trapped gas in the cylinder.

1. **HEAT ADDED.** If the heat reservoir has a higher temperature (hot reservoir) than the gas inside the cylinder, heat flows from the reservoir *into the cylinder*. Internal energy increases, $+\Delta U$. Therefore, heat added is considered positive, $+Q$.

2. **HEAT REMOVED.** If the heat reservoir has a lower temperature (cold reservoir) than the trapped gas inside the cylinder, heat flows *out of the cylinder* into the reservoir. Internal energy decreases, $-\Delta U$. Therefore, heat removed is considered negative, $-Q$.

IF YOU SEE
work

mechanical energy transfer

On the gas: $+W$, volume decreases.

By the gas: $-W$, volume increases.

Isometric: $W = 0$, volume does not change.

IF YOU SEE
heat

thermal energy transfer

Heat added: $+Q$

Heat removed: $-Q$

Adiabatic: No heat added or removed: $Q = 0$

3. ADIABATIC PROCESS. This is a thermodynamic process where *no heat is added or removed*. This occurs when the heat reservoir and the trapped gas are in thermal equilibrium (same temperature). Heat transfer requires a temperature difference. So, no net heat flow occurs between objects in thermal equilibrium, $Q = 0$.

ENERGY MODEL SUMMARIZED

Figure 19.2 summarizes the interactions among internal energy, work, and heat. The signs on each variable are linked to their effect on the internal energy, ΔU, of the system (the gas trapped in the cylinder). Table 19.2 summarizes additional information and trends.

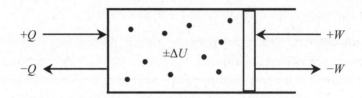

Figure 19.2 Thermodynamics energy model

Table 19.2 Interactions Among Internal Energy, Work, and Heat

If You See . . .	Result	Related Events	Result
Increase in internal energy	$+\Delta U$	Increase in temperature	$+\Delta T$
Decrease in internal energy	$-\Delta U$	Decrease in temperature	$-\Delta T$
Constant internal energy	$\Delta U = 0$	Isothermal: constant temperature	$\Delta T = 0$
Work done on the gas	$+W$	Volume of gas compresses	$-\Delta V$
Work done by the gas	$-W$	Volume of gas expands	$+\Delta V$
No work is done	$W = 0$	Isometric: constant volume	$\Delta V = 0$
Heat added	$+Q$	Heat reservoir is hotter	
Heat removed	$-Q$	Heat reservoir is colder	
No heat added or removed	$Q = 0$	Adiabatic: no heat transfer	

IF YOU SEE any thermodynamic quantities

1st law of thermodynamics

$\Delta U = Q + W$

FIRST LAW OF THERMODYNAMICS

The **first law of thermodynamics** is a statement of conservation of energy for thermal processes. Work, W, and heat, Q, are the only two processes that can affect the internal energy, U, of a gas. Their effect can be summarized as follows.

- For a system where internal energy comprises only thermal energy, the change in the internal energy of the system is equal to the energy transferred into or out of the system by work and heat.
- The previous statement can be summarized as an equation:

$$\Delta U = Q + W$$

Examples 19.1 and 19.2 demonstrate how the first law of thermodynamics is applied.

EXAMPLE 19.1

First Law of Thermodynamics

During a thermodynamic process, 200 joules of heat are added to a gas while 300 joules of work are done by the gas. Determine the change in internal energy.

WHAT'S THE TRICK?

Apply the first law of thermodynamics. Heat added is positive, $+Q$. Work done by the gas pushes the piston outward, expanding the volume. To accomplish this, the gas loses energy and work is negative, $-W$.

$$\Delta U = Q + W$$
$$\Delta U = (200 \text{ J}) + (-300 \text{ J})$$
$$\Delta U = -100 \text{ J}$$

EXAMPLE 19.2

First Law of Thermodynamics

(A) During an isothermal process, 1,000 joules of heat are removed from a trapped gas. Determine the change in internal energy of the gas.

WHAT'S THE TRICK?

You need to know the definition of the word "isothermal." In an isothermal process, the temperature remains constant, $\Delta T = 0$, and the internal energy does not change, $\Delta U = 0$.

(B) Determine the work done on or by the gas.

WHAT'S THE TRICK?

Apply the first law of thermodynamics, including the information in part A, $\Delta U = 0$. The heat removed is negative, $-Q$.

$$\Delta U = Q + W$$
$$(0) = (-1,000 \text{ J}) + W$$
$$W = 1,000 \text{ J}$$

(C) Is this work done on the gas or by the gas?

WHAT'S THE TRICK?

You need to know the difference between work done on the gas and work done by the gas. Positive work means that the internal energy of the gas is increasing. The environment must compress the gas in order to increase the internal energy. This is work done on the gas.

HEAT ENGINES

A **heat engine** is a device that converts thermal energy into other forms of energy. Heat engines operate between a high temperature (hot), T_H, and a low temperature (cold), T_C. Heat Q_H is added into the engine at a high temperature. Heat Q_C is removed from the engine at a low temperature. The heat transferred into the engine is greater than the heat removed from the engine, $Q_H > Q_C$. The difference in energy can be used to do useful work, W.

$$W = Q_H - Q_C$$

An analogy would be using a waterfall to turn a paddle wheel. Figure 19.3 compares the waterfall on the left to a heat engine on the right.

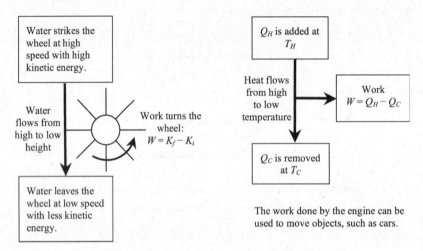

Figure 19.3 Energy flow in heat engines

Efficiency

IF YOU SEE

efficiency

$e = \left| \dfrac{W_{net}}{Q_H} \right|$

and

$e = \dfrac{T_H - T_C}{T_H}$

Expressed as a percentage. The second formula solves for a theoretical best-case efficiency.

The heat added to an engine, Q_H, can be thought of as thermal fuel. The heat removed from the engine, Q_C, can be thought of as thermal exhaust. Engine exhaust is not ideal, especially in a world concerned with pollution and dwindling resources. However, no heat engine can operate without expending some waste heat, Q_C, at low temperature, T_C. The warm engine of a car after it has been running is an example of low-temperature (T_C) waste heat (Q_C). This is the heat from gasoline that was not used to propel the car forward. Engine **efficiency** is the relationship between the total (net) work, W_{net}, produced by an engine to the heat added, Q_H. Efficiency measures useful output compared with input. Miles per gallon is a type of efficiency. It is a ratio of the car's useful output (miles traveled) to input (fuel). The efficiency, e, of a heat engine is also a ratio of output (net work) to input (heat added). Efficiencies are calculated using absolute values, and they are expressed as percentages.

$$e = \left| \frac{W_{net}}{Q_H} \right|$$

Heat engines have a theoretical maximum efficiency. This is calculated using the temperature extremes, T_H and T_C, between which the engine operates. The best possible efficiency of a heat engine is calculated as follows:

$$e = \frac{T_H - T_C}{T_H}$$

Calculations with this formula require Kelvin temperatures. Real engines have efficiencies that are lower than this theoretical best-case efficiency.

EXAMPLE 19.3

Efficiency of a Heat Engine

An engine operates between 27°C and 127°C. Determine its theoretical efficiency.

WHAT'S THE TRICK?

When temperatures are given, work in the Kelvin scale and use the following formula:

$$e = \frac{(400)-(300)}{(400)} = 0.25 = 25\%$$

ENTROPY

The natural flow of heat is from high temperature to low temperature. Systems with different temperatures contain particles that are moving at different average speeds. When these systems mix, all types of collisions occur. However, mathematical probability drives the direction of heat transfer. The warmer system has more fast-moving particles, while the cooler system has more slow-moving particles. Therefore, collisions between fast particles and slow particles are more likely. In these collisions, energy is transferred from the fast particles to the slow particles. Energy continues to transfer until both systems have the same average particle speeds. Therefore, the equilibrium state is the statistically most probable state that can occur.

Entropy is a way to quantify the probability of finding a system in a particular state. Figure 19.4(a), below, shows two gases separated by a movable wall. The gas in the left compartment initially has more molecules than the gas in the right compartment. Figures 19.4(b) and 19.4(c) show two possible states that the system can be found in after the wall has been removed.

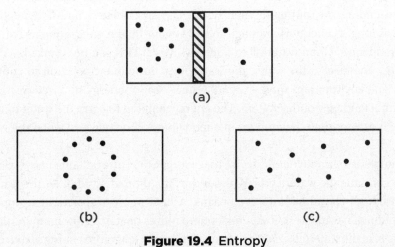

(a)

(b)   (c)

Figure 19.4 Entropy

Figures 19.4(b) and 19.4(c) are snapshots of the system at an instant in time. The gas particles are in random motion, and both of the resulting diagrams are actually possible. However, the highly organized pattern in Figure 19.4(b) is about as likely as winning millions in a lottery. It has a low probability, and therefore it has low entropy. Random states like the one seen in Figure 19.4(c) are statistically more likely to occur. They are more probable and thus have higher entropy. Probability drives systems toward random and disordered equilibrium states with higher entropy. Therefore, entropy is associated with messiness or randomness. The more random a system is, the greater its entropy. In addition, the natural trend is toward equilibrium and toward greater entropy.

Look again at Figure 19.4(c), the most probable result when the wall was removed. What is the likelihood that the molecules in Figure 19.4(c) will spontaneously separate back into their original compartments, returning to their locations in Figure 19.4(a), and then remain in those positions without a wall to hold them? This scenario is not likely and does not take place. This means that removing the wall is an irreversible step. Entropy drives thermodynamic processes toward equilibrium. It ensures that isolated (no environmental interference) thermal processes are irreversible.

Heat engines increase entropy. When gasoline is added into a car, the gasoline is actually large, organized molecules that are, in liquid form, stored in the car's gas tank. During combustion the gasoline is converted into many small molecules of water vapor and carbon dioxide. Now in gaseous form, they exit the car at the exhaust pipe and quickly mix with other molecules in the atmosphere. When the gas tank is empty it is very unlikely that these exact molecules of water and carbon dioxide will join back together to create liquid gasoline in the car's gas tank. Combustion increases messiness, and it is irreversible, resulting in an increase in entropy.

SECOND LAW OF THERMODYNAMICS

The **second law of thermodynamics** addresses entropy, the drive toward equilibrium and its irreversible nature.

- The entropy of an isolated system cannot decrease.
- The entropy of isolated systems always increases until the system reaches equilibrium.
- Once at equilibrium, the entropy of the system remains constant.

An **isolated system** is a system that follows natural tendencies and does not interact with the surrounding environment. When systems are not isolated, natural tendencies may be reversed as long as energy is supplied to the system by the environment. For example, the natural tendency is for a waterfall to flow downward. This can be reversed. As the Sun's radiant energy is added to the water, the water rises through evaporation to continue the water cycle. The environment must expend a great deal of energy in an effort to interfere with the natural tendency of the system. The energy required to carry the water to the top of the waterfall is greater than the amount of energy that will be released when the water falls on its own.

A **heat pump** is a thermodynamic device that acts like the Sun's radiant energy, evaporating the water to continue the water cycle. Without the Sun, ultimately all of Earth's water would flow down from all elevations into the oceans. A heat pump transfers heat opposite the natural direction of natural heat flow. A heat pump moves heat from low temperature to high temperature. Like the waterfall analogy, a heat pump must add more energy to move heat the wrong direction than would be transferred if heat moved from a region of high temperature to a region of low temperature on its own.

The following general trends are consequences of entropy and the second law of thermodynamics.

- The natural tendency is for systems to move to equilibrium and for entropy (disorder) to increase.
- When systems with different temperatures come into contact, heat flows spontaneously from the high-temperature region to the low-temperature region until thermal equilibrium is reached.
- Heat engines can never be 100 percent efficient.
- Heat pumps reverse entropy and move heat from a low-temperature region to a high-temperature region. This requires the addition of energy from the environment. The energy that must be added is greater than the energy that would be released if heat flowed normally.

IF YOU SEE

entropy

Disorder, messiness

The natural tendency is for entropy to increase as systems seek equilibrium.

SUMMARY

1. **INTERNAL ENERGY IS THE TOTAL ENERGY OF A SYSTEM.** The change in internal energy, ΔU, is more important than the actual amount of internal energy. The change in internal energy of a thermodynamic system is due only to changes in thermal energy, which is the total of the microscopic kinetic energies of the particles in the system. In thermodynamic systems, all other energies are constant and do not change.

2. **THE SYSTEM'S ENERGY CAN BE CHANGED THROUGH WORK AND HEAT.** Work is a mechanical change in energy that changes the volume of the gas by physically moving a piston. Heat is the microscopic transfer of thermal energy by touching the system with hot or cold heat reservoirs.

3. **THE FIRST LAW OF THERMODYNAMICS IS A STATEMENT OF CONSERVATION OF ENERGY.** The change in the internal energy of a thermodynamic system is equal to the energy transferred into or out of the system by heat and work: $\Delta U = Q + W$.

4. **HEAT ENGINES CAN NEVER BE 100 PERCENT EFFICIENT.** A heat engine adds heat (Q_H) at high temperature (T_H) and expends heat (Q_C) at low temperature (T_C). The difference between the heat added and the heat removed is an amount of net work (W_{net}). The efficiency of the heat engine is the ratio of the net work (output) to the energy added at high temperature (input).

5. **ENTROPY IS ASSOCIATED WITH MESSINESS AND DISORDER.** Entropy quantifies the likelihood that a system will be found in a particular state. Highly organized systems are less likely and have low entropy. In contrast, random, disorganized systems are more likely and have high entropy. When systems interact, statistical probability drives the entire system toward the most likely state, which is equilibrium. Therefore, the natural trend is for entropy to increase. It can be reversed, but this requires the addition of energy from outside the system.

6. **THE SECOND LAW OF THERMODYNAMICS SETS THE NATURAL DIRECTION OF HEAT FLOW FROM HIGH TEMPERATURE TO LOW TEMPERATURE.** This flow of heat is driven by entropy. Reversing the flow of heat is possible but requires the addition of energy from the environment. The quantity of energy that must be added will be greater than the energy that would be released if the system were to run in the natural direction. A device reversing natural heat flow is known as a heat pump.

If You See	Try	Keep in Mind
Internal energy	Change in internal energy measures the change in the system's total energy. It is tied to temperature changes.	U increases: $+\Delta U$ and $+\Delta T$ U decreases: $-\Delta U$ and $-\Delta T$ Isothermal: constant temperature $\Delta U = 0$ and $\Delta T = 0$
Work	Mechanical change in energy involving changes in the volume of the gas.	On the gas: gas compressed $+W$ and $-\Delta V$ By the gas: gas expands $-W$ and $+\Delta V$ Isometric (isochoric): constant volume $W = 0$ and $\Delta V = 0$
Heat	Microscopic thermal energy transfer	Heat added: $+Q$ Heat removed: $-Q$ Adiabatic: no heat added or removed $Q = 0$
Problems involving internal energy, work, and heat	First law of thermodynamics $\Delta U = Q + W$	Use the variable trends and signs discussed above.
Efficiency	$$e = \left| \frac{W_{\text{net}}}{Q_H} \right|$$ $$e = \frac{T_H - T_C}{T_H}$$	T_H and T_C must be in kelvin. The second formula solves for a theoretical best-case efficiency.
Entropy	The most natural state for a system is one in which the particles are randomly disordered. This is when entropy is high. It is associated with messiness or randomness.	The natural tendency is for the entropy of a system to increase until it reaches equilibrium.
Second law of thermodynamics	For an isolated system, the natural direction of heat flow is from high temperature to low temperature.	Heat engines can never be 100 percent efficient. Heat pumps can make heat flow backward, but this requires a lot of energy input from the environment.

PRACTICE EXERCISES

1. In an isothermal process

   (A) the temperature is zero
   (B) the volume change is zero
   (C) the work is zero
   (D) the heat added is zero
   (E) the internal energy change is zero

2. In an adiabatic process, there is no

   (A) change in pressure
   (B) change in volume
   (C) change in temperature
   (D) change in internal energy
   (E) heat added or removed

3. The first law of thermodynamics is essentially a statement of

   (A) entropy
   (B) heat transfer
   (C) conservation of energy
   (D) thermal energy
   (E) internal energy

4. During a thermodynamic process, 500 joules of heat are removed from a gas while 300 joules of work are done on the gas. Determine the change in internal energy.

   (A) −800 J
   (B) −200 J
   (C)   0 J
   (D) 200 J
   (E) 800 J

5. During an adiabatic process, the internal energy of the gas increases by 1,600 joules. Which statement is correct?

   (A) 1,600 J of work are done on the gas, and the temperature increases.
   (B) 1,600 J of work are done by the gas, and the temperature increases.
   (C) 1,600 J of work are done on the gas, and the temperature decreases.
   (D) 1,600 J of work are done by the gas, and the temperature decreases.
   (E) No work is done on the gas, and the temperature increases.

6. During an isometric process, 450 joules of heat are removed from a trapped gas. Which statement is true?

   (A) The volume of the gas is increasing.
   (B) The volume of the gas is decreasing.
   (C) The temperature of the gas is increasing.
   (D) The temperature of the gas is decreasing.
   (E) Work = 450 J

7. A heat engine operates between 25°C and 100°C. The theoretical efficiency is most nearly

   (A) 10%
   (B) 20%
   (C) 50%
   (D) 70%
   (E) 80%

8. A heat engine absorbs 160 J of heat and exhausts 120 J to a cold reservoir. What is the efficiency of this engine?

   (A) 25%
   (B) 40%
   (C) 50%
   (D) 75%
   (E) 80%

9. The entropy of isolated systems

   (A) is zero
   (B) is one
   (C) decreases
   (D) remains constant
   (E) increases

10. Which of these is a consequence of the second law of thermodynamics?

   (A) The entropy of isolated systems always increases.
   (B) The natural direction of heat flow is from hot to cold.
   (C) No heat engine can ever be 100 percent efficient.
   (D) A heat pump requires energy from outside the system to operate.
   (E) All of these

ANSWERS EXPLAINED

	Key Words	Needed for Solution	Now Solve It
1. **(E)**	Isothermal	Knowledge/definitions	In an isothermal process, the temperature remains constant and the change in internal energy is zero. Answer A may catch the eye, but it is wrong. It states that the temperature is zero, not that the change is zero.
2. **(E)**	Adiabatic	Knowledge/definitions	In an adiabatic process, no heat is added to or removed from a system.
3. **(C)**	First law	Knowledge/definitions	The first law of thermodynamics is a statement of conservation of energy for systems involving thermal energy.
4. **(B)**	Heat removed . . . work done on the gas . . . change in internal energy	First law of thermodynamics $$\Delta U = Q + W$$	Heat removed is negative. Work done on the gas is positive, and the gas is compressed. $$\Delta U = Q + W$$ $$\Delta U = (-500 \text{ J}) + (300 \text{ J})$$ $$\Delta U = -200 \text{ J}$$
5. **(A)**	Adiabatic . . . internal energy	Knowledge/definitions First law of thermodynamics $$\Delta U = Q + W$$	In an adiabatic process, no heat is added or removed. $$\Delta U = Q + W$$ $$(1{,}600 \text{ J}) = (0 \text{ J}) + W$$ $$W = 1{,}600 \text{ J}$$ Positive work is done on the gas. Since internal energy increased, the temperature increased.
6. **(D)**	Isometric . . . heat removed	Knowledge/definitions First law of thermodynamics $$\Delta U = Q + W$$	In an isometric process, volume is constant and work is zero. $$U = Q + W$$ $$U = (-450 \text{ J}) + (0 \text{ J})$$ $$U = -450 \text{ J}$$ When internal energy decreases, temperature decreases.
7. **(B)**	25°C and 100°C . . . efficiency	$$e = \frac{T_H - T_C}{T_H}$$	This requires Kelvin temperatures. $$e = \frac{(373 \text{ K}) - (298 \text{ K})}{(373 \text{ K})} = 0.20 = 20\%$$ Note that answer E uses the wrong temperature scale.

	Key Words	Needed for Solution	Now Solve It
8. **(A)**	Heat engine absorbs . . . exhausts . . . efficiency	$W_{net} = Q_H - Q_C$ $e = \left\| \dfrac{W_{net}}{Q_H} \right\|$	Net work is the difference between heat absorbed and heat exhausted. $W_{net} = Q_H - Q_C = (160 \text{ J}) - (120 \text{ J}) = 40 \text{ J}$ $e = \left\| \dfrac{W_{net}}{Q_H} \right\| = \left\| \dfrac{40 \text{ J}}{160 \text{ J}} \right\| = 0.25 = 25\%$
9. **(E)**	Entropy	Knowledge/definitions	Isolated systems follow natural trends. The environment cannot interfere with them. Entropy for an isolated system always increases.
10. **(E)**	Second law	Knowledge/definitions	All of the answers are consequences of the second law of thermodynamics.

Atomic and Quantum Phenomena 20

→ **DEVELOPMENT OF THE ATOMIC THEORY**

→ **ENERGY-LEVEL TRANSITIONS**

→ **IONIZATION ENERGY/WORK FUNCTION**

→ **PHOTOELECTRIC EFFECT**

The theory that matter is composed of indivisible units, known as atoms, dates back to the ancient Greeks. At the turn of the nineteenth century, a series of experiments led scientists to modify many of their previous ideas about the composition and behavior of atoms. This chapter will discuss several of the experiments leading to these modifications and explain the current understanding of atoms and their quantum behavior. Emphasis will be placed on key experiments, experimenters, and the atomic model that is most likely to be on the SAT Subject Test in Physics. Table 20.1 lists the variables that will be used in this chapter. The topics covered in this chapter will do the following:

- Review atomic-model history through key experiments and experimenters.
- Define light quantum (photons) and its role in transmitting energy.
- Explain energy-level diagrams in illustrating quantum phenomena.
- Define the photoelectric effect.

Table 20.1 Variables Used for Atomic and Quantum Phenomena

New Variables	Units
K_{max} = maximum kinetic energy	J (joules)
ϕ = work function	J (joules)
h = Planck's constant	J • s (joule • seconds)

DEVELOPMENT OF THE ATOMIC THEORY

The ancient Greeks first proposed the word *atom* as the name for an indivisible unit of matter. Several observations and prominent experiments have led to a greater understanding of atoms. The following sections include major highlights and scientists involved in the development of the atomic theory.

J. J. Thomson

In 1897, J. J. Thomson discovered that even atoms themselves were divisible when he discovered the electron. Thomson knew that matter had an overall neutral charge, so he theorized that an atom of matter would be a mixture of positive and negative components.

His model, often referred to as the "raisin cake model" or "plum-pudding model," visualized the atom as containing positive and negative charges that were distributed throughout the interior of the atom.

Figure 20.1 Plum-pudding model of an atom

Ernest Rutherford

Ernest Rutherford, a student of Thomson, began to experiment with what appeared to be charged rays emanating from crystals of uranium. He named these rays alpha and beta. He also determined that they consist of streams of particles. Rutherford determined that the alpha particle is, in fact, a doubly charged positive ion, which we now know to be a helium nucleus consisting of two protons and two neutrons. The beta ray consists of a negative particle.

To investigate the interior of the atom, Rutherford used a radioactive source to fire alpha particles through a thin sheet of gold foil. A screen sensitive to alpha particles surrounded the gold foil to record the strikes of the alpha particles. If Thomson's model was correct, the even distribution of positive and negative components of gold atoms should not greatly affect the path of the positive alpha particles. As a result, Rutherford expected the fairly heavy alpha particles to pass through the gold atoms with little deflection.

After the experiment was concluded, Rutherford was surprised at the result. Although most of the alpha particles passed through the gold foil as expected, some of the alpha particles were deflected at extreme angles. In fact, a few nearly reversed direction completely. Rutherford likened this to shooting a cannonball at a piece of tissue paper and watching it bounce back. These results prompted Rutherford to propose that the positive region inside the gold atoms (protons were not yet discovered) was actually concentrated in a very tiny nucleus at its center, as illustrated in Figure 20.2.

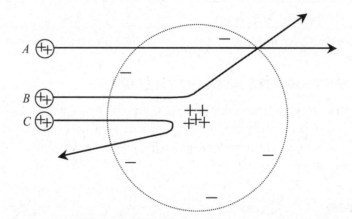

Figure 20.2 Rutherford's gold-foil experiment

Most alpha particles, like the one labeled *A* in Figure 20.2, passed through the gold foil without being deflected. A small number of positive alpha particles were deflected at extreme angles, such as particle *B* in Figure 20.2. These particles had to pass close to a dense positive region in order to be deflected in this manner. The most amazing results were those similar to *C* in Figure 20.2. These alpha particles nearly bounced back and must have encountered a very dense positive region head on. Rutherford was able to use the particle traces to map the atom. He proposed that the atom was mostly empty space, allowing the majority of particles through the gold atoms without interference. The deflected traces revealed a very small but extremely positive region at the center of the atom. The results led Rutherford to propose the following characteristics for atoms.

- Atoms are mostly empty space.
- The majority of the mass of an atom is concentrated in a tiny central nucleus.
- The central nucleus is positively charged.
- The electrons orbit the nucleus in a manner similar to planets orbiting the Sun.

Rutherford's observations provided new insights into the structure of the atom. However, many questions needed to be resolved. For example, how can the nucleus stay together with the electrostatic repulsive forces that must be acting on the positive charges? How can an orbiting electron be sustained indefinitely around the nucleus of the atom? Theoretically, an electron should lose energy and then fall into the positive nucleus. Obviously, this wasn't happening. If it did, then all matter would have disappeared long ago. More research was needed.

Spectroscopy

Spectroscopy is the study of the light spectrum emitted by luminous objects, such as the Sun, and the light emitted by gas discharge tubes. A gas discharge tube emits light by using a high voltage to shoot electrons through atoms in gaseous form. The electrons collide with the atoms, adding energy to the atoms. Eventually, the excess energy is lost. In the process, light with specific wavelengths is emitted. Each element emits a unique spectrum of light consisting of exact wavelengths. When viewed through a diffraction grating, the colors of the emitted light appear as a series of discrete lines. Each line in the pattern has a single wavelength and color. The patterns formed by these lines differ for each element, and they act like a fingerprint or bar code identifying each element.

Max Planck

In 1900, Max Planck attempted to explain the color spectrum seen when substances were heated to the point of glowing. The light emitted was due to atomic oscillations. The patterns seen could be explained only if Planck assumed that the atomic oscillations had very specific quantities of energy. He suggested that the oscillations were quantized (came in specific quantities). He was able to develop a mathematical relationship for these oscillations. He determined that it was based on a constant, now known as Planck's constant, h.

$$h = 6.63 \times 10^{-34} \text{ joule} \bullet \text{seconds} = 4.14 \times 10^{-15} \text{ electron volt} \bullet \text{seconds}$$

IF YOU SEE
Ernest Rutherford

Gold-foil experiment

The atom is mostly empty space with a tiny region, the nucleus, containing all the positive charge. Electrons orbit at a great distance from the nucleus.

Electron Volts

Planck's constant may be given in units of joule-seconds and/or units of electron volt–seconds. An **electron volt** (eV) is an alternate unit of energy. Atoms are incredibly small, so working with joules of energy is not ideal. It is like measuring the length of a pen with a mile stick. The electron volt is a unit that is scaled to match the size of an atom. The conversion between joules and electron volts involves the same numerical value as the charge on an electron.

$$1 \text{ electron volt} = 1.6 \times 10^{-19} \text{ joules}$$

Albert Einstein

In 1905, Albert Einstein applied Planck's idea of quantization to electromagnetic radiation. He suggested that light is quantized and that it consists of massless, particle-like packets that have a specific quantum (quantity) of energy. These packets of light came to be known as **photons**. Einstein clarified the relationship suggested by Max Planck. Einstein determined that the energy of a photon, E, is the product of frequency, f, and Planck's constant, h.

$$E = hf$$

Chapter 15, "Waves," showed that wave speed was a function of wavelength and frequency and that the speed of light in a vacuum is c.

$$v = f\lambda$$
$$c = f\lambda$$

The second formula can be rearranged to solve for frequency, $f = c/\lambda$. Then, substitute this into the equation $E = hf$. The result is a useful equation that relates the energy of a photon of light to its wavelength.

$$E = \frac{hc}{\lambda}$$

IF YOU SEE
photon,
energy,
frequency,
wavelength

$E = hf$

$E = \dfrac{hc}{\lambda}$

Frequency is directly proportional to energy. Wavelength is inversely proportional to energy.

EXAMPLE 20.1

Energy of Photons

How are the frequency and wavelength affected when the energy of photons is doubled?

WHAT'S THE TRICK?

The frequency of light is directly proportional to energy. High-frequency photons have more energy. The wavelength of light is inversely proportional to energy. Short wavelengths of light have more energy.

$E = hf$	and	$E = \dfrac{hc}{\lambda}$
$f = \dfrac{E}{h}$	and	$\lambda = \dfrac{hc}{E}$
$(2f) = \dfrac{(2E)}{h}$	and	$\left(\dfrac{1}{2}\lambda\right) = \dfrac{hc}{(2E)}$

Doubling photon energy doubles frequency and halves wavelength.

Niels Bohr

In 1913, Niels Bohr used Einstein's light quanta to suggest a model of the atom that explained why electrons do not fall into the nucleus and why the light emitted from excited atoms produces the observed emission spectra. Bohr theorized that the electrons of atoms could occupy only exact **energy levels**. An energy level is a specific energy state with an exact quantum (quantity) of energy.

The **absorption** and **emission** of specific wavelengths of light are related to the energy difference between these energy levels. When atoms absorb light, the absorbed photons combine with the electrons. The energy of the photons adds to the energy of the electrons. Electrons with this added energy are said to be **excited** and must occupy a higher energy level. When atoms emit light, the excited electrons lose energy by emitting photons. The less energetic electrons must now occupy lower energy levels. The light that is absorbed and emitted is restricted by the energy levels in the atom. Only photons with quanta (quantities) of energy matching the exact difference between energy levels can be absorbed and emitted by an atom. Every atom has unique energy levels, and the difference between energy levels varies from atom to atom. Photons emitted from different elements experience different energy changes, resulting in unique wavelengths and colors. As a result, each element emits a unique color spectrum.

Bohr borrowed elements of Einstein's idea of light quanta and merged them with atomic spectra to propose a quantum mechanical model of atomic structure. The addition of exact energy levels provided the stability that the Rutherford model lacked. The next section details the absorption and emission of light according to the Bohr model of the atom.

IF YOU SEE
Niels Bohr

energy levels

Electrons occupy exact energy levels. Absorbed and emitted photons of light move electrons between energy levels.

ENERGY-LEVEL TRANSITIONS

Niels Bohr's research involved the simplest atom possible, the hydrogen atom. This atom consists of a single proton and a single electron. A partial energy-level diagram of the hydrogen atom is shown in Figure 20.3.

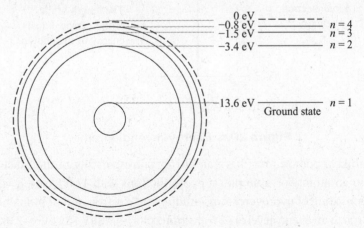

Figure 20.3 Energy levels of an atom

Although the actual model of the atom is not as simple as that shown in Figure 20.3, the Bohr model is still used in energy-level problems. The left side of the sketch shows a simplified Bohr model of a hydrogen atom. It shows the energy levels as circles similar to the orbits of planets in the solar system. The right side of the diagram shows the corresponding **energy-level diagram** of the atom. In an energy-level diagram, horizontal lines are used to portray the energy levels. The **ground state** is the lowest level an electron can occupy. It is numbered as the first energy level ($n = 1$). All the higher energy states are known collectively as the excited states. They are numbered from the ground state to the edge of the atom, $n = 2$, $n = 3$, etc. Be careful with the excited states. Since the ground state is $n = 1$, the first excited state is $n = 2$ and the second excited state is $n = 3$. The energy levels typically have negative values and are often measured in electron volts. The edge of the atom has a value of zero electron volts. The ground state for hydrogen has an energy of –13.6 electron volts. Moving deeper into the atom is similar to taking an elevator ride below ground. The floor numbers become larger the farther down the elevator travels, and the elevator is moving in the negative direction.

Absorption

Absorption occurs when a photon of light with the correct amount of energy enters an atom and is absorbed by an electron. The energy of the photon adds to the energy of the electron, creating an excited electron called a **photoelectron**. The high-energy photoelectron must move to a higher energy level. If the hydrogen atom is radiated by light with 12.1 electron volts of energy, the electron will absorb the photon of light and their energies will add.

$$E_{\text{electron initial}} + E_{\text{photon}} = E_{\text{electron final}} = (-13.6 \text{ eV}) + (12.1 \text{ eV}) = -1.5 \text{ eV}$$

IF YOU SEE

absorption

Add the photon energy to the initial electron energy to find the new energy level.

Absorption is drawn as an upward arrow in an energy-level diagram.

Absorptions are indicated with upwardly drawn arrows in energy-level diagrams. The absorption calculated above is depicted in Figure 20.4. The photoelectron formed in this absorption moves upward 12.1 electron volts from the ground state to the third energy level, $n = 3$ (the second excited state).

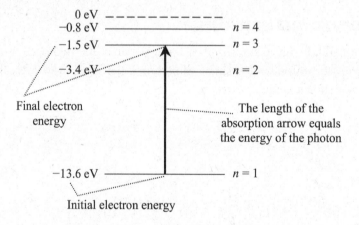

Figure 20.4 Energy-level diagram

To be absorbed, a photon must have an energy corresponding to the exact difference in energy levels in an atom. For example, suppose photons with 11.0 electron volts of energy radiate a gas consisting of hydrogen atoms. Adding the electron and photon energies should move the electron to an energy level of –2.6 electron volts (–13.6 eV + 11 eV = –2.6 eV). However, this energy level does not exist. The absorption cannot take place, and these mismatched photons pass through the hydrogen atoms.

Emission

Nature prefers low-energy states. So, electrons in high-energy levels will spontaneously move to lower energy levels until they finally reach the ground state. When electrons drop to a lower energy level, they lose energy. The energy they lose is given off as a photon of light. An

emission is the light given off by an atom when electrons drop to lower energy levels. If an electron in the hydrogen atom at energy level two ($n = 2$) drops to the ground state ($n = 1$), it will lose 10.2 electron volts of energy.

$$E_{photon} = E_2 - E_1 = (-3.4 \text{ eV}) - (-13.6 \text{ eV}) = 10.2 \text{ eV}$$

The transitions to lower energy levels are haphazard. An electron in the third energy level might return all the way to the ground state in a single step. In a different atom, another electron in the third energy level might first drop to the second energy level and then drop to the ground state. Energy-level problems involve samples that contain vast quantities of atoms. All the possible drops between the energy levels take place in many different atoms simultaneously. Emissions of light are pictured as downward arrows in energy-level diagrams. Figure 20.5(a) shows all the possible emissions due to energy-level drops from energy level three. Figure 20.5(b) shows the possible emissions due to energy drops from energy level four.

IF YOU SEE

emission

All possible energy drops to the ground state can occur. The energy of emitted photons is the difference between energy levels.

Emissions are drawn as downward arrows in an energy-level diagram.

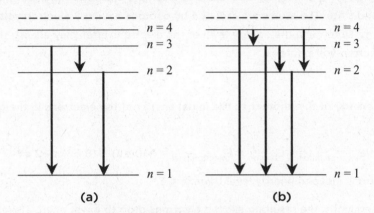

Figure 20.5 Emission of light

Think of the energy-level diagram as a series of stairs and the electron as a ball on the stairs. When the electron is given enough energy to move to the high stair, such as the fourth energy level ($n = 4$), it can then fall to the bottom floor ($n = 1$) in a variety of ways. The ball can fall from the fourth stair to the first without hitting any of the other stairs. Or, it can hit any combination of the stairs along the way. Each drop releases a photon with an amount of energy equal to the energy difference between the stairs.

To find the energy of a photon, simply subtract the energy levels between which the electron is moving. Once the energy of the photon (E) is found, it can be used in the following formulas to determine the frequency (f) and wavelength (λ) of the emitted light.

$$E = hf \quad \text{and} \quad E = \frac{hc}{\lambda}$$

Each wavelength of light has a unique color, which produces the variety of colors seen in color spectra. The Sun is composed of many types of atoms with electrons dropping through all kinds of energy differences. As a result, a continuous spectrum of light is emitted from the Sun. However, an excited gas made of one type of atom will emit a discrete spectrum containing light matching only its energy-level differences. Keep in mind that many emitted wavelengths of light are not visible to the human eye. The hydrogen atom has two well-known light series that are produced when electrons drop down to a specific energy level.

- Balmer series: Transitions to $n = 2$, creating visible light.
- Lyman series: Transitions to $n = 1$, creating ultraviolet light (not visible).

EXAMPLE 20.2

Energy-Level Diagrams

$$0\ eV\ ----------$$
$$-1\ eV\ \underline{\hspace{4cm}}\quad n = 4$$
$$-2\ eV\ \underline{\hspace{4cm}}\quad n = 3$$
$$-5\ eV\ \underline{\hspace{4cm}}\quad n = 2$$
$$-10\ eV\ \underline{\hspace{4cm}}\quad n = 1$$

(A) The energy-level diagram above shows a sample of atoms initially in the ground state. The atoms are radiated by photons having 8 electron volts of energy. Determine the energy level of the electrons after they absorb the 8-electron-volt photons.

WHAT'S THE TRICK?

Add the energy of the photons to the initial energy of the electrons in the ground state.

$$E_{\text{electron initial}} + E_{\text{photon}} = E_{\text{electron final}} = (-10\ eV) + (8\ eV) = -2\ eV$$

The electrons move to energy level three, $n = 3$.

(B) Subsequently, the resulting excited electrons drop to lower energy levels, emitting photons of light. Determine all the possible energies of the emitted photons.

WHAT'S THE TRICK?

Determine every possible drop that can occur from $n = 3$.

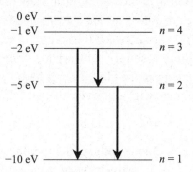

Now, subtract the different energy levels for each drop.

$$E_{\text{photon}} = E_3 - E_1 = (-2\ eV) - (-10\ eV) = 8\ eV$$
$$E_{\text{photon}} = E_3 - E_2 = (-2\ eV) - (-5\ eV) = 3\ eV$$
$$E_{\text{photon}} = E_2 - E_1 = (-5\ eV) - (-10\ eV) = 5\ eV$$

IONIZATION ENERGY/WORK FUNCTION

In energy-level problems, electrons receive just enough energy to reach a higher energy level inside the atom. However, if the energy of the incoming photons is greater than the energy difference between the ground state and the edge of the atom, then the electrons are ejected from the atom. In this process, the atom becomes a positive ion. The minimum energy required to accomplish this is known as the **ionization energy**. For atoms with electrons in the ground state, the ionization energy is equal to the absolute value of the ground-state energy. For hydrogen gas with a ground state of –13.6 electron volts, the ionization energy is equal to 13.6 electron volts. The ionization energy is the minimum energy needed to eject an electron and ionize an atom. The ionization energy is also known as the **work function**, ϕ.

PHOTOELECTRIC EFFECT

The photoelectric effect involves the ionization of atoms by striking them with photons that exceed the energy difference between the ground state and the edge of the atom. Figure 20.6 shows a hypothetical atom with a ground state of –10 electron volts that is radiated by photons with 12 electron volts of energy.

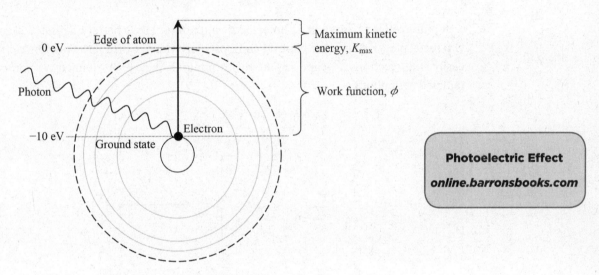

Figure 20.6 The photoelectric effect

If these energies are added in the same manner as in the previous section, the result is a positive energy instead of a negative energy.

$$E_{\text{electron initial}} + E_{\text{photon}} = E_{\text{electron final}} = (-10 \text{ eV}) + (12 \text{ eV}) = 2 \text{ eV}$$

Electrons with positive energies have been ejected from the atom. This electron will leave the atom. When it does, the electron will have 2 electron volts of energy. The ejected electron will be moving with this excess energy, which is known as the **maximum kinetic energy**, K_{max}, of the ejected electron. In order to be ejected, the electron first had to move from the ground state, –10 electron volts, to the edge of the atom, 0 electron volts. This required the addition of 10 electron volts of energy. The energy to move from the ground state to the edge of the atom is known as the work function, ϕ. The work function is essentially the absolute value of the ground-state energy. For the atom in Figure 20.6, the work function is:

$$\phi = |E_{\text{ground state}}| = |-10 \text{ eV}| = 10 \text{ eV}$$

The excess energy of the ejected electrons can be determined as follows:

$$K_{max} = E_{photon} - \phi$$

The energy of a photon is related to its frequency, $E_{photon} = hf$. This expression can be substituted into the previous equation to complete the equation for the photoelectric effect.

$$K_{max} = hf - \phi$$

Both versions of this formula may be encountered. In Figure 20.6, the energy of the incident photon was given. So, the first formula, $K_{max} = E_{photon} - \phi$, is used.

$$K_{max} = (12 \text{ eV}) - (10 \text{ eV}) = 2 \text{ eV}$$

What is the significance of moving ejected electrons? When certain metallic substances with low work functions are radiated with high-energy photons, countless electrons are ejected. These electrons are in motion. A lot of moving electrons make up a current. As a result, this phenomenon is a way to generate an electric current using photons of light. It is known as the **photoelectric effect** since it converts photon energy to electric energy. This is how electricity is generated using sunlight. The photoelectric effect is at the heart of solar power.

A **photocell** is a battery-like photoelectric apparatus. Like a battery, a photocell consists of two metal plates. One of the plates is composed of a metal with a low work function that will easily emit electrons, e, when radiated with photons. Figure 20.7 depicts a photocell being radiated with photons.

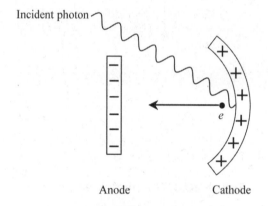

Incident photon

e

Anode   Cathode

Figure 20.7 Photocell

Each atom in the plate radiated by the photons ejects electrons, which move with kinetic energy, K_{max}, toward the opposite plate. The plate radiated with photons loses electrons and becomes positive. The plate receiving the excess electrons becomes negative. This creates a potential difference, V, between the plates. Energy is conserved during this process. The kinetic energy of the electrons, K_{max}, is converted into electric potential energy, U_E.

$$U_E = K_{max}$$

When a wire is connected between the plates, the potential difference provides the pressure to move electrons from the negative plate to the positive plate. Note that usually current is regarded as the flow of positive charges. However, the photoelectric effect focuses on the released electrons and tracks the actual electron flow, which is technically a negative current.

The **photoelectric-effect experiment** consists of a photocell, an ammeter, and a variable power supply connected in series, as shown in Figure 20.8.

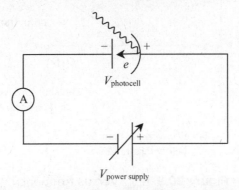

Figure 20.8 Photoelectric-effect experiment

The variable power supply (battery symbol with an arrow running through it) creates a second potential difference in addition to the potential difference produced in the photocell. The power supply is wired into the circuit backward, so its electric potential can cancel the electric potential generated by the photocell. This experiment is not about creating solar power. Instead, it is designed to measure and test the properties of the photoelectric effect. The ammeter records the amount of current flowing in the circuit. In the experiment, light, with different frequencies (energy) and intensities (brightness) are used to stimulate the photocell. When the variable power supply is adjusted so that the ammeter reads zero, no current is flowing in the circuit. This occurs when the potential (electric pressure) of the photocell, which is pushing electrons counterclockwise in the circuit, is equal to the potential of the power supply, which is pushing electrons clockwise in the circuit. The voltage displayed by the adjustable power supply is known as the stopping potential, V_s, since it stops current from flowing. The stopping potential of the power supply equals the potential induced in the photocell when the current is no longer flowing. When the ammeter reads zero:

$$V_{\text{photocell}} = V_s$$

Several key observations were made as the frequency of the photons incident on the photocell was steadily increased.

1. Photons with very low frequencies created no potential in the photocell, indicating that no electrons were ejected.
2. As the frequency of the photons was steadily increased, a threshold frequency was encountered that induced a potential in the photocell, indicating that electrons were starting to flow.
3. Increasing the frequency of the photons above the threshold frequency increased the potential of the photocell and the energy of the emitted electrons. The increase in the energy of the emitted electrons was linear and matched the following equation:

$$K_{\text{max}} = hf - \phi$$

The results of the photoelectric-effect experiment are summarized in the graph of electron energy versus photon frequency shown in Figure 20.9.

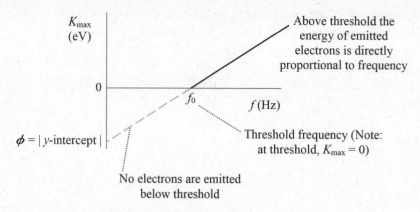

Figure 20.9 K_{max} **versus frequency**

The graph of the photoelectric effect is a linear graph of the kinetic-energy equation. If it is compared with the equation for a line, several key facts stand out.

Photoelectric effect: $K_{max} = hf - \phi$

Equation of a line: $y = mx + b$

IF YOU SEE

photoelectric effect

Light energy is converted into electric energy

$K_{max} = hf - \phi$

$U_E = K_{max}$

Increasing the photon frequency increases the energy of emitted electrons and electric potential.

Increasing the intensity of light increases the number of electrons emitted and the current.

- The slope of the graph is Planck's constant, h.
- The work function, ϕ, has the same value as the y-intercept but has the opposite sign. The y-intercept of the photoelectric effect is negative, so the work function is the absolute value of the y-intercept.
- The threshold frequency occurs at a point where K_{max} is equal to zero. This fact can be used to determine the work function, ϕ, when given the threshold frequency, f_0.

$$K_{max} = hf - \phi$$
$$(0) = hf_0 - \phi$$
$$\phi = hf_0$$

In another experiment, the photon frequency was held at a constant value. That value was capable of ejecting photons. Instead, the intensity of light was varied. This experiment yielded different results.

1. Increasing the intensity of light increases the number of photons incident on the photocell. When more photons strike the photocell, more electrons are ejected.
2. Increasing the light intensity does not change the voltage of the photocell or the energy of the ejected electrons.
3. If the power supply is adjusted to zero, the voltage of the photocell will cause a current to flow. Increasing the intensity of light increases the current flowing in the circuit.

The observations recorded in the photoelectric-effect experiment indicate that light acts as a particle rather than as a wave.

SUMMARY

1. **THEORIES ON ATOMIC STRUCTURE AND ITS QUANTUM NATURE WERE ADVANCED IN A SERIES OF EXPERIMENTS AT THE TURN OF THE TWENTIETH CENTURY.** Rutherford's gold-foil experiment demonstrated that atoms were a small, positive, central nucleus with a great deal of empty space surrounded by negative electrons. Bohr's quantum ideas integrated Einstein's light-quanta theories to support Rutherford's planetary model of atoms with stable (quantum) states known as energy levels.

2. **LIGHT QUANTA ARE DISCRETE PACKETS OF ENERGY.** Einstein proposed that light energy is transmitted through the vacuum of space in discrete quantities. The photon, as it later came to be known, led to an understanding of the wave-particle nature of light. The frequency of a photon is directly proportional to the energy it transmits.

3. **ENERGY-LEVEL DIAGRAMS ILLUSTRATE THE ABSORPTION AND EMISSION OF LIGHT.** When an atom absorbs light, the energy of the photon is added to the energy of the electron. The now-excited electron occupies a higher energy level. Electrons prefer to be in the ground state and spontaneously lose energy, moving to lower energy levels. The lost energy leaves the atom as photons of light. The energy of the photons emitted is equal to the energy drop experienced by the electrons.

4. **LIGHT CAUSING AN ELECTRICAL DISCHARGE IS KNOWN AS THE PHOTO-ELECTRIC EFFECT.** Light with a minimum threshold frequency can be used to induce the emission of electrons from a piece of metal. The emitted electrons can be collected on a second piece of metal, creating a potential difference between the two plates. This battery-like apparatus is known as a photocell, which will create a current when wired into a circuit.

5. **THE PHOTOELECTRIC EXPERIMENT REVEALED KEY OBSERVATIONS REGARDING THE PHOTOELECTRIC EFFECT.** Adjusting the frequency of light results in the release of electrons only above a threshold frequency. Above this frequency, the energy of emitted electrons is proportional to frequency. Adjusting the intensity of light affects the number of emitted electrons but does not affect their energy. The photoelectric-effect experiment is evidence that light has a particle characteristic.

If You See	Try	Keep in Mind
Ernest Rutherford	Knowledge/definitions Fired alpha particles (helium nuclei) at a thin piece of gold foil	Atoms are mostly empty space. All the positive charge is located at their center, the nucleus.
Photon energy, frequency, and/or wavelength	$E = hf$ $E = \dfrac{hc}{\lambda}$	Frequency is directly proportional to energy, and wavelength is inversely proportional to energy.
Niels Bohr	Proposed energy levels. Used Einstein's quantized view of light to explain light spectra and why electrons do not fall into the nucleus.	Electrons occupy exact energy levels. Photons of light absorbed and emitted by atoms have exact energies matching the difference between energy levels.
Absorption of photons	Add the energy of the photon to the energy of the electron	Absorptions are drawn as upward arrows in energy-level diagrams.
Emission of photons	Electrons can experience a variety of drops on their way to the ground state. Each drop releases a different amount of energy and a different photon of light.	Emissions are drawn as downward arrows in energy-level diagrams.
Photoelectric effect	Light energy is converted into electric energy $K_{\max} = hf - \phi$ $U_E = K_{\max}$	Increasing photon energy above a threshold increases the energy of emitted electrons and the electric potential of photocells. Increasing light intensity increases the number of emitted electrons and current.

PRACTICE EXERCISES

1. An alpha particle is the nucleus of which atom?

   (A) Carbon-14
   (B) Hydrogen
   (C) Helium
   (D) Uranium
   (E) Plutonium

2. The Rutherford gold-foil experiment demonstrated

   I. the plum-pudding model
   II. atoms are mostly empty space
   III. electrons occupy specific energy levels

   (A) I only
   (B) II only
   (C) III only
   (D) I and III only
   (E) II and III only

3. In the Bohr model of the atom,

   I. electrons occupy energy levels with exact quantities of energy
   II. the photoelectric effect is described
   III. the absorption and emission of light spectra are predicted

   (A) I only
   (B) II only
   (C) III only
   (D) I and III only
   (E) II and III only

Questions 4–5
Use the energy-level diagram for the atoms of a substance to answer the next two questions.

```
 0 eV  – – – – – – – – – – –
–1 eV  ————————————  n = 4
–2 eV  ————————————  n = 3


–5 eV  ————————————  n = 2




–10 eV ————————————  n = 1
```

4. Which electron transition will result in the emission of a photon with the longest wavelength?

   (A) $n = 4$ to $n = 3$
   (B) $n = 4$ to $n = 2$
   (C) $n = 4$ to $n = 1$
   (D) $n = 3$ to $n = 2$
   (E) $n = 3$ to $n = 1$

5. What would be the frequency of a photon created when an electron moves from quantum state $n = 4$ to quantum state $n = 2$?

   (A) $\dfrac{h}{5}$

   (B) $\dfrac{h}{4}$

   (C) $\dfrac{2}{h}$

   (D) $\dfrac{4}{h}$

   (E) $\dfrac{5}{h}$

6. The photoelectric effect provided experimental evidence that light

   (A) has a wave characteristic
   (B) is a transverse wave
   (C) can diffract
   (D) can constructively and destructively interfere
   (E) has a particle characteristic

7. In a photoelectric experiment, the frequency of light is steadily increased. Which statement below is NOT correct?

(A) Below the threshold frequency, no electrons are emitted.
(B) Above the threshold frequency, electrons are emitted.
(C) Increasing the frequency of light increases the energy of the emitted electrons.
(D) Increasing the frequency of light increases the potential difference of the photocell.
(E) Increasing the frequency of light increases the induced current.

8. In a photoelectric experiment where photons are being emitted, the intensity of light is steadily increased. Increasing the intensity of light striking the photocell will

(A) increase the threshold frequency, making it more difficult to eject photons
(B) decrease the threshold frequency, making it easier to eject photons
(C) increase the energy of individual emitted electrons
(D) decrease the energy of individual emitted electrons
(E) increase the number of electrons emitted

Questions 9–10

The graph below depicts the maximum kinetic energy of emitted electrons as a function of frequency for a photoelectric process.

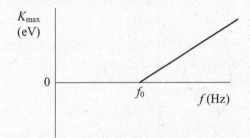

9. What does the slope of the graphed function represent?

(A) Speed of light, c
(B) Electron volt, eV
(C) Planck's constant, h
(D) Electric potential, V
(E) Wavelength, λ

10. Determine the work function, ϕ, for this metal.

(A) $\dfrac{h}{f_0}$

(B) hf_0

(C) $\sqrt{2}\,hf_0$

(D) f_0

(E) $\dfrac{f_0}{h}$

ANSWERS EXPLAINED

	Key Words	Needed for Solution	Now Solve It
1. **(C)**	Alpha particle	Knowledge/definitions	An alpha particle is a helium nucleus consisting of 2 protons and 2 electrons. It is an ion with a +2 charge.
2. **(B)**	Rutherford	Knowledge/definitions	The atom is mostly empty space. II is true. Answer I is Thomson's atomic model, and answer III is a characteristic of Bohr's atomic model.
3. **(D)**	Bohr	Knowledge/definitions	Answers I and III are elements of the Bohr model. The Bohr model was concerned with the interior structure of the atom, not electrons emitted by the photoelectric effect.
4. **(A)**	Emission . . . longest wavelength	Photon energy is the difference between energy levels: $$E = \frac{hc}{\lambda}$$	Be careful. Students see long wavelength and they think of big energy drops. However, energy is inversely proportional to wavelength. Long wavelengths result from small energy changes. The smallest energy change occurs when electrons move from $n = 4$ to $n = 3$, a drop of only 1 eV.
5. **(D)**	Frequency	Photon energy is the difference between energy levels: $$E = hf$$	When the electron transitions from $n = 4$ to $n = 2$, it loses 4 eV of energy, which is released as a photon. $$E_{photon} = E_4 - E_1 = (-5 \text{ eV}) - (-1 \text{ eV}) = 4 \text{ eV}$$ The frequency of the photon can be determined by rearranging $E = hf$. $$f = \frac{E}{h} = \frac{4}{h}$$
6. **(E)**	Photoelectric effect . . . experimental evidence	Knowledge/definitions	The photoelectric effect is one of several experiments indicating that light has a particle characteristic. (Note that Young's double-slit experiment is evidence of light's wave characteristic.)
7. **(E)**	Photoelectric experiment	Distinguishing between the effects of increasing photon frequency and increasing intensity	Only answers A–D are consistent with increasing the frequency of light. Answer E has to do with intensity (see the next question).

	Key Words	Needed for Solution	Now Solve It
8. **(E)**	Photoelectric experiment	Distinguishing between the effects of increasing photon frequency and increasing intensity	Increasing the intensity of light increases the rate at which photons strike the photocell. This does not affect the photocell's threshold frequency. Increasing the intensity does not change the energy of each photon and therefore does not change the energy of the individual electrons emitted. Increasing the rate of photons emitted does increase the number of electrons emitted.
9. **(C)**	Graph . . . maximum kinetic energy . . . frequency	$K_{max} = hf - \phi$ $y = mx + b$	The slope of this function is Planck's constant, h.
10. **(B)**	Work function	$K_{max} = hf - \phi$	The answers to this question all contain the threshold frequency. At the threshold frequency, the maximum kinetic energy is zero, $K_{max} = 0$. $$K_{max} = hf - \phi$$ $$(0) = hf_0 - \phi$$ $$\phi = hf_0$$

Nuclear Reactions

21

- → QUARKS
- → NUCLEONS
- → SUBATOMIC PARTICLES
- → ISOTOPES
- → THE STRONG FORCE
- → MASS-ENERGY EQUIVALENCE
- → RADIOACTIVE DECAY
- → FISSION AND FUSION

Nuclear reactions involve particles in the nucleus of an atom. Nuclear reactions encompass a variety of reactions that lead to changes in the composition of the nucleus. Altering the number of protons in the nucleus of an atom causes the atom to become an entirely different element, a process known as **transmutation**. This chapter will review the following concepts and particles related to nuclear reactions:

- Introduce quantities used to understand nuclear reactions.
- Identify subatomic particles involved in nuclear reactions.
- Define isotopes and examine their importance in nuclear reactions.
- Examine the nature of the forces acting on nuclear particles.
- Understand the process of radioactive decay.
- Differentiate between the key nuclear reactions of fission and fusion.

QUARKS

Quarks are the elementary particles that compose protons, neutrons, and other subatomic particles. Quarks possess mass, charge, and spin. As a result, they are subject to all the fundamental forces: gravitation, electromagnetism, strong nuclear force, and weak nuclear force. However, quarks possess fractional, non-integer charges. Protons comprise two up quarks, each with a +2/3 charge, and one down quark, with a –1/3 charge. Neutrons are composed of two down quarks and one up quark. Quarks must combine with one another and never exist by themselves. A variety of quarks combining in different ways create all the subatomic particles described by the **Standard Model** of particle physics.

NUCLEONS

Nuclear reactions are concerned with the nucleus of the atom and the particles it contains. The subatomic particles contained in the nucleus of an atom are known as **nucleons**. They consist of protons and neutrons. Both the mass and the number of these particles are important when analyzing nuclear reactions.

Atomic Mass Units

A specialized unit of mass known as the atomic mass unit (u) was devised to make working with the mass of fundamental particles easier. The mass of both the proton and neutron were originally thought to be the same. They were each assigned a mass of 1.0 atomic mass unit for simplicity. It has since been determined that the masses are very similar but that the neutron has a slightly greater mass. The modern definition of an atomic mass unit is $\frac{1}{12}$ the mass of a carbon-12 atom. By using this scale, a proton has a mass of 1.00728 atomic mass units, while a neutron has a mass of 1.00866 atomic mass units. For rough calculations these masses are still both rounded off to 1.0 atomic mass unit. Atomic mass units are a more convenient scale to measure mass when working with fundamental particles.

Atomic Number and Mass Number

When the symbol for an element is used in a nuclear reaction, the atomic number and the mass number are written as subscripts and superscripts preceding the element's symbol, as shown in Figure 21.1.

$$\text{Mass number} \longrightarrow \quad \text{Atomic number} \longrightarrow {}^{12}_{6}C$$

Figure 21.1 Atomic number and mass number

The **atomic number** is the number of protons in an atom. The number of protons defines an element. For example, all carbon atoms have 6 protons, so the atomic number of all carbon atoms is always 6. If a carbon atom gains or loses protons, it is no longer a carbon atom. Changing the number of protons causes a transmutation of the atom into a completely different type of element.

The **mass number** provides three different and important numerical values.

1. The mass number is the *number of protons plus neutrons.* In Figure 21.1, the mass number is 12. So, there are 12 protons and neutrons in a carbon-12 atom. Since the atomic number is the number of protons only, the number of neutrons can be deduced by subtracting the atomic number from the mass number. The carbon atom in Figure 21.1 has 6 neutrons (12 − 6 = 6).
2. The mass number is also the *atomic mass* (mass of a single atom) measured in atomic mass units. The carbon atom in Figure 21.1 has a mass of 12 atomic mass units.
3. In addition, the mass number is also the *molar mass* (mass of one mole of atoms) measured in grams per mole. This is widely used in chemistry but is not a factor in this chapter.

SUBATOMIC PARTICLES

Table 21.1 lists the fundamental particles most likely to be encountered in nuclear reactions. Knowing the mass and atomic numbers of these basic particles will be an asset when analyzing nuclear reactions. In addition, the charged particles will interact with electric and magnetic fields. You must know which particles are charged and whether that charge is positive or negative. Occasionally, questions ask how these particles move in electric and magnetic fields. Although you do not need to know the exact mass of the particles listed in the table, you should be able to list them in order of their masses.

Table 21.1 Subatomic Particles

Name	Symbol	Charge	Rest Mass (u)
Proton	$_1^1\text{p}$	+1	1.00728
Electron	$_{-1}^0\text{e}$	−1	0.00055
Neutron	$_0^1\text{n}$	0	1.00866
Neutrino	ν_e	0	Nearly zero
Antineutrino	$\bar{\nu}_e$	0	Nearly zero
Alpha particle (helium nucleus)	$_2^4\alpha$ or $_2^4\text{He}$	+2	4.00150
Beta particle (an electron emitted from the nucleus)	$_{-1}^0\beta$ or $_{-1}^0\text{e}$	−1	0.00055
Gamma radiation	$_0^0\gamma$	0	0

In addition to protons, neutrons, and electrons, three new and important subatomic particles plus gamma radiation have been listed in the table.

Neutrino

Neutrinos are often a product of radioactive decay and nuclear reactions. A neutrino, ν_e, is a neutral particle with very little (nearly zero) mass. Neutrinos are more common in the universe than electrons and protons. However, neutrinos do not interact well with matter, and they have insignificant mass. Neutrinos also have an antiparticle variant known as the antineutrino, $\bar{\nu}_e$. They have been included here as their symbols may be encountered in nuclear-reaction formulas.

Alpha Particle

An **alpha particle** is simply the nucleus of a helium atom without any electrons. The alpha particle has a mass number of 4. This means it contains 4 nucleons (protons + neutrons), and it has a mass of 4.0 atomic mass units. Of the particles listed in Table 21.1, the alpha particle is the most massive. The atomic number is 2, indicating that the alpha particle contains 2 protons. As a result, the alpha particle must also contain 2 neutrons. Since the alpha particle contains two protons and no electrons, it is positively charged. The charge of an alpha particle is equal to the charge of two protons ($+3.2 \times 10^{-19}$ coulombs). As a result, it interacts with electric and magnetic fields as would any positive charge.

Beta Particle

A **beta particle** is an electron produced when a neutron undergoes a transmutation to become a proton. In the process, the positive and negative charges cancel and the neutron becomes slightly more massive than the proton. Under the right conditions, a neutron may spontaneously divide and become a proton and an electron (plus an antineutrino, $\bar{\nu}_e$, which can be ignored).

$$_0^1\text{n} \rightarrow {}_1^1\text{p} + {}_{-1}^0\text{e} + \bar{\nu}_e$$

When this happens, the electron originates in the nucleus and not in the energy levels surrounding the nucleus, where ordinary electrons are found. This electron, known as a beta particle, is ejected from the atom with high energy.

Electrons and beta particles have too little mass to affect the atomic mass of an atom noticeably. Think of electrons as adding as much mass to an atom as eyelashes add to the mass of a person. The mass of electrons is therefore not included in the atomic mass. The mass number for an electron has a value of 0 as expected. However, the atomic number (number of protons) is shown as −1. Essentially, an electron is the negative of a proton. If this is the case, shouldn't all stable atoms have an atomic number of 0 since the protons and electrons cancel each other? The atomic number is used for nuclear reactions and for particles that are inside the nucleus. Electrons in the energy levels surrounding the nucleus are not inside the nucleus, and their atomic number is ignored. However, when electrons such as beta particles engage in nuclear reactions, their atomic number (−1) is important to balance the particles during the reaction. The electrical charge of a beta particle is the same as that of any electron (-1.6×10^{-19} coulombs). A beta particle interacts with electric and magnetic fields in the same manner as does an electron.

Gamma Ray (Gamma Radiation)

Just like electrons, protons and neutrons can move between energy levels within the nucleus. Nuclear reactions can excite nucleons to higher energy levels. When the nucleons subsequently drop to lower energy levels, they emit photons. The photons emitted when nucleons drop to lower energy levels are incredibly energetic compared to those emitted when orbiting electrons change energy levels. These energetic photons are known as **gamma rays**, γ. Gamma rays are a form of electromagnetic radiation. They do not have mass or charge, and they are not influenced by electric or magnetic fields.

ISOTOPES

Although an atom must have a specific number of protons (a set atomic number) to remain a specific element, an atom does not need to contain a definite number of neutrons. Each element may have several combinations of neutrons that allow atoms of that element to exist in slightly different variations. As an example, carbon can be found in nature as $^{12}_{6}C$, $^{13}_{6}C$, and $^{14}_{6}C$, with 6, 7, and 8 neutrons, respectively. This means that although the atomic number of carbon is set, the mass number may vary. The forms of carbon with different numbers of neutrons are the possible **isotopes** of the carbon atom. Isotopes are the same chemical element but have different numbers of neutrons. The isotopes of an element are often reported with the element name followed by the mass number (carbon-12, carbon-13, and carbon-14). There is no need to report the atomic number (6) since it is a known fact that all carbon atoms have six protons.

Isotopes vary in both mass and their atomic stability. The difference in mass between isotopes is obvious. Carbon-14 has a mass of 14 atomic mass units, while carbon-12 has a mass of 12 atomic mass units. The mass is important in balancing nuclear reactions and when calculating the energy involved in nuclear reactions. However, the instability of certain isotopes, such as carbon-14 and uranium-235, allows nuclear reactions to occur.

THE STRONG FORCE

Students rarely ask why the protons in the nucleus cluster together when they should repel each other due to electrostatic forces. As it turns out, another force is operating in the nucleus, the **strong force**. The strong force attracts nucleons to one another. It attracts protons to protons, neutrons to neutrons, and protons to neutrons. While the electrostatic force is trying to separate the protons, the strong force holds them together. The magnitude of the strong force is greater than that of the electrostatic force. However, the strong force operates only at very small distances, such as those inside the nucleus. If a proton in the nucleus is moved away from the center of the atom, the strong force weakens with increasing distance. At a certain distance, the repulsion due to the electrostatic force will exceed the strong force. When this occurs, the electrostatic force will accelerate the proton out of the atom.

Adding neutrons to the nucleus helps hold the nucleus together. The neutrons add to the strong force, helping to hold the repelling protons near each other. In addition, the neutrons have no charge, so they do not contribute to the electrostatic repulsive force that acts to tear apart the nucleus. However, every atom has an optimum mix of protons and neutrons. When the number of protons and neutrons falls outside of an optimum range, the geometry of the nucleus weakens the strong force. This allows the electrostatic force either to eject a small portion of the nucleus or to tear apart the entire nucleus. Carbon-14 and uranium-235 are classic examples of unstable nuclei. When these atoms undergo nuclear reactions their nuclei experience changes in compositions. As a result, these elements undergo a transmutation into entirely new elements.

MASS-ENERGY EQUIVALENCE

During a nuclear reaction, the mass of the reactants at the start of the reaction does not equal the mass of the products produced by the reaction. The difference in mass between the products and the reactants is known as the **mass defect**, Δm. The mass defect is associated with the energy involved in a nuclear reaction. The amount of energy, E, associated with the mass defect can be calculated using the speed of light in a vacuum, c, and Albert Einstein's famous equation:

$$E = (\Delta m)c^2$$

This equation demonstrates **mass-energy equivalence**. During nuclear reactions, matter may be converted into energy or energy may be converted into matter.

Nuclear reactions release energy by converting a small amount of the original mass into energy. In nuclear reactions where energy is released, the reactants (ingredients) have more mass than the products formed during the reaction. To balance and account for all the original mass during the reaction, the mass defect must be added to the product side of the reaction.

$$\text{Reactants} \rightarrow \text{Products} + \Delta m$$

For example, a neutron splitting to form a proton and an electron can be shown as follows.

$$^1_0\text{n} \rightarrow\ ^1_1\text{p} +\ ^0_{-1}\text{e} + \bar{\text{v}}_e + \Delta m$$
$$1.00866\ \text{u} = 1.00728\ \text{u} + 0.00055\ \text{u} + \Delta m$$
$$\Delta m = 0.00083\ \text{u}$$

IF YOU SEE
nuclear
reactions

A small amount of mass (mass defect) is converted into energy, $E = mc^2$

The energy created in these reactions is released as a product of the reaction.

When the symbol for each particle is replaced by the appropriate mass in atomic mass units, it becomes apparent that the product side of the reaction is missing mass. The symbol for the mass defect is added to the side with the least amount of mass. So, the magnitude of the mass defect can be calculated.

The mass defect, Δm, is the small amount of mass that is converted into energy, $E = (\Delta m)c^2$, during the reaction. The actual calculation involves values and conversions that would be difficult to complete without a calculator and is unlikely to be tested. The values are shown here to assist you in understanding the concept of converting matter into energy.

$$\Delta m = 0.00083 \text{ u} \left(\frac{1.66 \times 10^{-27} \text{ kg}}{1 \text{ u}} \right) = 1.38 \times 10^{-30} \text{ kg}$$

$$E = (\Delta m)c^2 = (1.38 \times 10^{-30} \text{ kg})(3 \times 10^8 \text{ m/s})^2 = 1.24 \times 10^{-13} \text{ J}$$

This amount of energy is released when a neutron becomes a proton and an electron. Energy is a product of this reaction, and the mass defect can be replaced with energy in the reaction equation.

$$^1_0\text{n} \rightarrow ^1_1\text{p} + ^0_{-1}\text{e} + \bar{v}_e + \text{Energy}$$

Note that reactions can also run in reverse. If the above example were reversed, combining a proton and an electron would create a neutron. In the reverse reaction, the reactants would have less mass, so the mass defect would be added to the left side of the equation. In this process, energy is a reactant and energy must be converted into matter to allow the reaction to progress.

RADIOACTIVE DECAY

Radioactive decay occurs when an unstable isotope spontaneously loses energy by emitting particles from its nucleus. The decay processes were named in their order of discovery by using the first three letters of the Greek alphabet: alpha (α), beta (β), and gamma (γ). After their initial discovery, it was determined that an alpha particle was in fact the nucleus of a helium atom (^4_2He), a beta particle was actually an electron ($^0_{-1}\text{e}$), and gamma radiation was not a particle at all. Instead, it is a high-frequency photon ($^0_0\gamma$). Elements that naturally decay are said to be radioactive. Radioactive substances have a critical imbalance between the number of protons and neutrons in the nucleus. As atoms become larger, more neutrons are needed to maintain stability. Uranium-235 has 92 protons and 143 neutrons. The three common forms of natural radioactivity—alpha, beta, and gamma radiation—are discussed below.

The SAT Subject Test in Physics is not a chemistry exam. Although knowing the mass and atomic numbers of the fundamental particles, such as protons, electrons, neutrons, alpha particles, and beta particles, is important, you do not need to know the chemical symbols, names, mass numbers, and atomic numbers of all the elements in the periodic table. The element names and symbols will be given in the questions. Instead, test questions will focus on balancing the mass numbers and atomic numbers in nuclear reactions, as shown in the example problems in the following sections.

Alpha Decay

In an alpha decay, an alpha particle, $_2^4$He, is spontaneously ejected from the nucleus of an atom. If an alpha particle leaves the nucleus, the mass number of the atom is reduced by 4 while the atomic number is reduced by 2. Changing the atomic number causes a transmutation into a new element. The ejected alpha particle is the least dangerous form of radioactive decay. Although it is harmful if digested, an alpha particle can be stopped by both paper and skin.

IF YOU SEE
alpha decay

Subtract 4 from the mass number and 2 from the atomic number.

Conversely, if the mass and atomic numbers decrease by 4 and 2, respectively, then an alpha decay occurred.

EXAMPLE 21.1

The Product of an Alpha Decay

An isotope of uranium, $_{92}^{238}$U, undergoes alpha decay. In the process, the atom becomes an isotope of thorium. Which of the following elements is the result of this transmutation?

(A) $_{90}^{234}$Th   (B) $_{92}^{237}$Th   (C) $_{92}^{239}$Th   (D) $_{93}^{238}$Th   (E) $_{94}^{242}$Th

WHAT'S THE TRICK?

The particle resulting from the decay leaves the nucleus and is subtracted. Subtract the mass number (4) and atomic number (2) of the alpha particle from the original nucleus.

$$_{92}^{238}\text{U} \rightarrow {}_{92-2}^{238-4}\text{Th} + {}_2^4\alpha$$

$$_{92}^{238}\text{U} \rightarrow {}_{90}^{234}\text{Th} + {}_2^4\alpha$$

The resulting thorium nucleus has a mass number of 234 and an atomic number of 90. The answer is A.

EXAMPLE 21.2

Determining the Type of Radioactive Particle

An isotope of thorium, $_{90}^{227}$Th, undergoes a transmutation into an isotope of radium, $_{88}^{223}$Ra. What type of decay process caused this transmutation?

WHAT'S THE TRICK?

Determine the difference between the mass numbers and the atomic numbers.

$$_{90}^{227}\text{Th} \rightarrow {}_{88}^{223}\text{Ra} + {}_{90-88}^{227-223}\ ?$$

$$_{90}^{227}\text{Th} \rightarrow {}_{88}^{223}\text{Ra} + {}_2^4\alpha$$

The mass number has decreased by 4, and the atomic number has decreased by 2. These are the mass number and atomic number for an alpha particle.

The mass number is constant, and the atomic number increases by 1.

Conversely, if the mass number is constant and the atomic number decreases by 1, then beta decay occurred.

Beta Decay

A beta particle is released when a neutron in the nucleus of an atom decays into a proton and an electron. This electron originates in the nucleus and is known as a beta particle. Beta particles move at greater speeds than alpha particles and can be stopped with a thin sheet of metal, such as aluminum.

EXAMPLE 21.3

The Product of Beta Decay

An isotope of carbon, $^{14}_{6}C$, undergoes beta decay. In the process, the atom becomes an isotope of nitrogen. Which of the following is the result of this transmutation?

(A) $^{10}_{4}N$ (B) $^{13}_{6}N$ (C) $^{15}_{6}N$ (D) $^{14}_{7}N$ (E) $^{18}_{8}N$

WHAT'S THE TRICK?

The particle resulting from the decay leaves the nucleus and is subtracted. Subtract the mass number (0) and atomic number (–1) of the beta particle from the original nucleus.

$$^{14}_{6}C \rightarrow \, ^{14-0}_{6-(-1)}N + \, ^{0}_{-1}\beta$$

$$^{14}_{6}C \rightarrow \, ^{14}_{7}N + \, ^{0}_{-1}\beta$$

The resulting nitrogen nucleus has a mass number of 14 and an atomic number of 7. The answer is D.

EXAMPLE 21.4

Determining the Type of Radioactive Particle

An isotope of thorium, $^{231}_{90}Th$, undergoes a transmutation into an isotope of neptunium, $^{231}_{91}Np$. What type of decay process caused this transmutation?

WHAT'S THE TRICK?

Determine the difference between the mass numbers and atomic numbers.

$$^{231}_{90}Th \rightarrow \, ^{231}_{91}Np + \, ^{231-231}_{90-91} \, ?$$

$$^{231}_{90}Th \rightarrow \, ^{231}_{91}Np + \, ^{0}_{-1}e$$

The mass number is unchanged, and the atomic number decreased by –1. This is the mass number and atomic number for a beta particle.

Gamma Rays

Gamma rays are the result of a number of radioactive decay reactions, including alpha and beta decay. Gamma rays travel at the speed of light and have greater penetration than alpha or beta radiation. Very dense materials such as lead are needed to stop gamma rays. The release of gamma rays (high-energy photons) does not affect the atomic number or mass number of the atoms.

Decay Rate

The rate at which radioactive decay occurs is often measured in what is known as a **half-life**. One half-life is the time interval needed for half of a sample of radioactive atoms to decay. Carbon-14, for example, has a half-life of 5,740 years (1 half-life = 5,740 years). At the end of this time period, only half of the original sample of carbon-14 remains. The rest has undergone a transmutation into nitrogen-14. Table 21.2 shows how a sample consisting of 16 grams of carbon-14 would progress through several half-lives.

Table 21.2 Decay of Carbon-14

Half-life	Time in Years	Carbon-14	Nitrogen-14
0	0	16 g	0 g
1	1(5,740) = 5,740 years	(½)(16) = 8 g	8 g
2	2(5,740) = 11,480 years	(½)(8) = 4 g	12 g
3	3(5,740) = 17,220 years	(½)(4) = 2 g	14 g
4	4(5,740) = 22,960 years	(½)(2) = 1 g	15 g

Carbon-14 is created in Earth's upper atmosphere by cosmic rays bombarding nitrogen gas atoms. Carbon-14 is absorbed by plants and animals, and it appears in these organisms at the same concentration levels found in the atmosphere. When an organism dies, it no longer absorbs carbon-14. The amount of carbon-14 in the organism will begin to beta decay into nitrogen. Measuring the levels of carbon-14 in the fossilized remains of dead organisms can help determine the time period when extinct species existed on Earth.

EXAMPLE 21.5

Half-Life

A 120-gram sample of iodine-131 has a half-life of 8.0 days. How much of the original sample remains after 24 days?

WHAT'S THE TRICK?

Divide the time interval by the length of time of one half-life to determine how many half-lives have passed.

$$(24 \text{ days})\left(\frac{1 \text{ half-life}}{8.0 \text{ days}}\right) = 3 \text{ half-lives}$$

During each half-life, the sample of iodine is halved. If 3 half-lives have passed, then the original 120-gram sample will be reduced by half 3 times.

$$(120 \text{ g})\left(\frac{1}{2}\right)\left(\frac{1}{2}\right)\left(\frac{1}{2}\right) = 120 \text{ g} \left(\frac{1}{2}\right)^3 = 15 \text{ g}$$

IF YOU SEE
fission or
fusion

Fission:
splitting a
large nucleus
into smaller
nuclei.

Fission is used
in nuclear
reactors.

Fusion:
creating a
larger nucleus
from smaller
nuclei.

Fusion occurs
in stars, such
as the Sun.

FISSION AND FUSION

A **fission** reaction is the splitting of a large atom into smaller atoms. A **fusion** reaction involves combining smaller atoms to make a larger atom. Spontaneous fission and fusion reactions involve the release of energy.

Fission

Fission of larger atoms into smaller ones is typically induced by the bombardment of the larger atom with free neutrons. The addition of a free neutron temporarily creates a larger, unstable nucleus. The attractive strong force is no longer able to hold the protons together. The repulsive electrostatic force tears apart the nucleus, forming two smaller nuclei. This process also releases several additional free neutrons and energy in the form of gamma rays. A common fission reaction occurring in nuclear power plants involves the splitting of the uranium-235 atom as shown in the following reaction.

$$^{235}_{92}U + ^{1}_{0}n \rightarrow ^{92}_{36}Kr + ^{141}_{56}Ba + 3(^{1}_{0}n) + \text{Energy}$$

In the example, the uranium-235 atom undergoes a transmutation into two distinct atoms, krypton and barium. It also releases three more free neutrons as well as energy. The energy released is used to heat water until it becomes steam. The steam is then used to rotate a coil of wire in a magnetic field, generating electrical energy. The three free neutrons are able to bombard three more uranium-235 atoms, causing additional fission reactions, which release even more neutrons. In power plants, the reaction is controlled. However, the number of free neutrons and the subsequent fission reactions have the potential to grow exponentially in what is known as a **chain reaction**.

The main difference between radioactive decay and fission is that fission requires activation and produces free neutrons to continue the reaction. Radioactive decay occurs spontaneously and produces no free neutrons. Both reactions release energy.

EXAMPLE 21.6

Fission

How many neutrons are created in the following nuclear reaction?

$$^{235}_{92}U + ^{1}_{0}n \rightarrow ^{90}_{38}Sr + ^{143}_{54}Xe + ?\,n + \text{Energy}$$

WHAT'S THE TRICK?

The mass numbers and atomic numbers must remain constant. The arrow separating the reactants and products can be treated as an equal sign.

Mass numbers: 235 + 1 = 90 + 143 + ?
Atomic numbers: 92 + 0 = 38 + 54 + ?

The mass numbers on the product side of the reaction are missing three atomic mass units. Each neutron has a mass number of one atomic mass unit, so this fission reaction must produce three free neutrons.

$$^{235}_{92}U + ^{1}_{0}n \rightarrow ^{90}_{38}Sr + ^{143}_{54}Xe + 3(^{1}_{0}n) + \text{Energy}$$

Fusion

The fusion of two smaller atoms to become a larger atom requires a tremendous amount of activation energy to overcome the electrostatic repulsion of the protons. Unlike larger atoms, such as uranium-235 or uranium-238, which have a number of protons close together, making it relatively easy to induce them to break apart, bringing small atoms together to make larger ones is quite difficult. An example of a fusion reaction is the fusion of two isotopes of hydrogen atoms into a helium atom:

$$\,^2_1H + \,^2_1H \rightarrow \,^4_2He + Energy$$

The isotopes of hydrogen in the above equation are known as **deuterium** and consist of 1 proton and 1 neutron. A fusion reaction will release more energy, per the mass of the reactants, than a fission reaction. However, inducing fusion is quite difficult because of the tremendous amount of energy needed to overcome the electrostatic repulsion of the two smaller reactant nuclei.

The forces needed to drive these reactions occur naturally in stars, such as the Sun. The Sun is composed of an immense amount of gas (mostly hydrogen), and this huge amount of mass creates an enormous gravitational effect. As a result, these gases are under extreme temperature and pressure. In fact, they constitute a fourth state of matter known as **plasma**. These conditions provide the energy to initiate fusion reactions. Once running, fusion reactions produce significant amounts of excess energy.

EXAMPLE 21.7

Identifying Nuclear Reactions

Identify the nuclear process that is taking place in the following reaction.

$$\,^3_2He + \,^3_2He \rightarrow \,^4_2He + \,^1_1H + \,^1_1H$$

WHAT'S THE TRICK?

In fission reactions, the largest nucleus is on the reactant side (left side) of the reaction, and it splits into smaller nuclei. In fusion reactions, the largest nucleus is on the product side (right side) of the reaction. The largest nucleus is $\,^4_2He$, and it is on the product side. The reaction shown above is a fusion reaction. Note that this nucleus also appears to be an alpha particle. However, this is not alpha decay. Alpha decay involves a larger nucleus ejecting an alpha particle. This reaction involves two small nuclei fusing to create the larger helium atom.

SUMMARY

1. **THE MASS NUMBER AND ATOMIC NUMBER OF ELEMENTS ARE IMPORTANT QUANTITIES IN NUCLEAR REACTIONS.** The mass number is the number of nucleons (protons + neutrons) contained in the nucleus of an atom. The mass number is also the mass of a single atom in atomic mass units (u). The atomic number is the number of protons in an atom. The number of protons defines an element. If the number of protons changes, then an atom undergoes a transmutation into a different element.

2. **NUCLEAR REACTIONS INVOLVE SUBATOMIC PARTICLES.** These include protons, neutrons, electrons, alpha particles, and beta particles. Knowing the mass numbers and atomic numbers of these key fundamental particles is important when analyzing nuclear reactions. In addition, knowing the charge on these fundamental particles is important if they are interacting with electric and magnetic fields. Gamma rays (gamma radiation) are also a product of nuclear reactions, but they are not particles. Gamma rays are high-energy photons of light energy. Gamma rays do not have mass or charge.

3. **ISOTOPES ARE ATOMS WITH THE SAME NUMBER OF PROTONS BUT A DIFFERENT NUMBERS OF NEUTRONS.** In some isotopes, an imbalance between the number of protons and neutrons creates instability in the nucleus. This instability is responsible for many nuclear reactions.

4. **THE STRONG FORCE ATTRACTS NUCLEONS TO ONE ANOTHER AND HOLDS THE NUCLEUS TOGETHER.** The strong force operates over very short distances and attracts protons and neutrons to each other. However, in unstable isotopes, conditions may exist where the repulsive electrostatic force temporarily overcomes the strong force, allowing nuclear reactions to take place.

5. **RADIOACTIVE DECAY.** There are three main types of radioactive decay. An alpha particle is a helium nucleus with no electrons. A beta particle is an electron released from the nucleus. Gamma rays are high-energy photons. The decay rate of radioactive isotopes is quantified in half-lives.

6. **FISSION AND FUSION ARE NUCLEAR REACTIONS THAT CREATE LARGE AMOUNTS OF ENERGY.** In fission and fusion reactions, a quantity of mass, known as the mass defect, is lost during the reaction. This amount of mass is converted into energy and can be calculated using Einstein's mass-energy equivalence equation: $E = (\Delta m)c^2$. Fission is the splitting of large atoms into smaller ones by bombarding large atoms with free neutrons. The reaction releases additional free neutrons, creating a chain reaction. Fusion occurs when smaller atoms fuse into larger atoms. This process occurs naturally in stars.

If You See	Try	Keep in Mind
A nuclear reaction	Mass defect, Δm Mass-energy equivalence: $$E = (\Delta m)c^2$$	The products of a nuclear reaction usually have less mass than the reactants. The missing mass defect is converted into energy, which is released as a product.
Alpha decay	Alpha particle: ^4_2He	If an alpha particle is ejected from the nucleus, reduce the mass number by 4 and the atomic number by 2. Conversely, if a reaction is given and the mass and atomic numbers decrease by 4 and 2, then the reaction is an alpha decay.
Beta decay	Beta particle: $^0_{-1}\text{e}$	When a beta particle is ejected from the nucleus, the mass number remains constant and the atomic number increases by 1. Conversely, if a reaction is given and the mass number remains constant while the atomic number increases by 1, then beta decay has occurred.
Fission or fusion	Fission is the splitting of a large nucleus into smaller nuclei. The largest atom in the reaction will be on the left side of the equation. Fusion involves combining smaller nuclei to create a larger nucleus. The largest atom in the reaction will be on the right side of the equation.	Fission occurs in nuclear reactors. The energy created is used to generate electrical energy. Fusion occurs in stars, such as the Sun.

PRACTICE EXERCISES

1. Which answer correctly ranks the fundamental particles in order from most massive to least massive?

   (A) Neutron > proton > electron > alpha particle = beta particle
   (B) Alpha particle > beta particle > proton > neutron > electron
   (C) Alpha particle > proton > neutron > electron > beta particle
   (D) Alpha particle > neutron > proton > electron = beta particle
   (E) Neutron > proton > electron > alpha particle > beta particle

2. An alpha particle is ejected from the nucleus of an atom. Which of the following answers describes the change in the nucleus of the atom?

   (A) The atomic number increases by 1, and the mass number remains unchanged.
   (B) The atomic number increases by 1, and the mass number decreases by 4.
   (C) The atomic number increases by 1, and the mass number increases by 2.
   (D) The atomic number increases by 2, and the mass number increases by 4.
   (E) The atomic number decreases by 2, and the mass number decreases by 4.

3. A nucleus undergoes a transmutation from $^{234}_{84}$Po to become $^{234}_{85}$At. The nuclear reaction that has occurred to accomplish this is

   (A) alpha decay
   (B) beta decay
   (C) gamma decay
   (D) fission
   (E) fusion

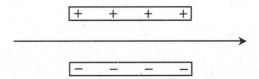

4. The product of a nuclear reaction passes between two charged plates. Which of the following entities would experience the motion shown in the diagram above?

   (A) Beta particle
   (B) Electron
   (C) Proton
   (D) Alpha particle
   (E) Gamma ray

5. A radioactive sample with a half-life of 3 days is analyzed after 15 days. The amount of remaining radioactive material as a fraction of the original sample is most nearly

   (A) $\frac{1}{32}$

   (B) $\frac{1}{16}$

   (C) $\frac{1}{8}$

   (D) $\frac{1}{4}$

   (E) $\frac{1}{2}$

6. A radioactive sample with a half-life of 4 days is discovered to have $\frac{1}{8}$ of its radioactive material remaining. How many days has the sample been experiencing radioactive decay?

   (A) 4 days
   (B) 6 days
   (C) 8 days
   (D) 12 days
   (E) 16 days

7. Which of the following is an isotope of $^{238}_{92}\text{U}$?

   (A) $^{141}_{56}\text{Ba}$

   (B) $^{234}_{90}\text{Th}$

   (C) $^{235}_{92}\text{U}$

   (D) $^{235}_{92}\text{Th}$

   (E) $^{238}_{90}\text{Th}$

8. Which statement regarding the forces in the nucleus is NOT correct?

   (A) The strong force attracts protons to protons.
   (B) The strong force attracts neutrons to both protons and other neutrons.
   (C) At short distances, the strong force is stronger than the electric force.
   (D) Adding neutrons to a nucleus adds to both the strong force and the electrostatic force.
   (E) When the ratio of neutrons to protons falls outside of an optimal range, the nucleus becomes unstable.

9. How many neutrons are liberated during the following nuclear reaction?

$$^{235}_{92}U + ^{1}_{0}n \rightarrow ^{90}_{38}Sr + ^{142}_{54}Xe + ? (^{1}_{0}n) + Energy$$

(A) 1
(B) 2
(C) 3
(D) 4
(E) 5

10. Categorize the following nuclear reaction:

$$^{1}_{1}H + ^{2}_{1}H \rightarrow ^{3}_{2}He + Energy$$

(A) Alpha decay
(B) Beta decay
(C) Gamma decay
(D) Fission
(E) Fusion

ANSWERS EXPLAINED

		Key Words	Needed for Solution	Now Solve It
1.	**(D)**	Ranks the fundamental particles . . . massive	Knowledge of fundamental particles $^4_2\alpha$, $^1_0 n$, $^1_1 p$, $^0_{-1}\beta$, and $^0_{-1} e$	Compare the mass numbers. Although the proton and neutron appear to have the same mass, the neutron is slightly more massive. Even though the beta particle originates from the nucleus, it is an electron and has the same mass as an electron.
2.	**(E)**	Alpha particle . . . change in the nucleus	Alpha decay $^4_2\alpha$	The alpha particle is ejected and leaves the nucleus. Subtract the atomic and mass numbers of an alpha particle from the original nucleus. This reduces the atomic number by 2 and the mass number by 4.
3.	**(B)**	Transmutation . . . nuclear reaction	Assess how the mass number and atomic number have changed.	The mass number remained constant, while the atomic number increased by 1. This is consistent with the removal of a beta particle from the nucleus. Removing (subtracting) the atomic number (–1) of the beta particle is the same as adding 1 to the atomic number.
4.	**(E)**	Product of a nuclear reaction . . . diagram	Knowledge/definitions	Only an uncharged entity would follow a straight path when passing through charged plates. Gamma rays are not particles. They are photons of light energy and have neither mass nor charge.
5.	**(A)**	Half-life of 3 days . . . remaining . . . after 15 days	During each half-life, the remaining sample of radioactive material is halved.	To find the number of half-lives that have taken place, divide the length of time that has passed by the length of one half-life. $$(15 \text{ days}) \left(\frac{1 \text{ half-life}}{3 \text{ days}} \right) = 5 \text{ half lives}$$ Halve the original sample 5 times. However, this problem is asking for a fraction of the original sample. Treat the original sample as having a value of 1. $$(1) \left(\frac{1}{2}\right)\left(\frac{1}{2}\right)\left(\frac{1}{2}\right)\left(\frac{1}{2}\right)\left(\frac{1}{2}\right) = \left(\frac{1}{2}\right)^5 = \frac{1}{32}$$
6.	**(D)**	Half-life of 4 days . . . $\frac{1}{8}$ of its radioactive material remaining	Determine how many times $\frac{1}{2}$ needs to be multiplied by itself to arrive at $\frac{1}{8}$.	$\frac{1}{2}$ raised to the third power is $\frac{1}{8}$. $$\left(\frac{1}{2}\right)\left(\frac{1}{2}\right)\left(\frac{1}{2}\right) = \left(\frac{1}{2}\right)^3 = \frac{1}{8}$$ Three half-lives have taken place. Each half-life is 4 days long. As a result, it takes 12 days for the sample to decay to $\frac{1}{8}$ of its original size.

	Key Words	Needed for Solution	Now Solve It
7. **(C)**	Isotope	Knowledge/definitions	An isotope has the same number of protons but a different number of neutrons than the original atom. The name (symbol) of the atom remains unchanged.
8. **(D)**	Forces in nucleus . . . NOT correct	Strong force versus the electrostatic force	Neutrons and protons attract each other with the strong force. However, neutrons have no charge, and they do not add to the electrostatic force that repels the protons in the nucleus.
9. **(D)**	How many neutrons . . . following nuclear reaction	Balance the mass numbers and atomic numbers.	Mass numbers: $$235 + 1 = 90 + 142 + ?$$ Atomic numbers: $$92 + 0 = 38 + 54 + ?$$ The missing mass number is 4, and the missing atomic number is 0. This reaction produces 4 additional neutrons: $4(_0^1 n)$.
10. **(E)**	Categorize . . . nuclear reaction	Knowledge/definitions	The largest nucleus is on the product (right) side of the reaction equation. Fusion is the process where smaller nuclei are combined to create a larger nucleus.

Relativity

22

→ **SPECIAL THEORY OF RELATIVITY**

→ **TIME, LENGTH, AND MASS**

In 1905, Albert Einstein wrote four scientific papers that would change physics forever. One of those papers, on the photoelectric effect, would win him the Nobel Prize in 1921. Another paper, on special relativity, used creative-thought experiments to understand the nature and behavior of the speed of light and objects traveling near such speeds. Although his ideas are complex and often seem counterintuitive, observations of testable phenomena have consistently agreed with his theory of special relativity. This chapter will review the following concepts regarding special relativity:

- Define the concept of special relativity and the laws of physics within a reference frame.
- Determine what will happen to time, length, and mass as objects approach the speed of light.

SPECIAL THEORY OF RELATIVITY

Until Einstein's paper in 1905, most physicists believed that the universe was filled with an invisible medium called ether. They reasoned that ether was necessary in order for light, considered by most physicists to be a wave, to propagate through space. However, experiments such as the famous Michelson-Morley experiment all failed to prove the existence of ether.

Einstein viewed the problem in a completely different manner and proposed an explanation for these failed experiments. He viewed light as a quantum particle (later named a photon) that traveled with a specific speed in a vacuum, $c = 3 \times 10^8$ meters per second. Einstein then suggested that the speed of light is the same for all **inertial reference frames**. An inertial reference frame is a frame of reference moving at a constant velocity. (Inertia is the tendency of objects to continue moving at constant velocity.) Einstein stated that all the laws of physics are the same for any inertial reference frame. This is known as Einstein's **first postulate of special relativity**.

In addition to the speed of light's being the same for all inertial reference frames, Einstein stated that even if light were emitted from a moving source, it would continue to have a velocity of c. This is known as Einstein's **second postulate of special relativity**. The second postulate is counterintuitive to the laws of motion that describe the relative motion of two moving objects. Consider a ball being thrown forward from a moving vehicle. The true speed of the ball is the speed of the throw plus the speed of the moving vehicle. This is not so with light. Light beams from a car's headlight travel at c regardless of the speed of the car. The true speed of light is always c when measured in any inertial reference frame.

**IF YOU SEE
a moving
light source**

$c = 3 \times 10^8$

The speed of light is the same for all observers, regardless of the motion of the light source.

Einstein's special relativity is a special case since inertial reference frames are not accelerating. In 1915, Einstein proposed a theory of general relativity that encompassed accelerating reference frames in addition to inertial reference frames. The general theory of relativity, however, is beyond the scope of the SAT Subject Test in Physics.

TIME, LENGTH, AND MASS

Using the assumption that light travels at a constant velocity of c for any inertial reference frame, several noticeable effects will happen to time, length, and mass at speeds approaching c. Note that the formulas in the following sections are provided for context. You will probably not have to solve them on the actual examination as doing so would require a calculator. The exam will instead ask generalized questions regarding the effects on time, length, and mass when objects are moving relative to an observer.

Time Dilation

Picture a person watching a moving object. An observer will usually interpret his or her own motion as being stationary and will perceive the other object as moving. Both the observer (stationary) and the object (moving at constant velocity) are equipped with clocks. The observer will see his or her own clock as running normally but will see the clock on the moving object as running slowly. This effect is known as **time dilation**. The amount that the clock appears to be running slowly can be calculated with the following equation:

$$t = \frac{t_0}{\sqrt{1 - v^2/c^2}}$$

Time, t_0, is the time on the clock of the stationary observer, which appears to be running normally. However, when the stationary observer looks at the clock on the moving object, it will have a different, slower time, t. The velocity of the moving object is v, and c is the speed of light. If the moving object has a small velocity, both clocks will appear to be running at the same time, $t \approx t_0$. The time difference between the two clocks will be negligible, and the effects of time dilation will go unnoticed.

$$t = \frac{t_0}{\sqrt{1 - v^2/c^2}} \approx \frac{t_0}{\sqrt{1 - 0}} \approx t_0$$

The speeds of manmade objects are too slow for dilation effects to appear. As a result, we are unaware that dilation actually occurs. However, as the speed of the object nears the speed of light, the effects of time dilation increase dramatically. When an object moves at 99.5 percent of the speed of light ($v = 0.995c$), the time difference in the clocks is substantial.

$$t = \frac{t_0}{\sqrt{1 - \frac{v^2}{c^2}}} \approx \frac{t_0}{\sqrt{1 - \frac{(0.995\cancel{c})^2}{(1.000\cancel{c})^2}}} \approx 10t_0$$

To an outside stationary observer, events in the moving frame (time t) appear to take 10 times as long as the same events in the observer's stationary frame (time t_0). The events in the moving frame appear to be running at one-tenth the speed of events in the observer's frame of reference. The clock and a person on the moving object would appear to be moving in slow motion. At the speed of light, $v = c$, the clock on the moving object would appear to stop entirely.

Since the incredible speeds needed to observe time dilation are unlikely to be attained by humans, evidence of time dilation has been found by examining tiny, high-speed particles. Muons are negatively charged elementary particles with a lifespan of about 2.2×10^{-6} seconds. They are formed by the collision of cosmic rays with the atmosphere. Even though they travel at nearly the speed of light ($0.99c$), the distance from the atmosphere to Earth is such that they should decay before they hit the ground. Yet they are detected at the surface of Earth. This indicates that muons are experiencing time dilation.

Length Contraction

When an object moves, the length of the object and anything moving with the object appears to contract in length. The observed length is calculated with the following equation:

$$L = L_0 \sqrt{1 - v^2/c^2}$$

You should note that the actual length affected by motion is the length that matches the direction of motion. If an object is moving to the right, only length along the x-axis is affected. The height of the object in the y-direction and its depth in the z-direction remain unchanged. In the formula above, length L_0 is the object's rest length. This is the length of the object measured when the object is at rest or by someone moving at the same speed as the object. Length L is the length measured by a stationary observer, someone not moving with the object. At low velocities, v, these lengths are nearly identical, $L \approx L_0$, and the length contraction is not noticeable.

$$L = L_0 \sqrt{1 - v^2/c^2} = L_0 \sqrt{1 - 0} = L_0$$

The length of a moving car is shorter by a distance that is smaller than the diameter of 1 atom. However, if a spacecraft were capable of near-light-speed travel, then an observer watching the spacecraft race by would perceive the length of the entire spacecraft and length of everything in the spacecraft to be smaller than expected. However, a person on the spacecraft (in the spacecraft's same inertial frame) would measure the spacecraft as having its normal length, L_0. At the speed of light, $v = c$, length would become 0, destroying three-dimensional space.

Momentum Effects

Momentum, like time and length, is also affected by the motion of objects. At slow speeds the momentum of an object is directly proportional to its velocity, $p = m_0 v$, where m_0 is an object rest mass. The rest mass is the mass of the object when it is stationary. However, as the speed of an object approaches c, the momentum of the object increases in a dramatic nonlinear manner.

$$p = \frac{m_0 v}{\sqrt{1 - v^2/c^2}}$$

IF YOU SEE an object moving near light speed

Length contraction

Length, in the direction of motion, will appear to decrease if viewed by a stationary observer.

As with time and length, the effects are unnoticed at the low speeds that dominate human experience.

$$p = \frac{m_0 v}{\sqrt{1 - v^2/c^2}} = \frac{m_0 v}{\sqrt{1 - 0}} = m_0 v$$

SUMMARY

1. **SPECIAL RELATIVITY APPLIES TO ALL INERTIAL REFERENCE FRAMES.** Einstein's first postulate of special relativity states that the laws of physics are the same in all inertial reference frames. Einstein's second postulate of special relativity states that the speed of light is a constant, c, regardless of the motion of the observer or the speed of the light source. Special relativity applies only to reference frames moving at constant velocities.

2. **AS AN OBJECT APPROACHES THE SPEED OF LIGHT, ITS TIME, LENGTH, AND MASS ARE AFFECTED.** Time slows down (time dilation), length contracts, and mass increases. Muon particles, which are created in Earth's upper atmosphere, are evidence supporting the theory of special relativity.

If You See	Try	Keep in Mind
A moving light source	The speed of light is the same for all observers in inertial frames of reference: $c = 3 \times 10^8$ m/s.	This is true even if the light source is moving.
An object moving at the speed of light	(1) Time dilation: A clock on the moving object will appear to run slowly.	The effects of time dilation, length contraction, and mass increase go unnoticed at the slow speeds that humans experience. The effects are dramatic at speeds near the speed of light.
	(2) Length contraction: The only length affected is the length in the direction of motion.	
	(3) Mass will increase: The moving object will appear larger than its rest mass.	

PRACTICE EXERCISES

1. A spaceship leaving Earth at a speed of $0.95c$ flashes a laser beam back at Earth. An observer on Earth will register this laser beam as having what speed?

   (A) $0.05c$

   (B) $0.5c$

   (C) $0.95c$

   (D) c

   (E) $1.05c$

2. A spacecraft with a speed of $0.99c$ passes by a stationary observer. The stationary observer makes observations regarding the speed of a clock on the spacecraft, the length of the spacecraft, and the mass of the spacecraft. Which set of observations correctly indicates how the motion of the spacecraft has altered these values?

	Clock	Length	Mass
(A)	Slower	Shorter	Decreased
(B)	Slower	Shorter	Increased
(C)	Same	Shorter	Decreased
(D)	Faster	Shorter	Increased
(E)	Faster	Longer	Decreased

3. A spaceship traveling at $0.8c$ passes parallel to Earth. A rod of length L is inside the spaceship. According to an astronaut inside the spaceship, what is the length of the rod?

   (A) $0.2L$

   (B) $0.8L$

   (C) L

   (D) $1.2L$

   (E) $1.8L$

$v = 0.85c$

4. A spacecraft is moving to the right with a speed of $0.85c$ as shown in the diagram above. The spacecraft is carrying a cube-shaped box as viewed from inside the spacecraft. Which of the following represents the appearance of the box as seen by a stationary observer watching the spacecraft pass by?

(A)

(B)

(C)

(D)

(E)

ANSWERS EXPLAINED

	Key Words	Needed for Solution	Now Solve It
1. **(D)**	Flashes a laser beam . . . observer . . . register . . . speed	Second postulate of special relativity	The second postulate of special relativity states that the speed of light is not dependent upon the motion of the light source. The speed of light will be measured as $c = 3 \times 10^8$ m/s in inertial reference frames.
2. **(B)**	Speed of a clock . . . length . . . mass	Time dilation, length contraction, and mass effect	A stationary observer will perceive the clock on a moving object to be slower, its length (in the direction of motion) to be shorter, and its mass to be larger.
3. **(C)**	Traveling at $0.8c$ . . . rod of length L . . . astronaut inside the spaceship	Length contraction	The astronaut inside of the spaceship will not see any change in the length of the rod. The astronaut is in the same inertial reference frame as the rod.
4. **(D)**	Speed of $0.85c$ . . . moving right; appearance of the box . . . stationary observer	Length contraction	Length contraction affects only the direction of motion. The spacecraft is moving in the x-direction. The box will appear shorter in the x-direction. The height and depth will remain unchanged.

Historical Figures and Contemporary Physics

23

→ **HISTORICAL FIGURES**
→ **CONTEMPORARY PHYSICS**

The SAT Subject Test in Physics occasionally contains questions where students need to associate a specific principle, law, or important experiment with the person who conducted that particular research. In addition, the exam occasionally mentions some aspects associated with contemporary physics. Whether the questions are about the work of an important historical figure or about contemporary physics, they are not very detailed. These questions do not require you to understand completely the work done or the physics principles involved. Instead, they require simple recognition and matching. As a result, this chapter simply matches scientists with their most important work and the contemporary physics principle with its main characteristics.

HISTORICAL FIGURES

Newtonian Mechanics

GALILEO GALILEI (1564–1642)

- Bodies dropped from the same height will all fall with the same acceleration, g. The distance they travel is proportional to the square of time, $y = \frac{1}{2}gt^2$.
- **Principle of inertia**: The natural state of motion is uniform constant velocity.

ISAAC NEWTON (1642–1727)

- **First law of motion** (law of inertia): Modifies Galileo's principle of inertia. The natural state of motion is constant velocity unless acted upon by an unbalanced force.
- **Second law of motion** ($\Sigma F = ma$): The acceleration of an object is directly proportional to the net force acting on the object. Acceleration is inversely proportional to the mass of the object.
- **Third law of motion**: When two objects interact (action-reaction pair), an equal and opposite force acts on each object. This causes an opposite reaction. However, the actual motion depends on the masses involved.
- **Law of gravity** ($F_g = Gm_1m_2/r^2$): Two masses, m_1 and m_2, at a distance apart of r attract each other. The magnitude of the attraction is proportional to the masses and inversely proportional to the square of the distance (inverse-square law).

JAMES WATT (1736–1819)

- **Power**: While working with steam engines to improve their efficiency, Watt developed the concept of power. The units of power are named after him.

JOHANNES KEPLER (1571–1630)

- **First law of planetary motion**: Planetary motion is elliptical.
- **Second law of planetary motion**: A line drawn from the central body (Sun) to an orbiting body (planet) will sweep equal areas of space in equal time intervals.
- **Third law of planetary motion**: The square of the period (time of one orbit) is proportional to the radius of the orbit, $T^2 \propto r^3$.

Electricity and Magnetism

CHARLES-AUGUSTIN DE COULOMB (1736–1806)

- **Coulomb's law** ($F_E = kq_1q_2/r^2$): Two charges, q_1 and q_2, at a distance of r, will attract/repel each other. The magnitude of the attraction/repulsion is directly proportional to the magnitude of the charges and is inversely proportional to the square of the distance (inverse-square law).

GEORG SIMON OHM (1789–1854)

- **Ohm's law**: Determined the relationship among potential, current, and resistance: $V = IR$.

MICHAEL FARADAY (1791–1867)

- **Electromagnetic fields**: Introduced a way to visualize electric and magnetic fields as lines extending through space.
- **Electromagnetic induction**: Discovered the principle of electromagnetic induction where an emf, $\mathcal{E} = \Delta\phi/t$, can be induced (created) in order to stimulate the flow of a current in a loop of conducting material. The induced emf is a potential. It is created by changing the flow of the magnetic field (changing the flux $\Delta\phi$) passing through the loop of conducting material.

HEINRICH LENZ (1804–1865)

- **Lenz's law**: Dictates the direction of an induced current in a closed loop of conducting material, based on conservation of energy.

JAMES CLERK MAXWELL (1831–1879)

- **Electromagnetic waves**: Mathematically demonstrated that light is an electromagnetic wave moving at the speed of light.

Waves and Optics

THOMAS YOUNG (1773–1829)

- **Young's double-slit experiment**: Shined monochromatic light through two narrow slits to create an interference pattern. The resulting pattern demonstrated that light is a wave phenomenon.

CHRISTIAN DOPPLER (1803–1853)

- **Doppler effect**: The shift in wavelength and frequency perceived in sound and light when the source is moving toward or away from an observer.

Thermal Physics

LORD KELVIN (1824–1907)

■ **Absolute zero**: Developed the concept of absolute zero and its associated temperature scale. The units of temperature for this scale are named after him.

JAMES JOULE (1818–1889)

■ **Heat and work equivalence**: Devised an experiment to show that the temperature of water could be increased by applying a flame or by doing mechanical work on it (rapidly stirring it). Showed that heat and work are both methods of adding energy to a system. This means that the energy associated with heat (calories) can simply be converted during the work. The units of work and energy are named after Joule.

Modern Physics

ALBERT MICHELSON (1852–1931) AND EDWARD MORLEY (1838–1923)

■ **Michelson-Morley experiment**: Designed an elaborate device known as an **interferometer** to detect the motion of Earth through the invisible ether. The ether was believed to be the medium that allowed light to travel through space. The experiment failed to prove the existence of the ether and opened the door for new lines of thinking, such as Einstein's special theory of relativity.

J. J. THOMSON (1856–1940)

■ **Discovered the electron**: While working with cathode rays passing through electric plates, he deduced the existence of the electron. He suggested that the atom is similar to a plum pudding. In this model, the atom is viewed as having an overall positive charge (pudding) and the negative electrons are distributed randomly (plums) throughout the atom.

MAX PLANCK (1858–1947)

■ **Founder of quantum theory**: While working with emitted light spectra, Planck suggested that light energy can be emitted only in multiples of specific quantities and are thus quantized. He derived a constant (Planck's constant) that determines the energy associated with the specific quantities he observed.

ALBERT EINSTEIN (1879–1955)

■ **Einstein's miracle year, 1905**: Published four papers that changed physics. Three of these are addressed in introductory physics courses.
■ **Photoelectric effect**: Suggested that light acts like a particle (photon) and that the energy of the photon is quantized. The energy of the photon can be calculated by multiplying Planck's constant by the frequency of the photon: $E = hf$.
■ **Special relativity**: All the laws of physics are the same in inertial frames (frames of reference that have constant velocity). The speed of light in a vacuum is a constant 3×10^8 m/s regardless of the motion of the light source.
■ **Mass-energy equivalence**: Matter can be converted into energy and vice versa according to the equation $E = mc^2$.

- **General relativity**: Objects with mass, such as the Sun, are regarded as distorting the geometry of spacetime. The perception of the Sun's gravity acting on other masses, such as Earth, is actually a response to the curvature of spacetime. The orbit of Earth is dictated by the curvature of spacetime, rather than by a pull of gravity.

ERNEST RUTHERFORD (1871–1937)

- **Gold-foil experiment**: Fired alpha particles (positive helium nuclei without their electrons) at a very thin piece of gold foil. The scattering of the alpha particles revealed that the atom is mostly empty space consisting of a dense positive nucleus surrounded by orbiting electrons.

NIELS BOHR (1885–1962)

- **Planetary model of the atom, including specific energy levels**: Combining the concepts of his predecessors, mentioned above, Bohr deduced a model of the atom that explains why electron orbits do not decay and fall into the nucleus. His calculated energy levels, which electrons must occupy, match completely with the observations of light spectra emitted from these atoms.

CONTEMPORARY PHYSICS

Astrophysics

Astrophysics is the physics of celestial objects such as stars, planets, galaxies, etc. Research is conducted with a variety of telescopes and arrays that receive and analyze electromagnetic radiation. These include ground-based and space-based telescopes, such as the Hubble Space Telescope, X-ray telescopes, and radio-wave arrays. Astrophysics seeks to resolve the origin of the universe and to explain its properties.

Chaos Theory

Chaos theory is a mathematical theory that attempts to explain the behavior of complex and chaotic systems. When a complex series of events is set in motion the results can vary drastically depending on small initial changes in the system. A popular phrase describing this is "the butterfly effect." Chaos theory is not limited to physics. It is involved in processes such as erosion and fluctuations in stock-market prices. In astrophysics, the universe we now live in was greatly influenced by the very initial conditions under which the universe was formed.

Dark Energy

A form of energy throughout the universe that is hypothesized to be accelerating the expansion of the universe that was observed by Edwin Hubble.

Dark Matter

The total mass of the universe appears not to match the gravitational effects observed throughout the universe. Scientists have hypothesized that a form of matter that cannot be seen with telescopes accounts for the missing mass. This matter is under investigation and has been called dark matter.

Gravity Waves

In 1916, Albert Einstein's general theory of relativity described gravity as a curvature of spacetime. He also predicted that when masses accelerate, the curvature of spacetime would be disturbed, resulting in gravity waves that propagate through space at the speed of light. Corroborating evidence of this theory came in 2015 from the Laser Interferometer Gravitational Wave Observatory (LIGO) project, which detected the existence of gravitational waves.

Microprocessor

A microprocessor is a complex single circuit consisting of many miniaturized components. Examples of microprocessors are the chips that run computers and smartphones. The entire central processing unit of a computer is often a single microprocessor. Microprocessors are based on semiconductor and transistor technology.

Semiconductor

A semiconductor is a material that can act as a conductor or as an insulator. It is the key to modern miniaturized circuit design. Silicon is the most widely used semiconductor and is used in the manufacture of transistors and microprocessors.

Superconductivity

A superconductor is a material that has zero electrical resistance when cooled below a critical temperature.

String Theory

String theory hypothesizes that the elementary particles making up matter are actually linear oscillations or strings. The theory attempts to explain how everything interconnects and hopes to eliminate inconsistencies among other earlier theories.

Transistor

A transistor is an electrical component used in integrated circuits of all modern electrical devices. A transistor can both amplify the electrical signal it receives and act as a switch. A transistor typically has three leads: the base, the collector, and the emitter. A small amount of current passing through the base can control a larger current at the collector. This determines the amount of current leaving the transistor at the emitter. The amplification characteristic is used in electronic amplifiers, such as those found in stereo equipment. Transistors are used in the computer/semiconductor industry as a switch to establish the on/off (1,0) binary code used by computers.

PRACTICE EXERCISES

1. The relationship among potential, current, and resistance in electrical circuits is described by

   (A) Coulomb's law
   (B) Faraday's law
   (C) Lenz's law
   (D) Ohm's law
   (E) Snell's law

2. The idea that light comes in packets and that these packets are quantized according to the equation $E = hf$ was postulated by

   (A) Albert Einstein
   (B) Albert Michelson
   (C) James Maxwell
   (D) Ernest Rutherford
   (E) J. J. Thomson

3. The link between mechanical work and heat was demonstrated by

   (A) Albert Einstein
   (B) James Joule
   (C) Lord Kelvin
   (D) James Maxwell
   (E) James Watt

4. Which of the following is concerned with mathematically modeling the behavior of complex and seemingly random systems?

   (A) Astrophysics
   (B) Chaos theory
   (C) Dark matter
   (D) String theory
   (E) The theory of relativity

5. Which of these materials has the properties of both a conductor and an insulator?

   I. Dark matter
   II. Semiconductor
   III. Superconductor

   (A) I only
   (B) II only
   (C) III only
   (D) I and II only
   (E) II and III only

ANSWERS EXPLAINED

	Key Words	Needed for Solution	Now Solve It
1. **(D)**	Potential, current, and resistance	Knowledge Problems in this chapter deal with specific facts that are either memorized or recognized	The relationship among these variables was described by Georg Simon Ohm and is known as Ohm's law, $V = IR$.
2. **(A)**	Light . . . packets . . . $E = hf$		This was suggested by Albert Einstein in his work on the photoelectric effect.
3. **(B)**	Link . . . work . . . heat		James Joule conducted the famous experiment demonstrating heat-work equivalence. Energy can be added to a system as either mechanical work or heat.
4. **(B)**	Complex . . . random systems		This is chaos theory.
5. **(B)**	Properties of both a conductor and an insulator		Semiconductors have the properties of both a conductor and an insulator. Semiconductors are widely used in all modern electronic devices.

Don't forget you have five full-length practice SAT Physics tests with this book—one Diagnostic, three at the end of this book, and one online. The link to the online practice test can be found on the inside front cover or by going to *online.barronsbooks.com*

PRACTICE TESTS

ANSWER SHEET
Practice Test 1

1. Ⓐ Ⓑ Ⓒ Ⓓ Ⓔ
2. Ⓐ Ⓑ Ⓒ Ⓓ Ⓔ
3. Ⓐ Ⓑ Ⓒ Ⓓ Ⓔ
4. Ⓐ Ⓑ Ⓒ Ⓓ Ⓔ
5. Ⓐ Ⓑ Ⓒ Ⓓ Ⓔ
6. Ⓐ Ⓑ Ⓒ Ⓓ Ⓔ
7. Ⓐ Ⓑ Ⓒ Ⓓ Ⓔ
8. Ⓐ Ⓑ Ⓒ Ⓓ Ⓔ
9. Ⓐ Ⓑ Ⓒ Ⓓ Ⓔ
10. Ⓐ Ⓑ Ⓒ Ⓓ Ⓔ
11. Ⓐ Ⓑ Ⓒ Ⓓ Ⓔ
12. Ⓐ Ⓑ Ⓒ Ⓓ Ⓔ
13. Ⓐ Ⓑ Ⓒ Ⓓ Ⓔ
14. Ⓐ Ⓑ Ⓒ Ⓓ Ⓔ
15. Ⓐ Ⓑ Ⓒ Ⓓ Ⓔ
16. Ⓐ Ⓑ Ⓒ Ⓓ Ⓔ
17. Ⓐ Ⓑ Ⓒ Ⓓ Ⓔ
18. Ⓐ Ⓑ Ⓒ Ⓓ Ⓔ
19. Ⓐ Ⓑ Ⓒ Ⓓ Ⓔ
20. Ⓐ Ⓑ Ⓒ Ⓓ Ⓔ

21. Ⓐ Ⓑ Ⓒ Ⓓ Ⓔ
22. Ⓐ Ⓑ Ⓒ Ⓓ Ⓔ
23. Ⓐ Ⓑ Ⓒ Ⓓ Ⓔ
24. Ⓐ Ⓑ Ⓒ Ⓓ Ⓔ
25. Ⓐ Ⓑ Ⓒ Ⓓ Ⓔ
26. Ⓐ Ⓑ Ⓒ Ⓓ Ⓔ
27. Ⓐ Ⓑ Ⓒ Ⓓ Ⓔ
28. Ⓐ Ⓑ Ⓒ Ⓓ Ⓔ
29. Ⓐ Ⓑ Ⓒ Ⓓ Ⓔ
30. Ⓐ Ⓑ Ⓒ Ⓓ Ⓔ
31. Ⓐ Ⓑ Ⓒ Ⓓ Ⓔ
32. Ⓐ Ⓑ Ⓒ Ⓓ Ⓔ
33. Ⓐ Ⓑ Ⓒ Ⓓ Ⓔ
34. Ⓐ Ⓑ Ⓒ Ⓓ Ⓔ
35. Ⓐ Ⓑ Ⓒ Ⓓ Ⓔ
36. Ⓐ Ⓑ Ⓒ Ⓓ Ⓔ
37. Ⓐ Ⓑ Ⓒ Ⓓ Ⓔ
38. Ⓐ Ⓑ Ⓒ Ⓓ Ⓔ
39. Ⓐ Ⓑ Ⓒ Ⓓ Ⓔ
40. Ⓐ Ⓑ Ⓒ Ⓓ Ⓔ

41. Ⓐ Ⓑ Ⓒ Ⓓ Ⓔ
42. Ⓐ Ⓑ Ⓒ Ⓓ Ⓔ
43. Ⓐ Ⓑ Ⓒ Ⓓ Ⓔ
44. Ⓐ Ⓑ Ⓒ Ⓓ Ⓔ
45. Ⓐ Ⓑ Ⓒ Ⓓ Ⓔ
46. Ⓐ Ⓑ Ⓒ Ⓓ Ⓔ
47. Ⓐ Ⓑ Ⓒ Ⓓ Ⓔ
48. Ⓐ Ⓑ Ⓒ Ⓓ Ⓔ
49. Ⓐ Ⓑ Ⓒ Ⓓ Ⓔ
50. Ⓐ Ⓑ Ⓒ Ⓓ Ⓔ
51. Ⓐ Ⓑ Ⓒ Ⓓ Ⓔ
52. Ⓐ Ⓑ Ⓒ Ⓓ Ⓔ
53. Ⓐ Ⓑ Ⓒ Ⓓ Ⓔ
54. Ⓐ Ⓑ Ⓒ Ⓓ Ⓔ
55. Ⓐ Ⓑ Ⓒ Ⓓ Ⓔ
56. Ⓐ Ⓑ Ⓒ Ⓓ Ⓔ
57. Ⓐ Ⓑ Ⓒ Ⓓ Ⓔ
58. Ⓐ Ⓑ Ⓒ Ⓓ Ⓔ
59. Ⓐ Ⓑ Ⓒ Ⓓ Ⓔ
60. Ⓐ Ⓑ Ⓒ Ⓓ Ⓔ

61. Ⓐ Ⓑ Ⓒ Ⓓ Ⓔ
62. Ⓐ Ⓑ Ⓒ Ⓓ Ⓔ
63. Ⓐ Ⓑ Ⓒ Ⓓ Ⓔ
64. Ⓐ Ⓑ Ⓒ Ⓓ Ⓔ
65. Ⓐ Ⓑ Ⓒ Ⓓ Ⓔ
66. Ⓐ Ⓑ Ⓒ Ⓓ Ⓔ
67. Ⓐ Ⓑ Ⓒ Ⓓ Ⓔ
68. Ⓐ Ⓑ Ⓒ Ⓓ Ⓔ
69. Ⓐ Ⓑ Ⓒ Ⓓ Ⓔ
70. Ⓐ Ⓑ Ⓒ Ⓓ Ⓔ
71. Ⓐ Ⓑ Ⓒ Ⓓ Ⓔ
72. Ⓐ Ⓑ Ⓒ Ⓓ Ⓔ
73. Ⓐ Ⓑ Ⓒ Ⓓ Ⓔ
74. Ⓐ Ⓑ Ⓒ Ⓓ Ⓔ
75. Ⓐ Ⓑ Ⓒ Ⓓ Ⓔ

Practice Test 1

Do not use a calculator. To simplify numerical calculations, use $g = 10 \text{ m/s}^2$.

PART A

Directions: In this section of the exam, the same lettered choices are used to answer several questions. Each group of questions is preceded by five lettered choices. When answering questions in each group, select the best answer from the available choices and fill in the corresponding bubble on the answer sheet. Each possible answer may be used once, more than once, or not at all.

Questions 1–3

   (A) Alpha decay
   (B) Beta decay
   (C) Gamma ray
   (D) Fission
   (E) Fusion

Select the term from above that identifies the nuclear reactions described in questions 1 to 3.

1. An isotope of bismuth, $^{214}_{83}\text{Bi}$, undergoes a transmutation into an isotope of polonium, $^{214}_{84}\text{Po}$.

2. $^{235}_{92}\text{U} + ^{1}_{0}\text{n} \rightarrow ^{90}_{38}\text{Sr} + ^{143}_{54}\text{Xe} + 3\text{n} + \text{Energy}$

3. An isotope of thorium, $^{227}_{90}\text{Th}$, undergoes a transmutation into an isotope of radium, $^{223}_{88}\text{Ra}$.

The following diagrams depict sound waves moving outward from a sound source.

(A)

(B)

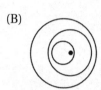

(C)

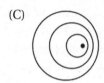

(D)

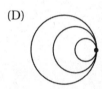

(E)

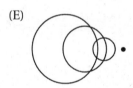

4. Which diagram depicts the sound waves created by a source moving with a speed v that is equal to the speed of sound?

5. Which diagram depicts the sound waves created by a source moving with a speed v that is less than the speed of sound?

6. Which diagram depicts the sound waves created by a source moving with a speed v that is greater than the speed of sound?

Questions 7–9

The following terms relate to circuits. Match the correct term with its definition.

    (A) Capacitance
    (B) Current
    (C) Power
    (D) Resistance
    (E) Voltage

7. The rate of charge flow.

8. The rate of energy dissipation in a circuit.

9. The potential difference between the terminals of the battery.

Questions 10–12

Match the correct image description with the optical instruments depicted in questions 10 to 12.

    (A) Real and inverted
    (B) Real and upright
    (C) No image forms, ray traces never intersect
    (D) Virtual and inverted
    (E) Virtual and upright

10.

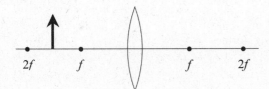

11.

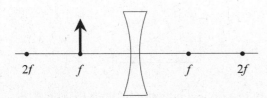

12.

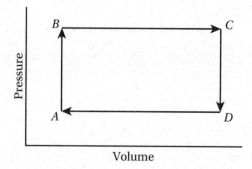

The pressure versus volume graph above represents four processes in a complete thermodynamics cycle of an ideal monoatomic gas in a closed container. The selections below represent the individual processes making up the cycle and are ordered to show the direction of each process. As an example, process AB begins at A and ends at B.

(A) Process AB
(B) Process BC
(C) Process CD
(D) Process DA
(E) Both process AB and process CD

13. According to the graph, which process terminates at the highest temperature?

14. During which process is work being done *on* the gas?

15. During which process is no work being done?

Questions 16–19

In an experiment a mass *m* is released from rest on two different inclines that have the same incline angle θ but have different surfaces. The first incline is frictionless. The second incline is rough, allowing the block to slide at constant velocity after it is given an initial push to overcome static friction.

(A) mg
(B) $mg \sin \theta$
(C) $\mu \, mg \sin \theta$
(D) $mg \cos \theta$
(E) Zero

16. For the trial involving the frictionless incline, what is the magnitude of the gravitational force acting on the mass?

17. For the trial involving the frictionless incline, what is the magnitude of the net force acting on the mass?

18. For the trial where the block moves at constant velocity, what is the magnitude of the friction force acting on the mass?

19. For the trial where the block moves at constant velocity, what is the magnitude of the net force acting on the mass?

Questions 20–22

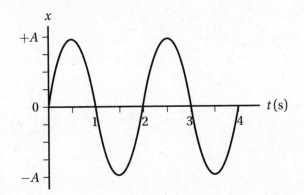

A mass is attached to the end of a vertically suspended spring. The mass is initially lowered to the equilibrium position, where the mass remains at rest when released. Define this position as zero potential energy. The mass is then displaced from equilibrium, released, and allowed to oscillate vertically with an amplitude of ±A. Air resistance is negligible. The graph above depicts the displacement, x, of the oscillation as a function of time.

(A) zero

(B) $\dfrac{A}{4}$

(C) $\dfrac{A}{2}$

(D) $\dfrac{A}{\sqrt{2}}$

(E) $\dfrac{A}{2\sqrt{2}}$

20. What is the magnitude of displacement when the mass experiences the greatest net force?

21. What is the magnitude of displacement when the mass has the greatest kinetic energy?

22. What is the magnitude of displacement when the mass's potential energy is equal to its kinetic energy?

> **Directions:** This section of the exam consists of questions or incomplete statements followed by five possible answers or completions. Select the best answer or completion, and fill in the corresponding bubble on the answer sheet.

Questions 23–25

Use the following speed-time graph for the motion of an object to solve questions 23 to 25.

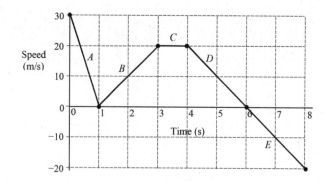

23. Determine the displacement during the first second (interval *A*).

(A) 10 m
(B) 15 m
(C) 20 m
(D) 25 m
(E) 30 m

24. During which interval is the object moving with a constant velocity?

(A) *A*
(B) *B*
(C) *C*
(D) *D*
(E) *A*, *B*, *D*, and *E*

25. Which of the following is true when $t = 6$ seconds?

   I. The object is accelerating.
   II. The object has an instantaneous speed of zero.
   III. The object has returned to the origin.

   (A) I only
   (B) II only
   (C) III only
   (D) I and II only
   (E) I, II, and III only

———————————————————————————

26. If a ball is thrown straight upward with an initial velocity of v, it will reach a height of h. If the initial speed of the ball is doubled, what will be the new maximum height?

   (A) $\sqrt{2}(h)$
   (B) $2h$
   (C) $2\sqrt{2}(h)$
   (D) $4h$
   (E) $4\sqrt{2}(h)$

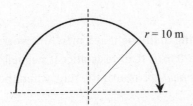

27. During a time of 5.0 seconds, an object moves through a half circle with a radius of 10 meters, as shown above. What is the magnitude of the object's average velocity during this motion?

   (A) 2 m/s
   (B) π m/s
   (C) 4 m/s
   (D) 2π m/s
   (E) 4π m/s

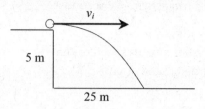

28. A ball is thrown horizontally from the edge of a 5-meter-tall building. It lands 25 meters from the base of the building, as shown in the diagram above. With what initial speed was the ball thrown?

    (A)   5 m/s
    (B) 10 m/s
    (C) 15 m/s
    (D) 20 m/s
    (E) 25 m/s

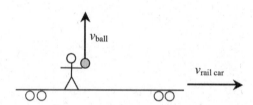

29. A flat railroad car is moving to the right at 5 m/s. A person standing on the car throws a ball straight upward at 20 m/s. If air resistance is negligible, where will the ball be *in relation to the person's new position* at the time when the ball returns to its original starting height?

    (A) The ball will land 20 meters in front of the person.
    (B) The ball will land 10 meters in front of the person.
    (C) The ball will land in the person's hand.
    (D) The ball will land 10 meters behind the person.
    (E) The ball will land 20 meters behind the person.

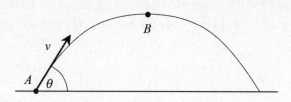

30. The diagram above depicts a projectile launched from point *A* with a speed of *v* at an angle of θ, above the horizontal. Determine the speed of the projectile when it reaches its maximum height at point *B*.

(A) 0

(B) v

(C) $\frac{1}{2}v$

(D) $v \cos \theta$

(E) $v \sin \theta$

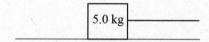

31. A 5.0-kilogram mass is pulled along a rough horizontal surface by a string, as shown above. The coefficient of kinetic friction between the surface and the object is 0.10. The tension in the string is 30 newtons. Determine the acceleration of the object.

(A) 1 m/s^2

(B) 2 m/s^2

(C) 3 m/s^2

(D) 4 m/s^2

(E) 5 m/s^2

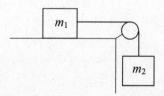

32. Two masses, $m_1 = 1$ kg and $m_2 = 3$ kg, are connected by a string that is draped over a massless, frictionless pulley, as shown above. Mass 1 is positioned on a frictionless horizontal surface, while mass 2 hangs freely. The masses are released from rest. Determine the acceleration of mass 2.

(A) 2.5 m/s^2

(B) 3.3 m/s^2

(C) 5.0 m/s^2

(D) 6.7 m/s^2

(E) 7.5 m/s^2

33. Earth ($m = 5.98 \times 10^{24}$ kilograms) pulls a 60-kilogram person toward it with a force of 600 N. With what amount of force does the person pull Earth toward themself?

(A) 0 N

(B) $\dfrac{1}{600}$ N

(C) 60 N

(D) 600 N

(E) 5.98×10^{24} N

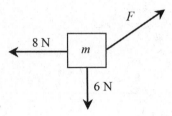

34. Three forces act on a mass, m, as shown in the diagram above. The mass remains at rest. Determine the magnitude of force F.

(A)   6 N
(B)   8 N
(C) 10 N
(D) 12 N
(E) 14 N

35. What is the apparent weight of a 60-kilogram astronaut who is experiencing a rocket launch from the surface of Earth with an acceleration of 40 meters per second squared?

(A)   600 N
(B) 1,200 N
(C) 1,800 N
(D) 2,400 N
(E) 3,000 N

Questions 36–37

A 1.0-kilogram mass is attached to the end of a 1.0-meter-long string. When the apparatus is swung in a vertical circle, the tension in the rope at the very bottom of the circle has a magnitude of 110 newtons.

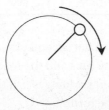

36. Determine the speed of the mass at the lowest point in the circle.

    (A)   5 m/s
    (B)   10 m/s
    (C)   25 m/s
    (D)   55 m/s
    (E)   110 m/s

37. Determine the minimum speed needed at the top of the loop in order for the mass to complete one cycle.

    (A) 1.0 m/s
    (B) 2.5 m/s
    (C) $\sqrt{10}$ m/s
    (D) $5\sqrt{10}$ m/s
    (E) 10 m/s

Questions 38–39

A 5.0-kilogram mass is moving in uniform circular motion with a radius of 1.0 meter and a frequency of 3.0 hertz.

38. Determine the tangential velocity of the mass.

    (A) $\frac{1}{6}\pi$ m/s

    (B) $\frac{2}{3}\pi$ m/s

    (C) $\frac{3}{2}\pi$ m/s

    (D) 3π m/s

    (E) 6π m/s

39. Determine the centripetal acceleration of the mass.

   (A) $3\pi^2$ m/s^2
   (B) $6\pi^2$ m/s^2
   (C) $12\pi^2$ m/s^2
   (D) $36\pi^2$ m/s^2
   (E) $72\pi^2$ m/s^2

_____

40. A car of mass m makes a turn on a flat section of road that has a radius of r. The coefficient of friction between the tires and the road is μ. The maximum speed at which the car can make the turn *without* skidding is v. If the mass of the car is doubled, what is the new maximum speed in the turn?

   (A) $\dfrac{1}{4}v$

   (B) $\dfrac{1}{2}v$

   (C) v
   (D) $2v$
   (E) $4v$

Questions 41–42

A 2.0-kilogram mass is pulled up a frictionless 30° incline at a constant speed of 0.5 meters per second.

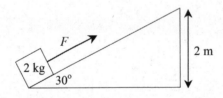

41. Determine the work done by force F to move the mass to a vertical height of 2.0 meters.

   (A) 0 J
   (B) 5 J
   (C) 10 J
   (D) 20 J
   (E) 40 J

42. Determine the power required to move the mass up the incline at constant speed.

   (A) 0 W
   (B) 2.5 W
   (C) 5.0 W
   (D) 10 W
   (E) 20 W

43. A 3.0-kilogram block is pressed against a spring that has a spring constant of 300 newtons per meter, as shown above. The block is moved to the left until the spring has been compressed 0.10 meters. The block and compressed spring are held in this stationary position for a brief amount of time. Finally, the block is released and the spring pushes the block to the right. What is the maximum speed reached by the block?

(A) 0 m/s
(B) 1 m/s
(C) 2 m/s
(D) 5 m/s
(E) 10 m/s

44. What amount of force is required to change the speed of a 1,500-kilogram car by 10 meters per second in a time of 5 seconds?

(A)   500 N
(B) 1,000 N
(C) 2,000 N
(D) 3,000 N
(E) 6,000 N

45. Which of these is true during an inelastic collision, in which no external forces act?

    I. Linear momentum is conserved.
   II. Kinetic energy is conserved.
  III. Mechanical energy is lost as heat.

(A) I only
(B) II only
(C) III only
(D) I and II only
(E) I and III only

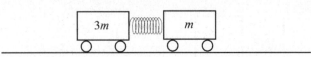

46. Two carts with masses $3m$ and m are placed on a horizontal track with a compressed spring positioned between them. The carts are released from rest. The $3m$ cart moves to the left with a speed of v. What is the speed of the cart on the right in the diagram above?

(A) $\frac{1}{3}v$

(B) $\frac{1}{2}v$

(C) v

(D) $2v$

(E) $3v$

47. A satellite of mass m orbits Earth at a height of h and a speed of v. What would the speed be for a satellite of mass $3m$ at a height of h?

(A) $\frac{1}{3}v$

(B) v

(C) $\sqrt{3}v$

(D) $3v$

(E) $9v$

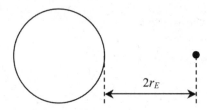

48. Determine the acceleration of gravity, in terms of g, at a point in space that is located a distance equal to two Earth radii ($2r_E$) above the surface of Earth, shown in the figure above.

(A) $\frac{1}{9}g$

(B) $\frac{1}{4}g$

(C) $\frac{1}{3}g$

(D) $\frac{1}{2}g$

(E) g

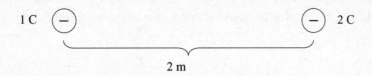

1 C $\quad$ 2 C

2 m

49. Determine the magnitude of the electric force acting on the 1-coulomb charge in the diagram above in terms of the Coulomb's law constant, k.

(A) $\frac{1}{4}k$

(B) $\frac{1}{2}k$

(C) k

(D) $2k$

(E) $4k$

50. The direction of an electric field is

(A) determined by Lenz's law
(B) determined by the right-hand rule
(C) the same as the direction of force acting on any type of charge
(D) the same as the direction of force acting on a negative charge
(E) the same as the direction of force acting on a positive charge

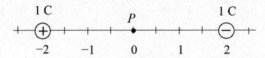

Distance in meters

51. In the diagram, a +1-coulomb charge is located 2 meters to the left of the origin. A −1-coulomb charge is located 2 meters to the right of the origin. Determine the electric potential, in terms of the Coulomb's law constant, k, at point P, located at the origin.

(A) zero

(B) $\frac{1}{4}k$

(C) $\frac{1}{2}k$

(D) k

(E) $2k$

52. A conducting sphere with a mass of 1.0 kilograms and a charge of 3.0 coulombs is initially at rest. Determine its speed after being accelerated through a 6.0-volt potential difference.

(A) 2.0 m/s
(B) 3.0 m/s
(C) 4.0 m/s
(D) 5.0 m/s
(E) 6.0 m/s

Questions 53–55

Use the following circuit diagram to answer questions 53 to 55.

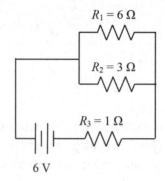

53. What current flows through the 1-Ω resistor?

(A) 0.5 A
(B) 1.0 A
(C) 2.0 A
(D) 3.0 A
(E) 4.0 A

54. What is the voltage drop across the 1-Ω resistor?

(A) 2 V
(B) 4 V
(C) 6 V
(D) 8 V
(E) 10 V

55. How much power is dissipated in the 1-Ω resistor?

(A) 4 W
(B) 10 W
(C) 16 W
(D) 20 W
(E) 32 W

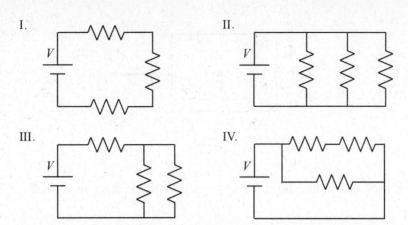

I.   II.

III.   IV.

56. In the circuit diagrams above, the resistors represent identical lightbulbs. Which circuit will have the brightest lightbulbs?

(A) I only
(B) II only
(C) III only
(D) IV only
(E) The brightness of the lightbulbs is the same in each circuit.

−z (into page)

× × × × ×
× × × × ×
q v
⊖ ——→ × × × × ×
× × × × ×
× × × × ×

57. Charge q moving with speed v in the +x-direction enters a uniform −z magnetic field, B, as shown in the diagram above. In what direction is the force of magnetism acting on the charge at the instant the charge first enters the magnetic field?

(A) +x
(B) +y
(C) −y
(D) +z
(E) −z

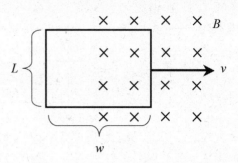

58. The induced emf in the loop at the instant shown in the diagram above is

    (A) 0
    (B) BLv
    (C) $\dfrac{BLv}{w}$
    (D) Bwv
    (E) $\dfrac{v}{BLw}$

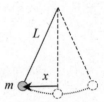

59. A pendulum is constructed with a string of length L and a mass m, as shown above. When the pendulum bob is displaced a distance x from equilibrium, its period of oscillation is T. What will be the new period if the mass, m, and the displacement from equilibrium, x, are both doubled?

    (A) $\dfrac{1}{4}T$
    (B) $\dfrac{\sqrt{2}}{2}T$
    (C) T
    (D) $\sqrt{2}\,T$
    (E) $4T$

60. A sound wave emitted by a source has a frequency f, a velocity v, and a wavelength λ.
     If the frequency is doubled, how will the speed and wavelength be affected?

(A) v, λ
(B) $2v, \lambda$
(C) $v, 2\lambda$
(D) $\frac{1}{2}v, \lambda$
(E) $v, \frac{1}{2}\lambda$

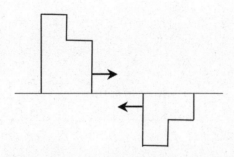

61. The two wave pulses shown above are moving toward one another. Which diagram
     depicts the waveform at the instant that the waves overlap and superimpose?

(A)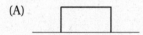

(B)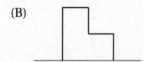

(C)

(D)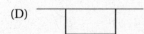

(E)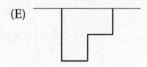

62. A guitar string vibrates in a manner resulting in a standing wave with a wavelength λ being formed in the string. A listener hears the fundamental frequency, f_1, for this particular string. The string is plucked a second time in a manner that produces the second harmonic for this string. How does the wavelength for the second harmonic compare with the wavelength at the fundamental frequency?

(A) $\frac{1}{2}\lambda$

(B) $\frac{\sqrt{2}}{2}\lambda$

(C) λ

(D) $\sqrt{2}\lambda$

(E) 2λ

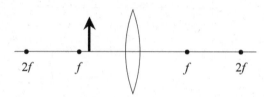

63. The object viewed by a convex lens is positioned just inside of the focus, as shown in the diagram above. Which of the following correctly describes the image?

(A) No image is formed
(B) Real and upright
(C) Real and inverted
(D) Virtual and upright
(E) Virtual and inverted

64. Which diagram below correctly illustrates the path of a light ray moving from point 1 in air (to the left of the solid line) to point 2 in glass (to the right of the solid line)?

(A)

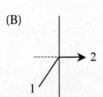

(B)

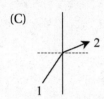

(C)

(D)

(E)

65. An image formed by a convex mirror is

(A) real and upright
(B) real and inverted
(C) virtual and upright
(D) virtual and inverted
(E) No image is formed by this mirror.

66. Monochromatic light with wavelength λ passes through two narrow slits that are a distance d apart. The resulting interference pattern appears as a series of alternating bright and dark regions on a screen located a length L meters behind the slits. How could the experiment be altered so that the spacing between bright regions on the screen is decreased?

(A) Use light with a shorter wavelength.
(B) Decrease the distance from the slits to the screen.
(C) Increase the distance between the slits.
(D) Perform the experiment under water.
(E) All of the above.

67. The bending of light as it passes through a narrow opening, or slit, is known as

(A) absorption
(B) diffraction
(C) interference
(D) polarization
(E) refraction

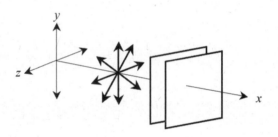

68. Unpolarized light is projected onto two polarizing filters, as shown above. The polarizing filters contain thin threads all oriented parallel to one another. The filters are rotated with respect to each other, and two key positions are identified. In one position, no light is transmitted through the filter. In the other position, the maximum amount of light is transmitted. How must the microscopic threads in the two filters be oriented so that these results are witnessed?

	No light transmitted	Maximum light transmitted
(A)	Parallel	Parallel
(B)	Parallel	Perpendicular
(C)	Perpendicular	Parallel
(D)	Perpendicular	Perpendicular

(E) No solution exists that allows for these observations.

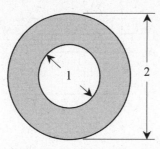

69. A metal washer is a flat, circular piece of metal with a hole through its center, as pictured above. What will be the effect of heating this washer?

(A) No change occurs as the effects on the diameters cancel.
(B) Diameter 1 will decrease, and diameter 2 will decrease.
(C) Diameter 1 will decrease, and diameter 2 will increase.
(D) Diameter 1 will increase, and diameter 2 will decrease.
(E) Diameter 1 will increase, and diameter 2 will increase.

70. The specific heat of a liquid is 2,000 joules/kilogram • kelvin. How much heat is required to raise the temperature of 3.0 kilograms of this liquid from 10°C to 30°C?

(A)   300 J
(B) 13,333 J
(C) 30,000 J
(D) 60,000 J
(E) 120,000 J

71. A heat engine operates between 100°C and 500°C. The theoretical efficiency is most nearly

(A) 10%
(B) 20%
(C) 50%
(D) 70%
(E) 80%

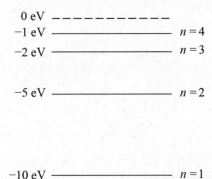

0 eV ‒ ‒ ‒ ‒ ‒ ‒ ‒ ‒ ‒ ‒ ‒
−1 eV ————————————— $n = 4$
−2 eV ————————————— $n = 3$

−5 eV ————————————— $n = 2$

−10 eV ————————————— $n = 1$

72. The energy-level diagram above shows a sample of atoms initially in the ground state. The atoms are radiated by photons having 9 electron volts of energy. After the absorption, photons are emitted by the sample of atoms. Several energies for the emitted photons are listed below. Which of these energies is NOT possible for photons emitted by the atom diagrammed above?

(A) 1 eV
(B) 2 eV
(C) 3 eV
(D) 5 eV
(E) 8 eV

73. In a photoelectric experiment, the frequency of light is steadily increased. Which statement below is NOT correct?

(A) Below the threshold frequency, no electrons are emitted.
(B) Above the threshold frequency, electrons are emitted.
(C) Increasing the frequency of light increases the energy of the emitted electrons.
(D) Increasing the frequency of light increases the potential difference of the photocell.
(E) Increasing the frequency of light increases the induced current.

74. A radioactive sample with a half-life of 25 days is analyzed after 100 days. The amount of remaining radioactive material as a fraction of the original sample is most nearly

(A) $\dfrac{1}{32}$

(B) $\dfrac{1}{16}$

(C) $\dfrac{1}{8}$

(D) $\dfrac{1}{4}$

(E) $\dfrac{1}{2}$

75. A spacecraft with a speed of $0.99c$ in the $+x$-direction passes by a stationary observer. The dimensions of the spacecraft will appear altered along which axis/axes?

(A) x only
(B) y only
(C) z only
(D) y and z only
(E) x, y, and z

ANSWER KEY
Practice Test 1

1.	B	26.	D	51.	A
2.	D	27.	C	52.	E
3.	A	28.	E	53.	C
4.	D	29.	C	54.	A
5.	C	30.	D	55.	A
6.	E	31.	E	56.	B
7.	B	32.	E	57.	C
8.	C	33.	D	58.	B
9.	E	34.	C	59.	C
10.	A	35.	E	60.	E
11.	E	36.	B	61.	A
12.	A	37.	C	62.	A
13.	B	38.	E	63.	D
14.	D	39.	D	64.	C
15.	E	40.	C	65.	C
16.	A	41.	E	66.	E
17.	B	42.	C	67.	B
18.	B	43.	B	68.	C
19.	E	44.	D	69.	E
20.	E	45.	E	70.	E
21.	A	46.	E	71.	C
22.	D	47.	B	72.	B
23.	B	48.	A	73.	E
24.	C	49.	B	74.	B
25.	D	50.	E	75.	A

DIAGNOSTIC CHART

Subject Area	Question Numbers	Questions Incorrect	Chapter(s) to Study
Mechanics	16, 17, 18, 19, 20, 21, 22, 23, 24, 25, 26, 27, 28, 29, 30, 31, 32, 33, 34, 35, 36, 37, 38, 39, 40, 41, 42, 43, 44, 45, 46, 47, 48, 59		1–9
Electricity and Magnetism	7, 8, 9, 49, 50, 51, 52, 53, 54, 55, 56, 57, 58		10–13
Waves and Optics	4, 5, 6, 10, 11, 12, 60, 61, 62, 63, 64, 65, 66, 67, 68		14–17
Heat and Thermodynamics	13, 14, 15, 69, 70, 71		18–19
Modern Physics	1, 2, 3, 72, 73, 74, 75		20–22

SCORING YOUR TEST

How to Determine Your Raw Score

Your raw score is the number of correctly answered questions minus the incorrectly answered questions multiplied by ¼. An incorrectly answered question is one that you bubbled in but was incorrect. If you leave the answer blank, it does not count as an incorrect answer.

Number of correctly answered questions: _____

Number of incorrectly answered questions: _____ $\times \frac{1}{4}$ = _____

$$\underline{\hspace{3cm}} - \underline{\hspace{3cm}} = \underline{\hspace{3cm}}$$

Number Correct 　　Number Incorrect $\times \frac{1}{4}$ 　　Raw Score

ANSWERS EXPLAINED

1. **(B)**	The atomic number (number of protons) increases from 83 to 84. For this to happen, a neutron must eject a beta particle (an electron) and become a proton. The overall mass of the atom remains virtually unchanged because the mass of the ejected electron is very small. As a result, the atomic mass remains unchanged.
2. **(D)**	Fission is the breaking apart of an atom with a large number of protons and neutrons into two smaller atoms. Free neutrons are often released along with a tremendous amount of energy.
3. **(A)**	An alpha particle is the nucleus of a helium atom, $_2^4$He. When two protons and two neutrons are released from thorium-227, the resulting isotope is radium-223. Adding together the atomic numbers of helium and radium-223 as well as adding together their atomic masses will reveal thorium-227.
4. **(D)**	The dot representing the sound source is located at the extreme boundary of the sound wave it has created. The sound waves are superimposing on each other at this point, creating the sound barrier.
5. **(C)**	The sound source is moving in the same direction as the sound waves it emits. As a result, the sound waves bunch up in front of the sound source, and they stretch out behind the moving source. Since the source and waves all travel at constant speeds the wavelengths in front of the moving source should be shorter and equal to one another, while the wavelengths behind the source should be longer and also equal to one another.
6. **(E)**	The dot representing the sound source is located in front of the sound wave it has created.
7. **(B)**	Current is a measure of the amount of charge, Q, flowing per time, t. It is measured in units of amps (amperes).
8. **(C)**	Power is the rate of energy dissipation in a circuit. It is measured in units of watts.
9. **(E)**	Voltage is synonymous with potential difference. It is measured in joules per coulomb.
10. **(A)**	The rays will converge beyond the focal point on the right side of the lens, forming an inverted and real image.
11. **(E)**	The rays will diverge. The negative back ray traces will form an upright, virtual image.
12. **(A)**	The rays will reflect off the mirror beyond the focal point on the left side of the mirror, creating an inverted, real image.
13. **(B)**	In a closed container the number of moles of the trapped gas, n, remains constant. Under these conditions the ideal gas law, $PV = nRT$, dictates that the temperature of the trapped gas is directly proportional to both the pressure and the volume of the gas. The highest temperature would occur at point C (the end of process BC) where both the pressure and the volume of the gas are at their greatest.
14. **(D)**	Work is done on a trapped gas when the gas is compressed and volume decreases. This occurs during process DA.

15. **(E)**	In order for work to be done, either on the gas or by the gas, the volume of the gas must change. No work is done during processes AB or CD, where volume remains constant.
16. **(A)**	The gravitational force is the object's weight, mg. While a force with a magnitude of $mg \sin \theta$ acts parallel to the incline, this is not the gravitational force. It is the component of the gravitational force acting parallel to the incline.
17. **(B)**	On a frictionless incline a mass will accelerate parallel to the incline due to the net force acting in the same direction, parallel to the incline. The net force is $mg \sin \theta$.
18. **(B)**	Objects move at constant velocity when forces are balanced, and the net force is zero. The force of friction must be opposite and equal to the force pulling the mass down the incline, which is $mg \sin \theta$.
19. **(E)**	As stated in answer 18 above, objects move at constant velocity when the net force is zero.
20. **(E)**	In an oscillation the greatest net force and greatest acceleration occur when the oscillating mass is displaced a maximum distance (amplitude) from its equilibrium position. Exercise caution when judging force and acceleration based on trends in velocity alone. While the mass is instantaneously at rest at maximum displacement, this is where the restoring force and acceleration are greatest.
21. **(A)**	The greatest speed and kinetic energy in an oscillation occur at the instant the oscillating mass passes through equilibrium.
22. **(D)**	At the instant an oscillating mass reaches its maximum displacement, its instantaneous speed and kinetic energy are both zero. As a result, the total energy of the oscillating mass at maximum displacement consists solely of potential energy. Amplitude, A, is equivalent to maximum displacement. Since energy is conserved in oscillations the total energy at maximum displacement will be equal to the total energy at the instant when the oscillating mass's kinetic and potential energies are equal, $K = U = \frac{1}{2} kx^2$. $$K_{\max x} + U_{\max x} = K_{\text{where } U \text{ equal to } K} + U_{\text{where } U \text{ equal to } K}$$ $$0 + \frac{1}{2} kA^2 = \frac{1}{2} kx^2 + \frac{1}{2} kx^2$$ $$\frac{1}{2} A^2 = x^2$$ $$x = \frac{A}{\sqrt{2}}$$
23. **(B)**	The area under a speed-time graph is the displacement. Finding the area under interval A requires solving for the area of a triangle. $$\text{Displacement} = \text{Area} = \frac{1}{2} (30 \text{ m/s})(1 \text{ s}) = 15 \text{ m}$$
24. **(C)**	During interval C, the object is moving at a constant speed of 20 m/s.

25. **(D)**	The slope of a speed-time graph is acceleration. At $t = 6$ seconds, there is a negative, nonzero slope, indicating an acceleration. At $t = 6$ seconds, the instantaneous speed is also zero. In order to return to the origin, the displacement must be zero. Displacement is the area under a speed-time graph. For the first six seconds, the accumulated area is positive, indicating that the object has moved away from the origin in the positive direction. The zero value at $t = 6$ seconds is the object's instantaneous speed, not its position.
26. **(D)**	Try $\vec{v}^2 = \vec{v}_0^2 + 2\vec{g}h$ rearranged to solve for height. $$h = \frac{v_f{}^2 - v_i{}^2}{2g}$$ At max height, the final velocity is zero. $$h = \frac{v_i{}^2}{2g}$$ Doubling the initial speed of the ball will increase the maximum height by 4. $$4(h) = \frac{(2v_i)^2}{2g}$$
27. **(C)**	Velocity is a vector quantity and is calculated using the displacement vector. Displacement is the final position minus the initial position. It is the shortest straight-line distance from the initial position to the final position. Although the object moved along a circular path, its displacement is the diameter of the circle and is equal to 20 m. So, the velocity equals 4 m/s. $$v = \frac{\Delta x}{t} = \frac{(20 \text{ m})}{(5.0 \text{ s})} = 4 \text{ m/s}$$
28. **(E)**	Horizontal motion does not affect vertical motion. The amount of time the ball takes to hit the ground can be determined as follows: $$y = \frac{1}{2}gt^2$$ $$t = \sqrt{\frac{2y}{g}} = \sqrt{\frac{2(5 \text{ m})}{10 \text{ m/s}^2}} = 1 \text{ s}$$ The horizontal motion of a projectile is constant velocity. $$v_{ix} = \frac{x}{t} = \frac{(25 \text{ m})}{(1 \text{ s})} = 25 \text{ m/s}$$ $$= (25 \text{ m/s})(1 \text{ s}) = 25 \text{ m}$$
29. **(C)**	Before throwing the ball vertically, both the person and the ball have a horizontal velocity of 5 meters per second. Even though the ball is thrown straight up at 20 meters per second in the vertical direction, it continues to move independently at 5 meters per second horizontally. The cart also maintains the same horizontal speed. As a result, the ball stays above the cart during its motion and lands where it started, in the thrower's hand.

30. **(D)**	The horizontal velocity remains constant during flight at $v \cos \theta$. At the top of the flight, the vertical velocity becomes zero. However, the horizontal velocity remains unchanged.
31. **(E)**	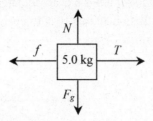

Orient the problem: *x*-direction   *y*-direction

Type of motion: accelerating   stationary

Sum the forces: $\Sigma F = T - f$ $\Sigma F = N - F_g$

Solve: $ma = T - \mu N$ $0 = N - F_g$

                              $ma = T - \mu(mg)$ $N = mg$

$$(5.0 \text{ kg})a = (30 \text{ N}) - (0.1)(5.0 \text{ kg})(10 \text{ m/s})$$
$$a = 5 \text{ m/s}^2$$

32. **(E)**	The direction of motion is parallel to the string. The sum of the forces acting on the system (both bodies simultaneously) will be the forces that are parallel to the string. The two equal and opposite tensions cancel.

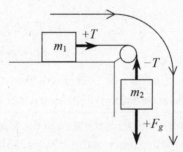

$$\Sigma F_{sys} = F_{g2}$$
$$(m_1 + m_2)a = (m_2)g$$
$$(1 \text{ kg} + 3 \text{ kg})a = (3 \text{ kg})(10 \text{ m/s}^2)$$
$$a = 7.5 \text{ m/s}^2$$

33. **(D)**	Newton's third law states that each force is balanced by an equal but opposite reactionary force. If Earth pulls on a person with a force of 600 N, then the person will also pull on Earth with a force of 600 N. The effect on the acceleration of the person, however, is much greater than the effect on the acceleration of the Earth. Given the same force, the more massive Earth does not accelerate as much as the less massive person.

34. **(C)**	In order for the mass that is acted upon by these forces to remain at rest, the sum of the forces must be zero. Force F must cancel the combined pull of the 8 N and 6 N forces. When the 8 N and 6 N forces are added using vector addition, they form the sides of a 3-4-5 triangle. Their vector sum is 10 N directed in the third quadrant. To cancel this 10 N force, force F must pull with 10 N in the opposite direction. 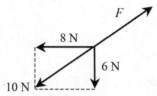
35. **(E)**	Humans sense weight by interacting with surfaces. When asked for the apparent (feeling of) weight of a person, solve for the normal force. 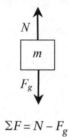 $$\Sigma F = N - F_g$$ $$N = F_g + \Sigma F$$ $$N = mg + ma$$ $$N = m(g + a) = (60 \text{ kg})(10 \text{ m/s}^2 + 40 \text{ m/s}^2) = 3{,}000 \text{ N}$$ Essentially, the acceleration of the spacecraft is added to the force of gravity.
36. **(B)**	The centripetal force, F_c, is the sum of the forces acting on the mass and is directed toward the center of the circular motion. When the mass reaches the bottom of the circle, tension is directed toward the center of the circle and is positive. Gravity acts downward, is directed away from the center of the circle, and is negative. $$F_c = T - F_g$$ $$m\frac{v^2}{r} = T - mg$$ $$(1.0 \text{ kg})\frac{v^2}{(1.0 \text{ m})} = (110 \text{ N}) - (1.0 \text{ kg})(10 \text{ m/s}^2)$$ $$v = 10 \text{ m/s}$$

| 37. **(C)** | At the top of the circle, both tension and gravity act downward toward the center of the circle, so both are positive.

$$F_c = F_g + T$$

$$m\frac{v^2}{r} = mg + T$$

The mass, radius, and gravity are all fixed variables that cannot be changed. However, as the speed of the circling mass is reduced, the tension in the string will decrease. The minimum speed occurs when the tension is zero at the exact instant that the mass is at the top of the loop.

$$m\frac{v^2}{r} = mg$$

$$v = \sqrt{rg}$$

$$= \sqrt{(1 \text{ m})(10 \text{ m/s}^2)}$$

$$= \sqrt{10} \text{ m/s}$$ |
|---|---|
| 38. **(E)** | $$T = \frac{1}{f} = \frac{1}{3.0 \text{ Hz}} = 1/3 \text{ s}$$

$$v = \frac{2\pi r}{T} = \frac{2\pi(1.0 \text{ m})}{(1/3 \text{ s})} = 6\pi \text{ m/s}$$

The mass does not affect the answer. Any object traveling under these conditions will have the same velocity. |
| 39. **(D)** | $$a_c = \frac{v^2}{r} = \frac{(6\pi \text{ m/s})^2}{(1.0 \text{ m})} = 36\pi^2 \text{ m/s}^2$$ |
| 40. **(C)** | The centripetal force, F_c, is the sum of the forces acting on the mass and is directed toward the center of the circular motion. For a car making a turn, the force acting toward the center of the turn is friction.

$$F_c = f$$

$$m\frac{v^2}{r} = \mu N$$

While the car is moving in the horizontal plane, it is not moving vertically. Therefore, the normal force and gravity are canceling in the y-direction. They must have equal magnitudes: $N = mg$.

$$m\frac{v^2}{r} = \mu mg$$

$$v_{max} = \sqrt{\mu rg}$$

Mass does not affect the outcome. It is not part of the equation. |
| 41. **(E)** | As there is no friction on the hill, the amount of work, W, done by force F pulling the block uphill is equal to the change in potential energy, ΔU_g, of the block regardless of the path taken by the block.

$$W = \Delta U_g = mg\Delta h$$

$$= (2 \text{ kg})(10 \text{ m/s}^2)(2 \text{ m})$$

$$= 40 \text{ J}$$ |

42. **(C)**	Power, P, is the rate of work: work divided by time. However, the time needed for the block to move up the hill has not been given. An alternate solution is needed. Work, W, is equal to the force, F, parallel to the motion of an object multiplied by the displacement, Δx, of the object. $$P = \frac{W}{t} = \frac{F\,\Delta x}{t}$$ The displacement divided by the elapsed time is velocity, v. $$P = Fv$$ $$= (10\ \text{N})(0.5\ \text{m/s})$$ $$= 5.0\ \text{W}$$ The force and velocity must be parallel to use this formula. If they are not parallel, use the component of force that is parallel to the velocity.
43. **(B)**	This involves conservation of energy. The potential energy of the spring is converted into the kinetic energy of the block: $$U_i + K_i = U_f + K_f$$ $$\frac{1}{2}kx^2 + 0 = 0 + \frac{1}{2}mv^2$$ $$\frac{1}{2}(300\ \text{N/m})(0.10\ \text{m})^2 = \frac{1}{2}(3.0\ \text{kg})v^2$$ $$v = 1\ \text{m/s}$$
44. **(D)**	Impulse, J, is defined as a change in momentum, Δp. Impulse can be calculated by multiplying the force, F, acting on an object by the elapsed time, t, that the force acts. The change in momentum is also equal to the mass, m, of an object multiplied by the change in velocity, Δv. $$J = \Delta p$$ $$Ft = (m\Delta v)$$ $$F(5\ \text{s}) = (1{,}500\ \text{kg})(10\ \text{m/s})$$ $$F = 3{,}000\ \text{N}$$
45. **(E)**	Linear momentum and total energy are always conserved, regardless if the collision is inelastic or elastic. However, in an inelastic collision, some of the kinetic energy is transformed into thermal energy, which is then lost as heat to the environment.
46. **(E)**	This is conservation of momentum. The total momentum of both objects added together must be the same before and after the collision. This is an explosion. Although the momentums of the two objects are in opposite directions, their magnitudes must be equal to conserve momentum. $$p_1 = p_2$$ $$m_1 v_1 = m_2 v_2$$ $$3mv = mv_2$$ $$v_2 = 3v$$

47. **(B)**	Orbits involve circular motion. The centripetal force, F_c, is the sum of the forces acting on the mass and is directed toward the center of the circular motion. For planets, this is the force of gravity. $$F_c = F_g$$ Two force-of-gravity equations can be substituted into this equation. Both are shown in the next step. Either one can be used to answer this question. $$m\frac{v^2}{r} = mg \quad \text{and} \quad m\frac{v^2}{r} = G\frac{mM}{r^2}$$ In both of these equations, the orbiting mass m cancels. The mass M in the right equation is the mass of the large central star. Since the mass of the orbiting body cancels, all planets at the same radius will have the same orbital speeds.
48. **(A)**	The measurement given is from the surface of Earth. The value needed for gravity calculations is the distance from the center of a planet. The point in space is located $3r_E$ from the *center* of Earth. The acceleration of gravity, g, at a point in space can be found using the gravity equation. $$g = G\frac{M_E}{r^2}$$ $$\left(\frac{1}{9}\right)g = G\frac{M_E}{(3r)^2}$$ Tripling the distance from the center of Earth will cause the gravity to be $\frac{1}{9}$ the value it is on the surface of Earth.
49. **(B)**	Coulomb's law can be expressed as $$F_E = k\frac{q_1 q_2}{r^2}$$ $$F_E = k\frac{(1\,\text{C})(2\,\text{C})}{(2\,\text{m})^2} = \frac{1}{2}k$$
50. **(E)**	By definition, the direction of an electric field is in the same direction as the force on a positive charge.
51. **(A)**	Electric potential, also called voltage, for a point charge can be determined by the following formula: $$V = k\Sigma\frac{q}{r} \qquad V = k\left(\frac{q_1}{|r_1|} + \frac{q_2}{|r_2|} + \dots\right)$$ When there is more than one point charge, the electric potential is the sum of the individual potentials. $$V = k\left(\frac{q_1}{|r_1|} + \frac{q_2}{|r_2|}\right) = k\left[\frac{(+1)}{(2)} + \frac{(-1)}{(2)}\right] = 0$$ The sign on the charge is included. However, the distance from each charge to the point in question is an absolute value.

52. **(E)**	This is conservation of energy. The electric potential energy is converted into kinetic energy. $$U_i + K_i = U_f + K_f$$ $$qV + 0 = 0 + \frac{1}{2}mv^2$$ $$(3\text{ C})(6.0\text{ V}) = \frac{1}{2}(1.0\text{ kg})v^2$$ $$v = 6\text{ m/s}$$
53. **(C)**	Add resistors R_1 and R_2 in parallel: $$\frac{1}{R_P} = \frac{1}{R_1} + \frac{1}{R_2} = \frac{1}{6\ \Omega} + \frac{1}{3\ \Omega} = \frac{1}{2\ \Omega}$$ $$R_{12} = R_P = 2\ \Omega$$ Add R_{12} and R_3 in series. $$R_S = R_{12} + R_3 = 2\ \Omega + 1\ \Omega = 3\ \Omega$$ $$R_{\text{total}} = R_S = 3\ \Omega$$ Use Ohm's law, $V = IR$, to determine the total current: $$I_{\text{total}} = \frac{V_{\text{battery}}}{R_{\text{total}}} = \frac{6\text{ V}}{3\ \Omega} = 2.0\text{ A}$$ Resistor R_3 is in series with the battery. In series, the current is constant: $$I_3 = I_{\text{total}} = 2.0\text{ A}$$
54. **(A)**	The current flowing through R_3 has been determined to be 2.0 A. Use Ohm's law, $V_3 = I_3 R_3$, to determine the voltage drop across R_3: $$V = (2.0\text{ A})(1\ \Omega) = 2\text{ V}$$
55. **(A)**	Any of these three power formulas can be used: $$P = IV = \frac{V^2}{R} = I^2 R$$ $$P_3 = I_3 V_3 = (2.0\text{ A})(2\text{ V}) = 4\text{ W}$$
56. **(B)**	Resistors in parallel receive the most current, dissipate the most power, and will also glow the brightest when they are lightbulbs.
57. **(C)**	The right-hand rule states that the thumb of the right hand points in the direction of motion of a charged particle, the extended fingers point in the direction of the magnetic field (into the page in this case), and the palm of the hand points in the direction of force (upward in this case). The right-hand rule applies to positively charged particles. Since this is a negatively charged particle, it will do the opposite and be forced downward in the $-y$ direction. This matches the answer if the left hand had been used instead of the right hand.
58. **(B)**	The induced emf, $\mathcal{E}$, is caused by a change in flux, ϕ. Flux can be determined by the area of the loop multiplied by the magnetic field passing through the loop. The amount of flux will change as the loop is moved with velocity v through the magnetic field, B. When a linear length of wire, L, enters a magnetic field, the emf generated in the wire can be determined with the following formula: $$\mathcal{E} = BLv$$

59. **(C)**	The period of a pendulum can be described as follows: $$T = 2\pi\sqrt{\dfrac{L}{g}}$$ The period is affected only by changes in the length, L, of a pendulum's string and the gravity field, g, it is in. Changes in the displacement, x, and mass, m, do not affect the period of a pendulum.
60. **(E)**	Wave speed is determined by the medium in which it travels. Since the medium has not changed, the wave speed must remain a constant v. The product of frequency and wavelength is the wave speed. Therefore, the wavelength must be reduced to $\dfrac{1}{2}\lambda$ when the frequency is doubled in order for wave speed to remain constant in the same medium.
61. **(A)**	When the two waves superimpose, they will add destructively since they are each on opposite sides of the axis of propagation. The wave pulse on the left is a larger and inverted version of the pulse on the right.
62. **(A)**	For strings and open tubes, the wavelengths of the harmonics can be found using the following formula, where n is the number assigned to the harmonic: $$\lambda_n = \dfrac{1}{n}\lambda_1$$ The second harmonic is $n = 2$. $$\lambda_2 = \left(\dfrac{1}{2}\right)\lambda_1 = \dfrac{1}{2}\lambda_1$$
63. **(D)**	For objects placed between the focus and a convex lens, the image will be magnified, virtual, and upright.
64. **(C)**	Light will refract toward a normal line drawn perpendicularly to the surface of the medium it is entering if that medium has a higher index of refraction than the one it is exiting. Glass has a higher optical density and a higher index of refraction than air.
65. **(C)**	A convex mirror is a divergent optical instrument. It can form only virtual images that are upright.
66. **(E)**	$$x_m = \dfrac{m\lambda L}{d}$$ In the above formula, x_m is a measure between the bright regions seen on the screen in Young's double-slit experiment. If the space between the bright regions decreases, then x_m will decrease, and vice versa. Decreasing the wavelength (λ), decreasing the distance from the slits to the screen (L), and increasing the slit spacing (d) will all decrease the x_m and the space between bright regions on the screen. If the experiment is performed under water, then the wave speed will decrease, causing a corresponding decrease in wavelength. All of these alterations will result in a closer spacing pattern.
67. **(B)**	This is the definition of diffraction.
68. **(C)**	When polarizing filters are oriented perpendicularly to each other, no light can pass through. When oriented in a parallel position, the maximum amount of light can pass through.

69. **(E)**	Heating a metal will cause the metal to expand. The entire ring, outer diameter, and inner diameter will all expand proportionally.
70. **(E)**	$$Q = mc \, \Delta T$$ $$Q = (3.0 \text{ kg})(2,000 \text{ J/kg} \cdot \text{K})(30° - 10°)$$ $$Q = 120,000 \text{ J}$$
71. **(C)**	$$e = \frac{T_H - T_C}{T_H}$$ $$= \frac{(500 + 273) - (100 + 273)}{(500 + 273)}$$ $$= \frac{400 \text{ K}}{773 \text{ K}} \approx 0.50$$ $$= 50\%$$ The temperatures must be in kelvins.
72. **(B)**	9 eV is enough energy to raise the electron from the ground state, $n = 1$, to an excited state at $n = 4$. The possible photon's energies emitted can be determined by taking the difference between any two energy levels on the electron's way back to the ground state. All of those energies are possible except for 2 eV.
73. **(E)**	Once the threshold frequency has been reached, the current begins. The current cannot increase with increasing the frequency. To increase the current, the intensity of the light must be increased. Increasing the intensity of the light increases the number of photons. This would cause more electrons to be ejected.
74. **(B)**	If each half-life lasts for 25 days, then 100 days is four half-lives. $$(100 \text{ days})\left(\frac{1 \text{ half-life}}{25 \text{ days}} \right) = 4 \text{ half-lives}$$ During each half-life the sample is reduced by half. $$\left(\frac{1}{2}\right)^4 = \left(\frac{1}{2}\right)\left(\frac{1}{2}\right)\left(\frac{1}{2}\right)\left(\frac{1}{2}\right) = \frac{1}{16}$$
75. **(A)**	As objects approach the speed of light, their length in the direction of motion will appear to decrease to an outside observer. Since this spacecraft is moving in the x-direction, only this dimension appears shorter.

ANSWER SHEET
Practice Test 2

1. (A) (B) (C) (D) (E)   21. (A) (B) (C) (D) (E)   41. (A) (B) (C) (D) (E)   61. (A) (B) (C) (D) (E)
2. (A) (B) (C) (D) (E)   22. (A) (B) (C) (D) (E)   42. (A) (B) (C) (D) (E)   62. (A) (B) (C) (D) (E)
3. (A) (B) (C) (D) (E)   23. (A) (B) (C) (D) (E)   43. (A) (B) (C) (D) (E)   63. (A) (B) (C) (D) (E)
4. (A) (B) (C) (D) (E)   24. (A) (B) (C) (D) (E)   44. (A) (B) (C) (D) (E)   64. (A) (B) (C) (D) (E)
5. (A) (B) (C) (D) (E)   25. (A) (B) (C) (D) (E)   45. (A) (B) (C) (D) (E)   65. (A) (B) (C) (D) (E)
6. (A) (B) (C) (D) (E)   26. (A) (B) (C) (D) (E)   46. (A) (B) (C) (D) (E)   66. (A) (B) (C) (D) (E)
7. (A) (B) (C) (D) (E)   27. (A) (B) (C) (D) (E)   47. (A) (B) (C) (D) (E)   67. (A) (B) (C) (D) (E)
8. (A) (B) (C) (D) (E)   28. (A) (B) (C) (D) (E)   48. (A) (B) (C) (D) (E)   68. (A) (B) (C) (D) (E)
9. (A) (B) (C) (D) (E)   29. (A) (B) (C) (D) (E)   49. (A) (B) (C) (D) (E)   69. (A) (B) (C) (D) (E)
10. (A) (B) (C) (D) (E)   30. (A) (B) (C) (D) (E)   50. (A) (B) (C) (D) (E)   70. (A) (B) (C) (D) (E)
11. (A) (B) (C) (D) (E)   31. (A) (B) (C) (D) (E)   51. (A) (B) (C) (D) (E)   71. (A) (B) (C) (D) (E)
12. (A) (B) (C) (D) (E)   32. (A) (B) (C) (D) (E)   52. (A) (B) (C) (D) (E)   72. (A) (B) (C) (D) (E)
13. (A) (B) (C) (D) (E)   33. (A) (B) (C) (D) (E)   53. (A) (B) (C) (D) (E)   73. (A) (B) (C) (D) (E)
14. (A) (B) (C) (D) (E)   34. (A) (B) (C) (D) (E)   54. (A) (B) (C) (D) (E)   74. (A) (B) (C) (D) (E)
15. (A) (B) (C) (D) (E)   35. (A) (B) (C) (D) (E)   55. (A) (B) (C) (D) (E)   75. (A) (B) (C) (D) (E)
16. (A) (B) (C) (D) (E)   36. (A) (B) (C) (D) (E)   56. (A) (B) (C) (D) (E)
17. (A) (B) (C) (D) (E)   37. (A) (B) (C) (D) (E)   57. (A) (B) (C) (D) (E)
18. (A) (B) (C) (D) (E)   38. (A) (B) (C) (D) (E)   58. (A) (B) (C) (D) (E)
19. (A) (B) (C) (D) (E)   39. (A) (B) (C) (D) (E)   59. (A) (B) (C) (D) (E)
20. (A) (B) (C) (D) (E)   40. (A) (B) (C) (D) (E)   60. (A) (B) (C) (D) (E)

Practice Test 2

Do not use a calculator. To simplify numerical calculations, use $g = 10 \text{ m/s}^2$.

PART A

Directions: In this section of the exam, the same lettered choices are used to answer several questions. Each group of questions is preceded by five lettered choices. When answering questions in each group, select the best answer from the available choices and fill in the corresponding bubble on the answer sheet. Each possible answer may be used once, more than once, or not at all.

Questions 1–3

The motion of five separate objects is shown in the following velocity-time graph.

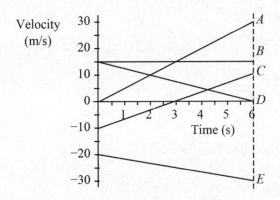

1. Which object is slowing the most during the entire 6-second elapsed time?

2. Which object has the greatest magnitude of acceleration?

3. Which object undergoes the greatest displacement during the 6-second time interval?

Questions 4–6

Choose the correct term from below to answer the following questions.

    (A) Kinetic energy
    (B) Potential energy
    (C) Total mechanical energy
    (D) Work
    (E) Power

4. The rate at which energy changes from one form to another or the rate at which energy is transferred into or out of a system.

5. The sum of the potential and kinetic energies of a system at a specific instant in time.

6. A quantity associated solely with the instantaneous position of an object.

Questions 7–9

The diagram below depicts an electron located between two charged plates. In addition, a uniform magnetic field is in the space between the plates. Use the diagram and the answers below to answer questions 7 to 9.

    (A) +x-direction
    (B) −x-direction
    (C) +y-direction
    (D) −y-direction
    (E) none, zero magnitude, and no direction

7. The direction of the electric field of the charged plates.

8. The direction of the electric force acting on the electron.

9. The direction of the magnetic force acting on the electron at the instant it begins to move.

Questions 10–12

Choose the correct term from below to answer questions 10 to 12.

  (A) Diffraction
  (B) Interference
  (C) Reflection
  (D) Refraction
  (E) Total internal reflection

10. The change in direction of light as it moves from one medium into another.

11. The spreading of waves that results when waves pass through an opening or encounter an obstacle.

12. The result of wave superposition.

Questions 13–15

Choose the correct experiment from below to answer questions 13 to 15.

  (A) Gold-foil experiment
  (B) Heat-work experiment
  (C) Michelson-Morley interferometer
  (D) Photoelectric-effect experiment
  (E) Young's double-slit experiment

13. Provided evidence that light has a particle characteristic.

14. Provided evidence proving that light has a wave characteristic.

15. A failed experiment, but its negative result indicated that the speed of light is constant regardless of the motion of the light source.

Questions 16–18

A gas is trapped in a cylinder with a movable piston. The choices listed below describe possible effects on the average kinetic energy of the gas particles, the pressure of the gas, and the volume of the gas as it undergoes a variety of thermodynamic processes.

Choose the answer from below that most correctly specifies the trends possible due to the aspects of the thermodynamic processes described in questions 16 to 18 below.

	Kinetic Energy	Pressure	Volume
(A)	Decrease	Increase	Constant
(B)	Constant	Increase	Decrease
(C)	Constant	Decrease	Increase
(D)	Increase	Constant	Increase
(E)	Increase	Increase	Constant

16. Work is done *on* a trapped gas, while temperature remains constant.

17. Work is done *by* a trapped gas, while temperature increases.

18. No work is done on a trapped gas, while temperature increases.

Questions 19–21

In this series of questions an object with a mass m and an initially velocity v_0 is moving to the right, in the positive X-direction, along a frictionless horizontal surface. The object is viewed from above, as shown in the diagram below. Questions 19 through 21 show possible force vectors that may be applied to the object in order to alter the motion of the object.

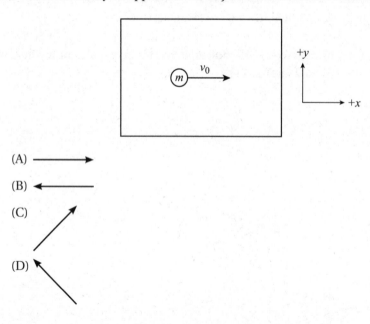

(E)

19. The application of this force vector will cause both the object's *x*-direction and its *y*-direction speeds to increase.

20. The application of this force vector will cause the object's *x*-direction speed to decrease while the *y*-direction speed increases.

21. The application of this force vector will slow the object but will not change the direction of the object.

22. Which of the following is NOT true for every possible motion involving a constant speed?

    (A) The magnitude of velocity is constant.
    (B) The object may be changing direction.
    (C) The acceleration is zero.
    (D) The magnitude of the net force acting on the object can be equal to or greater than zero.
    (E) The net work done on or by the object is zero.

23. A mass initially at rest experiences a constant acceleration a and is displaced a distance x, resulting in a final speed, v. What would be the final speed of the object if it moved the same distance x but with twice the acceleration, $2a$?

    (A) v
    (B) $\sqrt{2}(v)$
    (C) $2v$
    (D) $2\sqrt{2}(v)$
    (E) $4v$

24. Three objects, mass $1 = m$, mass $2 = 2m$, and mass $3 = 3m$, are taken to the top of a tall building. Mass 1 is thrown straight upward at a speed v, while mass 2 is thrown straight downward with a speed v, and mass 3 is released from rest. Air resistance is negligible. Which choice below correctly ranks the masses in order from least to greatest regarding both the flight time from launch at the top of the building to the ground level below and final speed at ground level?

Flight time	Final speed
(A) $2 < 1 < 3$	$3 < 2 < 1$
(B) $2 < 3 < 1$	$3 < 1 = 2$
(C) $2 < 3 < 1$	$1 < 3 < 2$
(D) $3 < 2 < 1$	$2 < 3 < 1$
(E) $3 < 2 < 1$	$1 = 2 = 3$

25. A projectile is launched with an initial speed v_0 at an angle θ above the horizontal. Air resistance is negligible. Which statement below is correct?

   (A) As the projectile moves upward acceleration is directed downward and its magnitude decreases until it reaches zero at maximum height. Then the projectile reverses direction and, while the acceleration continues to be directed downward, its magnitude increases as the object falls.

   (B) As the projectile moves upward acceleration is directed upward, and its magnitude decreases until it reaches zero at maximum height. Then the projectile reverses direction and the acceleration is now directed downward, and its magnitude increases as the object falls.

   (C) As the projectile moves upward acceleration is directed upward, and its magnitude increases until it reaches zero at maximum height. Then the projectile reverses direction and the acceleration is now directed downward, and its magnitude decreases as the object falls.

   (D) As the projectile moves upward acceleration is directed downward, and its magnitude increases until it reaches zero at maximum height. Then the projectile reverses direction and, while the acceleration continues to be directed downward, its magnitude decreases as the object falls.

   (E) As the projectile moves upward, reaches maximum height, reverses direction, and falls back to the ground the acceleration continually points downward and has the same constant magnitude the entire flight.

Questions 26–27

The object shown in the diagram below has mass m and weight w. The object is pulled to the right at constant velocity along a rough, horizontal surface by a string. The tension in the string has a magnitude T and is directed at an angle θ as measured from the horizontal.

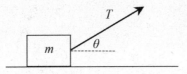

26. Determine the magnitude of the friction force.

   (A) 0
   (B) $T\cos\theta$
   (C) $T\sin\theta$
   (D) T
   (E) It cannot be determined without knowing the coefficient of friction.

27. Which of these correctly describes the relationship between the magnitude of the normal force, N, and the weight of the object, w?

   (A) $N = 0$
   (B) $N = m$
   (C) $N = w$
   (D) $N < w$
   (E) $N > w$

Questions 28–29

A 2.0-kilogram mass and a 1.0-kilogram mass are connected by a string. The masses are pulled along a horizontal surface by a 12-newton force.

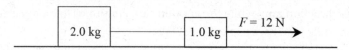

28. What is the acceleration of the 2.0-kilogram mass?

    (A) 1.0 m/s^2
    (B) 2.0 m/s^2
    (C) 3.0 m/s^2
    (D) 4.0 m/s^2
    (E) 6.0 m/s^2

29. Determine the magnitude of the tension in the string between the masses.

    (A) 2.0 N
    (B) 3.0 N
    (C) 4.0 N
    (D) 6.0 N
    (E) 8.0 N

---

30. A 60-kilogram person rides in an elevator that is accelerating upward at 1.0 meter per second squared. What is the apparent weight of the person?

    (A) 54 N
    (B) 66 N
    (C) 540 N
    (D) 600 N
    (E) 660 N

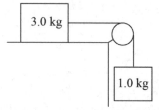

31. A 3.0-kilogram mass lies on a rough horizontal surface. It is attached to a 1.0-kilogram mass by a string draped over a pulley, as shown in the diagram above. What minimum coefficient of friction is needed in order for the blocks to remain at rest?

    (A) 0.25
    (B) 0.33
    (C) 0.50
    (D) 0.67
    (E) 0.75

32. A car on a flat section of road completes a turn that has a radius of 20 meters. The coefficient of friction between the tires and road is 0.50. What maximum speed can the car safely maintain in order to complete the turn without skidding?

(A) 5 m/s
(B) 10 m/s
(C) 15 m/s
(D) 20 m/s
(E) 25 m/s

33. A mass m is in uniform circular motion with a speed v and a radius r. How is the centripetal acceleration, a_c, affected if the radius is doubled while the speed remains constant?

(A) $\frac{1}{4}a_c$
(B) $\frac{1}{2}a_c$
(C) a_c
(D) $2a_c$
(E) $4a_c$

34. A roller coaster loop has a radius of 10.0 meters. What minimum speed is required at the top of the loop in order to complete the loop successfully?

(A) 2.5 m/s
(B) 5.0 m/s
(C) 7.5 m/s
(D) 10.0 m/s
(E) 12.5 m/s

Questions 35–36

An 8.0-kilogram mass is attached to a vertical spring. The mass is lowered 0.50 meters to equilibrium where it remains at rest. The stretching motion of the spring is plotted in the following force-displacement graph.

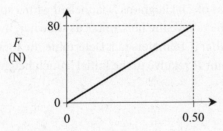

35. Determine the spring constant.

(A)   20 N/m
(B)   40 N/m
(C)   80 N/m
(D) 160 N/m
(E) 240 N/m

36. How much work is done on the spring to stretch it 0.50 meters?

(A) 10 J

(B) 20 J

(C) 40 J

(D) 80 J

(E) 160 J

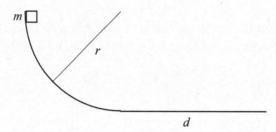

37. A mass m of 3.0 kilograms is released from rest on a surface that forms a quarter circle with a radius of 0.80 meters. When it reaches the bottom of the circular portion, the track is horizontal for a distance d of 1.2 meters. All surfaces are smooth and frictionless. Determine the speed of the mass when it reaches the end of the horizontal section of track.

(A) 2 m/s

(B) 4 m/s

(C) 6 m/s

(D) 12 m/s

(E) 16 m/s

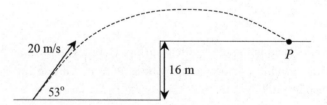

38. A projectile with a mass of 2.0 kilograms is launched with a speed of 20 meters per second at an angle of 53° above the horizontal, as shown in the figure above. It lands at point P on a building that is 16 meters tall. Determine the total mechanical energy the projectile strikes at point P, relative to the initial launch height.

(A) 100 J

(B) 200 J

(C) 400 J

(D) 800 J

(E) 1,600 J

39. Which of the following quantities is conserved in a perfectly elastic collision?

(A) Total velocity of the system
(B) Total linear momentum of the system
(C) Total kinetic energy of the system
(D) Both A and C
(E) Both B and C

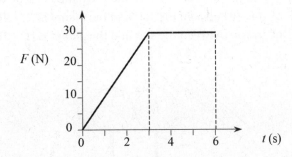

40. The graph above shows the force acting on an object during a 6-second time interval. What is the change in momentum during this elapsed time?

(A) 10 kg • m/s
(B) 60 kg • m/s
(C) 90 kg • m/s
(D) 135 kg • m/s
(E) 185 kg • m/s

Questions 41–42

A 3.0-kilogram mass moving with a speed of 2.0 meters per second is closing on a 2.0-kilogram mass moving with a speed of 1.0 meters per second, as shown in the diagram below. The masses collide and stick together.

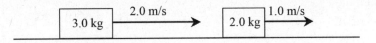

41. What is the speed of the combined masses after the collision?

(A) 1.2 m/s
(B) 1.6 m/s
(C) 1.8 m/s
(D) 2.0 m/s
(E) 2.4 m/s

42. What impulse is experienced by the 3.0-kilogram mass during the collision?

(A) −1.2 kg • m/s
(B) −1.6 kg • m/s
(C) −1.8 kg • m/s
(D) −2.0 kg • m/s
(E) −2.4 kg • m/s

43. A very large star of mass M and radius r undergoes a transformation into a neutron star. Assume that the mass remains constant while the radius of the star becomes 1/1,000 its original size, $(1 \times 10^{-3})r$. How does the change in the radius of the star affect the force of gravity, F_g, acting between the star and a planet that is orbiting the star?

(A) The force of gravity between the planet and the star remains unchanged.
(B) The new force of gravity between the star and the planet is $(1 \times 10^{-3})F_g$.
(C) The new force of gravity between the star and the planet is $(1 \times 10^{-6})F_g$.
(D) The new force of gravity between the star and the planet is $(1 \times 10^{-9})F_g$.
(E) The new force of gravity between the star and the planet is $(1 \times 10^{-12})F_g$.

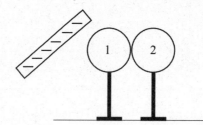

44. In the diagram above, two neutral, uncharged conducting spheres mounted on insulating stands are in contact with each other. A negatively charged rod is brought near the spheres but does not touch either sphere. While the negative rod is held in this position, the spheres are separated. As a result, sphere 1 now has a charge of Q. Which of these statements is true?

  I. Sphere 2 now has a $-Q$ charge.
 II. The spheres have been charged by induction.
III. If the spheres touch again, their total charge will be $2Q$.

(A) I only
(B) II only
(C) III only
(D) I and II only
(E) I, II, and III

Questions 45–46

In the diagram below, two charges, $q = -1$ coulomb and $Q = +4$ coulombs, are separated by a distance d of 1.0 meter.

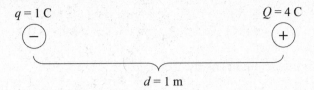

$q = 1$ C $Q = 4$ C

$d = 1$ m

45. Determine the location, as measured from charge q, where the electric field due to both charges has zero magnitude.

   (A) 0.25 m to the left of charge q
   (B) 0.33 m to the right of charge q
   (C) 0.33 m to the left of charge q
   (D) 1.00 m to the right of charge q
   (E) 1.00 m to the left of charge q

46. Charges q and Q attract one another with the electric force F_E. If the magnitude of both charges is doubled and the distance between the charges is also doubled, determine the new electric force acting on these charges.

   (A) F_E
   (B) $2F_E$
   (C) $4F_E$
   (D) $8F_E$
   (E) $16F_E$

Questions 47–49

Two charged plates each hold a charge Q of 3.0 coulombs, and they have a potential difference V of 6.0 volts. The plate spacing, d, is 20 centimeters. A positive charge, $q = 0.40$ coulombs, is located at the midpoint between the plates as shown in the diagram below.

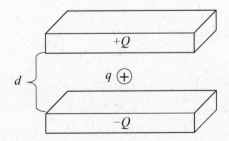

47. Determine the magnitude of the electric field between the plates.

   (A)   0.33 V/m
   (B)   1.5 V/m
   (C)   3.0 V/m
   (D) 30 V/m
   (E) 33 V/m

48. Determine the capacitance of the plates.

    (A) 0.33 F
    (B) 0.50 F
    (C) 2.0 F
    (D) 9.0 F
    (E) 18 F

49. What is the electric potential energy of charge q?

    (A) 0.20 J
    (B) 1.2 J
    (C) 1.5 J
    (D) 2.0 J
    (E) 2.4 J

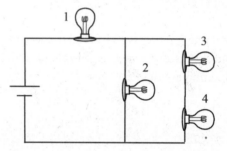

50. The lightbulbs connected in the circuit above all have identical resistances. Rank the lightbulbs in order from brightest to dimmest.

    (A) 1 > 2 > 3 = 4
    (B) 3 = 4 > 2 > 1
    (C) 1 > 3 = 4 > 2
    (D) 2 > 3 = 4 > 1
    (E) 1 = 2 > 3 = 4

Questions 51–53

Use the following diagram to answer questions 51 to 53.

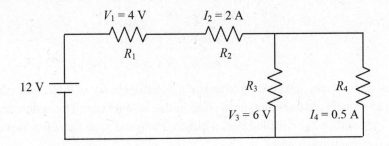

51. What is the resistance of R_1?

   (A) 0.5 Ω
   (B) 1 Ω
   (C) 2 Ω
   (D) 4 Ω
   (E) 16 Ω

52. What is the potential difference across R_2?

   (A) 1 V
   (B) 2 V
   (C) 4 V
   (D) 6 V
   (E) 12 V

53. What is the power consumption of R_3?

   (A) 1 W
   (B) 3 W
   (C) 6 W
   (D) 9 W
   (E) 12 W

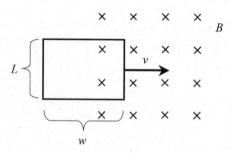

54. A sphere with a mass of 0.20 kilograms and a positive charge of 0.10 coulombs is moving at a speed of 10 meters per second in the +x-direction. The sphere enters a uniform 5.0-tesla magnetic field that is directed into the page (−z) as shown above. What is the resulting motion?

(A) The charge will circle clockwise with a 2.0 m radius.
(B) The charge will circle counterclockwise with a 2.0 m radius.
(C) The charge will circle clockwise with a 4.0 m radius.
(D) The charge will circle counterclockwise with a 4.0 m radius.
(E) The charge will circle clockwise with a 40 m radius.

55. Which of the following fields CANNOT change the speed of a charged mass that is acted upon by the field?

  I. Uniform gravity field
 II. Uniform electric field
III. Uniform magnetic field

(A) I only
(B) II only
(C) III only
(D) Both I and II
(E) Both II and III

56. A rectangular loop of wire with length L, width w, and resistance R is moved into a magnetic field, B, at a constant velocity, v. What is the magnitude of induced current in the loop as it enters the magnetic field?

(A) BLv

(B) $\dfrac{BLv}{R}$

(C) $BLvR$

(D) $\dfrac{R}{BLv}$

(E) $\dfrac{(BLv)^2}{R}$

57. A pendulum with a string length L oscillates with a period T. What does the string length need to be changed to in order to double the period?

(A) $\frac{\sqrt{2}}{2}L$

(B) $\sqrt{2}\,L$

(C) $2L$

(D) $2\sqrt{2}\,L$

(E) $4L$

58. Which graph correctly depicts the total energy during the oscillation of a frictionless spring-mass system?

(A)

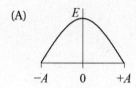

(B)

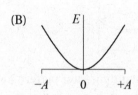

(C)

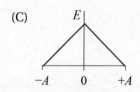

(D)

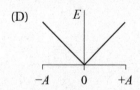

(E)

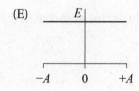

59. Sort the following electromagnetic waves in order from shortest to longest wavelength: infrared, gamma rays, microwaves, visible light, and radio waves.

(A) Microwaves, radio waves, visible light, infrared, gamma rays
(B) Gamma rays, infrared, visible light, radio waves, microwaves
(C) Gamma rays, visible light, infrared, microwaves, radio waves
(D) Radio waves, microwaves, visible light, infrared, gamma rays
(E) Microwaves, gamma rays, visible light, infrared, radio waves

60. Which term describes the separation of white light into separate colors in a glass prism due to small differences in the index of refraction for each wavelength?

(A) Diffraction
(B) Dispersion
(C) Interference
(D) Polarization
(E) Refraction

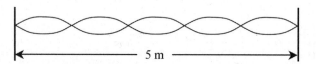

5 m

61. A 5-meter-long string is set into a vibration that creates the standing wave pattern in the diagram above. The speed of the wave in the string is 420 meters per second. What is the frequency of this vibration?

(A) 84 Hz
(B) 105 Hz
(C) 140 Hz
(D) 210 Hz
(E) 420 Hz

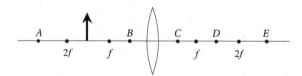

62. The object viewed by a convex lens is positioned outside of the focus, as shown in the diagram above. In which location will the image be formed?

(A) A
(B) B
(C) C
(D) D
(E) E

63. Light passes from glass (medium 1) into air (medium 2). The angle of the light in the glass (θ_1) is adjusted until total internal reflection is observed. Which of these represents the index of refraction of the glass, n_1?

(A) $\dfrac{1}{\sin \theta_1}$

(B) $\dfrac{1}{\sin \theta_2}$

(C) $\sin \theta_1$

(D) $\sin \theta_2$

(E) $(\sin \theta_1)(\sin \theta_2)$

64. An incident ray of white light passes from air into glass. Which color of light will experience the greatest refraction?

(A) Red
(B) Green
(C) Blue
(D) Violet
(E) None. All colors of light will be refracted at the same angle.

65. Monochromatic light is incident upon a single slit. The light passes through the slit and is projected on a screen behind the slit. Which of these describes the pattern seen on the screen?

(A) There is a single maximum consistent with a single slit.
(B) Light spreads out completely and fills the entire screen evenly.
(C) There is a large central maximum with faint secondary maxima.
(D) The pattern is identical to the results obtained using a double slit.
(E) The light converges into a single bright line.

66. When light passes from air into glass, how are its frequency, wavelength, and speed affected?

	Frequency	Wavelength	Speed
(A)	Decreases	Increases	Same
(B)	Decreases	Same	Increases
(C)	Same	Decreases	Decreases
(D)	Same	Decreases	Increases
(E)	Increases	Same	Increases

Questions 67–68

Mercury is used in thermometers to measure temperatures. The specific heat of mercury is 140 joules/kilogram • kelvin, and the coefficient of linear expansion for mercury is $60 \times 10^{-6}\,\text{K}^{-1}$.

67. How much heat must be added to increase the temperature of 0.002 kilograms of mercury by 10°C?

    (A) 0.028 J
    (B) 2.8 J
    (C) 4.2 J
    (D) 42 J
    (E) 420 J

68. Determine the change in length of a 0.20-meter column of mercury when the temperature increases by 10°C.

    (A) $1.2 \times 10^{-7}\,\text{m}$
    (B) $3.0 \times 10^{-5}\,\text{m}$
    (C) $1.2 \times 10^{-4}\,\text{m}$
    (D) $3.0 \times 10^{-4}\,\text{m}$
    (E) $3.4 \times 10^{-3}\,\text{m}$

---

69. A gas is trapped in a cylinder with a movable piston. During several thermodynamic processes, the pressure, P, and volume, V, of the gas are changed in order to increase the temperature of the gas from T to $6T$. Which of the following is NOT capable of creating this change in temperature?

    (A) Increase the pressure by a factor of 6 while holding the volume constant.
    (B) Triple the pressure, and double the volume.
    (C) Increase the pressure by a factor of 12 while reducing the volume to $\frac{1}{2}$.
    (D) Increase the pressure by a factor of 15 while reducing the volume to $\frac{1}{3}$.
    (E) Reduce the pressure to $\frac{1}{4}$ while increasing the volume by a factor of 24.

Questions 70–71

A photoelectric-effect experiment is conducted, and the results are graphed in the following kinetic energy versus frequency graph.

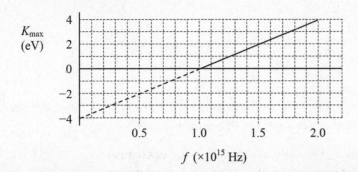

70. Which answer below correctly states the work function, ϕ, the threshold frequency, f_0, and the value of Planck's constant, h, as calculated from these experimental results?

	Work function	Threshold frequency	Planck's constant
(A)	4 eV	1.0×10^{15} Hz	4×10^{-15} eV • s
(B)	1.0×10^{15} Hz	4 eV	4×10^{-15} eV • s
(C)	4 eV	1.0×10^{15} Hz	8×10^{-15} eV • s
(D)	1.0×10^{15} Hz	4 eV	8×10^{-15} eV • s
(E)	–4 eV	2.0×10^{15} Hz	4×10^{-15} eV • s

71. If the photoelectric material is replaced with a new material that has a greater work function, how will the graph change?

(A) The graph will not change at all.
(B) The y-intercept will remain the same, but the slope of the graph will decrease.
(C) The y-intercept will remain the same, but the slope of the graph will increase.
(D) The y-intercept will change, and the slope of the graph will increase.
(E) The y-intercept will change, but the slope of the graph will remain the same.

72. An isotope of polonium, $^{218}_{84}$Po, undergoes beta decay. In the process, the atom becomes an isotope of astatine. Which of the following is the result of this transmutation?

(A) $^{214}_{82}$At

(B) $^{217}_{84}$At

(C) $^{219}_{84}$At

(D) $^{218}_{85}$At

(E) $^{222}_{86}$At

73. How many neutrons are liberated during the following nuclear reaction?

$$^{235}_{92}U + ^{1}_{0}n \rightarrow ^{90}_{38}Sr + ^{143}_{54}Xe + ?\left(^{1}_{0}n\right) + Energy$$

(A) 1
(B) 2
(C) 3
(D) 4
(E) 5

74. An object moving in the x-direction accelerates rapidly. As the object nears the speed of light, a stationary observer would report that the object's

(A) mass increases while its dimensions are unchanged
(B) mass decreases while its dimensions are unchanged
(C) mass increases and its length in the x-direction decreases
(D) mass decreases and its length in the x-direction decreases
(E) mass increases while its length, width, and depth all decrease

75. The mass of galaxies seems inconsistent with observed gravitational effects. Which of the following has been proposed to account for the missing mass in the galaxies?

(A) Chaos theory
(B) Dark matter
(C) String theory
(D) The general theory of relativity
(E) The special theory of relativity

ANSWER KEY
Practice Test 2

1.	D	26.	B	51.	C
2.	A	27.	D	52.	B
3.	E	28.	D	53.	D
4.	E	29.	E	54.	D
5.	C	30.	E	55.	C
6.	B	31.	B	56.	B
7.	B	32.	B	57.	E
8.	A	33.	B	58.	E
9.	D	34.	D	59.	C
10.	D	35.	D	60.	B
11.	A	36.	B	61.	D
12.	B	37.	B	62.	E
13.	D	38.	C	63.	A
14.	E	39.	E	64.	D
15.	C	40.	D	65.	C
16.	B	41.	B	66.	C
17.	D	42.	A	67.	B
18.	E	43.	A	68.	C
19.	C	44.	D	69.	D
20.	D	45.	E	70.	A
21.	B	46.	A	71.	E
22.	C	47.	D	72.	D
23.	B	48.	B	73.	C
24.	B	49.	B	74.	C
25.	E	50.	A	75.	B

DIAGNOSTIC CHART

Subject Area	Question Numbers	Questions Incorrect	Chapter(s) to Study
Mechanics	1, 2, 3, 4, 5, 6, 19, 20, 21, 22, 23, 24, 25, 26, 27, 28, 29, 30, 31, 32, 33, 34, 35, 36, 37, 38, 39, 40, 41, 42, 43, 57, 58		1–9
Electricity and Magnetism	7, 8, 9, 44, 45, 46, 47, 48, 49, 50, 51, 52, 53, 54, 55, 56		10–13
Waves and Optics	10, 11, 12, 14, 49, 60, 61, 62, 63, 64, 65, 66		14–17
Heat and Thermodynamics	16, 17, 18, 67, 68, 69		18–19
Modern Physics	13, 15, 70, 71, 72, 73, 74		20–22
Miscellaneous	75		23

SCORING YOUR TEST

How to Determine Your Raw Score

Your raw score is the number of correctly answered questions minus the incorrectly answered questions multiplied by ¼. An incorrectly answered question is one that you bubbled in but was incorrect. If you leave the answer blank, it does not count as an incorrect answer.

Number of correctly answered questions: _____

Number of incorrectly answered questions: _____ $\times \dfrac{1}{4}$ = _____

_____ $-$ _____ = _____

Number Correct   Number Incorrect $\times \dfrac{1}{4}$ Raw Score

ANSWERS EXPLAINED

1. **(D)**	To identify if an object is speeding up or slowing down, assess the trend in the speed (absolute value of velocity) of each object. Object A is speeding up from 0 to 30 m/s. Object B has a constant speed of 15 m/s. Object C is slowing during the first 3 seconds and then speeding up during the next three seconds. Object D is slowing during the entire interval from 15 m/s to zero and is therefore the correct answer. Object E does have negative velocity. However, its speed (absolute value of velocity) is increasing from 20 m/s to 30 m/s.
2. **(A)**	Acceleration is the slope of the velocity-time graph. The magnitude of acceleration is the absolute value of the slope. The greatest acceleration will have the steepest slope. Object A has the steepest slope and therefore the greatest acceleration, 5 m/s^2.
3. **(E)**	The area between the velocity-time graph and the x-axis is the displacement of an object. The area between the graph of object E and the x-axis represents a displacement of 135 m from the origin in the negative direction. Although negative, it is still the largest displacement.
4. **(E)**	Work is a change in energy. Power is the rate of work, which is also the rate of change in energy.
5. **(C)**	Total mechanical energy is the sum of the kinetic and potential energies of a system.
6. **(B)**	Potential energy is the energy associated with the instantaneous position of an object.
7. **(B)**	The direction of the electric field is the same as the direction of the force on a positive test charge. This means the electric-field lines are drawn away from positive plates, or charges, and toward negative plates or charges.
8. **(A)**	Negative charges move in the direction opposite that of electric-field lines.
9. **(D)**	The electric force described in answer 8 will begin to accelerate the electron to the right. Use the right-hand rule to find the force of magnetism on the moving charge. The thumb of the right hand points in the direction of motion of a charged particle, the extended fingers point in the direction of the magnetic field (into the page in this case), and the palm of the hand points in the direction of force (upward in this case). The right-hand rule applies to positively charged particles. Since this is a negatively charged particle, it will do the opposite and be forced downward in the $-y$-direction. As an alternative, you can use the left hand to solve for negative charges moving in magnetic fields.
10. **(D)**	Refraction is the bending of light, or any wave, as it moves from one medium into another.
11. **(A)**	Diffraction is the spreading of waves that results when waves pass through an opening or encounter an obstacle.
12. **(B)**	Wave superposition causes interference patterns in waves.

13. **(D)**	The photoelectric-effect experiment provided evidence that light behaves as a particle made up of discrete packets of energy known as quanta.
14. **(E)**	The interference patterns observed by passing light through two small slits in Young's experiment imitated the behavior of water waves passing through two small slits.
15. **(C)**	While trying to prove the existence of the ether, the medium in space in which light waves supposedly travel, the Michelson-Morley experiment failed to find the ether but did determine an accurate measurement for the speed of light.
16. **(B)**	In order for work to be done *on* a trapped gas the gas must be compressed, decreasing its volume. The average kinetic energy of gas particles is directly proportional to the temperature of the gas. If temperature is constant, then average kinetic energy is constant. While the question does not specify how pressure is affected, B is the only choice where kinetic energy is constant and volume decreases.
17. **(D)**	Volume expands when work is done *by* a trapped gas, which corresponds to answers C and D. If temperature increases, then the average kinetic energy of the gas particles must increase, which corresponds to answer D.
18. **(E)**	When no work is done *on* a gas the volume remains constant. This matches answers A and E. If temperature increases, then the average kinetic energy of the gas particles also increases. The faster-moving molecules would experience greater momentum changes as they collide with the surfaces of the container, causing an increase in pressure as well. This leaves only answer E as the correct choice.
19. **(C)**	The object is moving rightward in the +x-direction. For a force to increase its speed in this direction, there must be a component of force directed to the right. The object has zero initial velocity in the y-direction. For a force to increase its speed in this direction, there must be a component of force in the y-direction. Only the vector in answer C possesses both a rightward component of force and a component of force in the y-direction.
20. **(D)**	A force directed leftward, $-x$-direction, is needed to slow the rightward-moving object. For a force to increase the speed of the object in the y-direction there must be a component of force in the y-direction. Only the vector in answer D possesses both a leftward component of force and a component of force in the y-direction.
21. **(B)**	A force directed leftward, $-x$-direction, is needed to slow the rightward-moving object. The y-direction speed will be constant if no net force acts in this direction.

22. **(C)**	An object moving at a constant speed may be turning. When an object turns, it is changing direction and is therefore changing its velocity. If the velocity is changing, the object is accelerating. Uniform circular motion is an example. The magnitude of velocity is constant (A), the object is changing direction (B), and centripetal force is the net force acting on the object and is greater than zero (D). The work–kinetic energy theorem states that the net work is equal to a change in kinetic energy. If an object is moving at constant speed, then the change in kinetic energy and net work are both zero (E).
23. **(B)**	$$v_f^2 = v_i^2 + 2ax$$ The equation above simplifies to the following when $v_i = 0$ m/s: $$v = \sqrt{2ax}$$ Velocity is proportional to the square root of acceleration. $$\left(\sqrt{2}\right)v = \sqrt{2(2a)x}$$ Doubling acceleration increases the final velocity by $\sqrt{2}$.
24. **(B)**	All three masses experience the same acceleration of gravity acting downward. The value of mass is a distracter that can be ignored. Masses 2 and 3 both start at the same position, move through the same displacement, and experience the same acceleration. However, mass 2 already has an initial downward velocity. This means that it will have the shortest time of flight and arrive with the greatest speed. This narrows the choices to B and C. Mass 1 is thrown upward at speed v, rises to maximum height, reverses direction, and descends. When it reaches its initial height, mass 1 will have a downward speed v that is equal to its upward launch speed. This is the same downward speed that mass 2 possessed at the same location. As a result, mass 1 will have the same final speed as mass 2.
25. **(E)**	In the absence of air resistance, the net force acting on a projectile consists only of the force of gravity. Under these conditions, projectiles are subject only to the acceleration of gravity, which has a constant magnitude of 9.8 m/s^2 and constant downward direction.
26. **(B)**	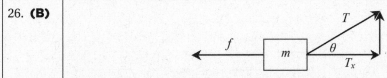 The friction force must be equal but opposite to the horizontal component of force, T_x, in order to maintain a constant velocity. $$f = T\cos\theta$$
27. **(D)**	The string is lifting the object by an amount equal to $T\sin\theta$. The sum of the normal force, N, and $T\sin\theta$ is equal to the weight of the object. Therefore, the normal force must be less than the weight of the object. $$N + T\sin\theta = w$$

28. **(D)**	Treat the blocks as a single system with a combined mass.
	$$\Sigma F_{sys} = F$$ $$(m_1 + m_2)a = F$$ $$a = \frac{F}{m_1 + m_2}$$ $$= \frac{(12\ N)}{(2.0\ kg) + (1.0\ kg)}$$ $$= 4\ m/s^2$$
29. **(E)**	The string between the two blocks provides the force accelerating the 2.0 kg block at the same rate as the acceleration of the system. $$\Sigma F_{2.0\ kg\ block} = T$$ $$ma = T$$ $$T = (2.0\ kg)(4.0\ m/s^2)$$ $$T = 8\ N$$
30. **(E)**	Apparent weight is equal to the normal force acting on the person. 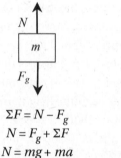 $$\Sigma F = N - F_g$$ $$N = F_g + \Sigma F$$ $$N = mg + ma$$ $$N = m(g + a) = (60\ kg)(10\ m/s^2 + 1\ m/s^2) = 660\ N$$
31. **(B)**	In order for the system to remain at rest, the opposing forces acting on the two masses must be equal and opposite. The tension in the string cancels. Set the magnitude of the friction force acting on mass 1 equal to the force of gravity acting on mass 2. 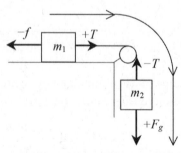 $$f_1 = F_{g2}$$ $$\mu m_1 g = m_2 g$$ $$\mu = \frac{m_2}{m_1} = \frac{(1.0\ kg)}{(3.0\ kg)} = 0.33$$

32. **(B)**	The friction force, f, holds the car in the turn and creates the net centripetal force, F_c, acting toward the center of the circular motion.$$F_c = f$$$$m\frac{v^2}{r} = \mu mg$$$$v = \sqrt{\mu rg}$$$$= \sqrt{(0.5)(20\text{ m})(10\text{ m/s}^2)}$$$$= 10\text{ m/s}$$
33. **(B)**	$$a_c = \frac{v^2}{r}$$Centripetal acceleration is equal to the square of the velocity divided by the radius. If the radius were doubled, the acceleration would be halved.$$\left(\frac{1}{2}\right)a_c = \frac{v^2}{(2r)}$$
34. **(D)**	At the top of the loop, both the normal force to the track and the force of gravity act downward toward the center of the circle. Both forces are positive.$$F_c = F_g + N$$$$m\frac{v^2}{r} = mg + N$$The mass, radius, and gravity are all fixed variables that cannot be reduced. However, as the speed of the roller coaster is reduced, the normal force of the track decreases. The minimum speed occurs when the normal force reaches zero at the top of the loop.$$m\frac{v^2}{r} = mg$$$$v = \sqrt{rg}$$$$= \sqrt{(10\text{ m})(10\text{ m/s}^2)}$$$$= 10\text{ m/s}$$
35. **(D)**	The spring constant is the slope of a force-displacement graph.$$k = \frac{80\text{ N} - 0\text{ N}}{0.50\text{ m} - 0\text{ m}} = 160\text{ N/m}$$
36. **(B)**	The work is the area underneath a force-displacement graph.$$W_s = \frac{1}{2}(80\text{ N})(0.50\text{ m}) = 20\text{ J}$$

37. **(B)**	This is conservation of energy. The potential energy at the top of the curved section of track is transformed into kinetic energy. The initial height is equal to the radius of the curved section of track, $h = r$. $$U_i + K_i = U_f + K_f$$ $$mgr + 0 = 0 + \frac{1}{2}mv^2$$ $$v = \sqrt{2gr}$$ $$= \sqrt{2\left(10 \text{ m/s}^2\right)(0.8 \text{ m})}$$ $$= 4 \text{ m/s}$$
38. **(C)**	Even though the question asks for the total energy at point P, you do not need to solve for projectile motion and then determine the landing velocity. Total mechanical energy is conserved and is the same at the beginning and at the end of the problem. It is easier to solve for the total mechanical energy at the beginning where it consists entirely of kinetic energy. $$\Sigma E_P = \Sigma E_i = KE = \frac{1}{2}mv^2$$ $$= \frac{1}{2}(2.0 \text{ kg})(20 \text{ m/s})^2$$ $$= 400 \text{ J}$$
39. **(E)**	Linear momentum is conserved during both elastic and inelastic collisions. In perfectly elastic collisions, kinetic energy is conserved as well.
40. **(D)**	Change in momentum is equal to the area underneath a force-time graph. $$\Delta p = \frac{1}{2}(30 \text{ N})(3 \text{ s}) + (30 \text{ N})(3) = 135 \text{ kg} \bullet \text{m/s}$$
41. **(B)**	This is conservation of momentum. This collision is perfectly inelastic. The mass sticks together to form one large, combined mass. $$m_1 v_{1i} + m_2 v_{2i} = (m_1 + m_2)v_f$$ $$(3.0 \text{ kg})(2.0 \text{ m/s}) + (2.0 \text{ kg})(1.0 \text{ m/s}) = (3.0 \text{ kg} + 2.0 \text{ kg})v_f$$ $$v_f = 1.6 \text{ m/s}$$
42. **(A)**	Impulse is equal to the change in momentum. This question is asking for the change in momentum of only the 3.0 kg mass. $$J = \Delta p = mv_f - mv_i = m(v_f - v_i)$$ $$J = (3.0 \text{ kg})(1.6 \text{ m/s} - 2.0 \text{ m/s}) = -1.2 \text{ kg} \bullet \text{m/s}$$
43. **(A)**	The universal law of gravitation, which can be expressed as: $$F_g = G\frac{m_1 m_2}{r^2}$$ The radial distance, r, between the two masses is a line drawn from the center of one mass to the center of the other mass. The actual radius of the masses is not a factor in the equation. Since the star did not lose any mass as it collapsed, the force between the star and planet remains unchanged.

44. **(D)**	The negative rod near sphere 1 causes negative charges to be repelled from sphere 1 onto sphere 2. As a result, sphere 1 has a greater positive charge while sphere 2 has a greater negative charge. If the spheres are separated, the spheres become charged. Since the rod did not touch either sphere, this was accomplished by induction. If sphere 1 receives a charge of Q, then due to conservation of charge sphere 2 must have an equal and opposite charge of $-Q$. If the spheres are again touched, they will neutralize, and neither sphere will then have a charge.
45. **(E)**	For the total electric field to equal zero, the individual electric-field vectors of the two charges must be opposite and equal to one another. The electric-field vector due to q points toward q, and the electric-field vector due to Q points away from Q. The only locations where these two field vectors are opposite each other is to the left of charge q and to the right of charge Q. However, in order for the magnitudes of the vectors to be equal, the zero point has to be closer to the smaller charge q. Therefore, the only location where the electric fields can cancel is to the left of charge q. This narrows the choices to A, C, or E. The magnitude of the electric field of each point charge can be found using the following equations: $$E_1 = k\frac{q}{r_1^2} \qquad \text{and} \qquad E_2 = k\frac{Q}{r_2^2}$$ r_1 and r_2 are the distances from each charge to the zero point. 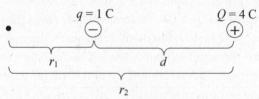 In order to cancel each other, E_1 and E_2 must be equal. So, the location of the zero point can be found by setting the two equations equal to each other where $r_2 = r_1 + d$. $$k\frac{q}{r_1^2} = \frac{Q}{(r_1+d)^2} \qquad \text{where } k \text{ cancels}$$ Since this is a multiple-choice question, a guess-and-check approach may be more efficient than solving the equations. Substitute the value for r_1 and set r_2 equal to $r_1 + d$. $$\frac{(1\text{ C})}{(1\text{ m})^2} = \frac{(4\text{ C})}{(1\text{ m} + 1\text{ m})^2}$$
46. **(A)**	$$F_E = k\frac{qQ}{r^2}$$ Doubling the values for q, Q, and r would yield: $$F_E = k\frac{(2q)(2Q)}{(2r)^2}$$ This is equivalent to the original expression.

47. **(D)**	$$V = Ed$$ $$E = \frac{V}{d} = \frac{(6.0\ \text{V})}{(0.20\ \text{m})} = 30\ \text{V/m}$$
48. **(B)**	$$C = \frac{Q}{V} = \frac{(3.0\ \text{C})}{(6.0\ \text{V})} = 0.50\ \text{F}$$
49. **(B)**	The electric potential energy, U_E, of charge, q, located a distance, d, from the charged plate that it is attracted to can be determined as follows: $$U_E = qEd = (0.40\ \text{C})(30\ \text{N/C})\,(0.10\ \text{m}) = 1.2\ \text{J}$$ The charge is located at the midpoint between the plates. Its energy is related to its position, not the distance between the plates. If the positive charge had been located initially on the positive plate, then the distance between the plates would have been used.
50. **(A)**	Brightness is determined by the power dissipated in the bulb. Power can be determined by using the following formula: $$P = I^2 R$$ All of the bulbs have the same resistance, R, so their brightness is determined by the amount of current flowing through them. All of the current must pass through bulb 1. After flowing past bulb 1, the current reaches a junction. Some current must go down the path leading to bulb 2, and the rest of the current must go down the path leading to bulbs 3 and 4. Current will take the path of least resistance, so more current will flow down the path leading to bulb 2 than the path leading to bulbs 3 and 4. However, the exact same amount of current will pass through both bulbs 3 and 4 because they are in series and receive the same current. As a result, bulbs 3 and 4 will have the same brightness.
51. **(C)**	The voltage of R_1 is shown to be 4 V. The current flowing through R_2 is shown to be 2 A. This same current must also flow through R_1 because they are in series with each other. Using Ohm's law, $V = IR$, the resistance of R_1 can be determined: $$R_1 = \frac{V_1}{I_1} = \frac{(4\ \text{V})}{(2\ \text{A})} = 2\ \Omega$$
52. **(B)**	The components in any loop of the circuit must use the voltage supplied by the battery. A loop exists containing R_1, R_2, and R_3. These must add up to the voltage of the battery. $$V_{\text{battery}} = V_1 + V_2 + V_3$$ $$(12\ \text{V}) = (4\ \text{V}) + V_2 + (6\ \text{V})$$ $$V_2 = 2\ \text{V}$$

53. **(D)**	Power can be determined using the following formula: $$P = IV$$ The voltage drop across R_3 is shown to be 6 V. The current across R_3 is equal to the total current flowing in the circuit *minus* the current flowing through the parallel resistor, R_4. The total current of 2 A is flowing through R_2 and splits up between R_3 and R_4. Since the current flowing through R_4 is 0.5 A, the current flowing in R_3 must be 2.0 A – 0.5 A = 1.5 A. $$P = IV = (1.5\ \text{A})(6\ \text{V}) = 9\ \text{W}$$
54. **(D)**	The right-hand rule states that the thumb of the right hand points in the direction of motion of a charged particle, the extended fingers point in the direction of the magnetic field (into the page in this case), and the palm of the hand points in the direction of the force (upward in this case). The right-hand rule applies to positively charged particles. The charge will circle counterclockwise. The magnitude of its radius can be determined by setting the centripetal force, F_c, equal to the force of the magnetic field on a moving charge. $$F_c = F_B$$ $$m\frac{v^2}{r} = qvB$$ $$r = \frac{mv}{qB} = \frac{(0.2\ \text{kg})(10\ \text{m/s})}{(0.1\ \text{C})(5.0\ \text{T})} = 4.0\ \text{m}$$
55. **(C)**	Magnetic fields apply a force in a direction that is perpendicular to the motion of the object. As a result, the forward speed is not changed; only the velocity is changed. Velocity changes because there is a change in direction but not in magnitude.
56. **(B)**	The induced emf, $\mathcal{E}$, is caused by a change in flux, ϕ. Flux can be determined by the area of the loop multiplied by the magnetic field passing through the loop. The amount of flux will change as the loop is moved with velocity v through the magnetic field, B. The equation to describe this is $\mathcal{E} = BLv$. The $\mathcal{E}$ is essentially an induced voltage. The current can be found by applying Ohm's law: $\mathcal{E} = V = IR$. $$I = \frac{V}{R} = \frac{BLv}{R}$$
57. **(E)**	$$T_P = 2\pi\sqrt{\frac{L}{g}}$$ In order to double the period T, the length must be quadrupled. $$(2T_P) = 2\pi\sqrt{\frac{(4L)}{g}}$$
58. **(E)**	During the oscillation, the total amount of energy will not change. The kinetic energy and potential energy will transform between each other, but their sum will remain constant.
59. **(C)**	This is the order of the electromagnetic spectrum listed from shortest to longest wavelength. It is also the electromagnetic spectrum listed from highest to lowest frequency.

60. **(B)**	Dispersion is caused by the slightly different indexes of refraction within a particular medium depending upon the wavelength. White light is made up of multiple wavelengths that experience dispersion in a prism.
61. **(D)**	There are 2.5 wavelengths visible in the drawing: $2.5\lambda = 5$ m.$$\lambda = 2 \text{ m}$$$$v = f\lambda$$$$f = \frac{v}{\lambda} = \frac{(420 \text{ m/s})}{(2 \text{ m})} = 210 \text{ Hz}$$
62. **(E)**	An object placed between the focal point, f, and twice the focal point, $2f$, will form a real and inverted image beyond the $2f$ point on the opposite side of a convex lens.
63. **(A)**	Snell's law can be expressed as:$$n_1 \sin \theta_1 = n_2 \sin \theta_2$$When total internal reflection occurs, θ_2 will equal 90°, and the sine of 90° is 1. Air, n_2, has an index of refraction of 1.$$n_1 \sin \theta_1 = (1)(1)$$$$n_1 = \frac{1}{\sin \theta_1}$$
64. **(D)**	Different wavelengths of light have slightly different indexes of refraction in glass. The shorter the wavelength, the greater the refraction resulting in the dispersion of the colors of light.
65. **(C)**	This pattern is caused by the interference of the light upon itself as it passes through the single slit and is evidence of the wave nature of light.
66. **(C)**	When light passes from one medium into another, the frequency does not change. Light moving from air, which has nearly the same index of refraction as a vacuum, will slow down.$$v = f\lambda$$When the frequency is constant, wavelength is directly proportional to wave speed. If the speed decreases, then the wavelength also decreases.
67. **(B)**	Heat, Q, is equal to the mass, m, multiplied by the specific heat, c, and the change in temperature, ΔT:$$Q = mc\,\Delta T$$$$Q = (0.002 \text{ kg})(140 \text{ J/kg} \cdot \text{K})(10 \text{ K}) = 2.8 \text{ J}$$The coefficient of linear expansion is not needed for this part. When in doubt, use kelvins. However, if the formula involves a change in temperature, ΔT, then either the Celsius or the Kelvin scale can be used; 1° is equal to 1 K.
68. **(C)**	The change in length, ΔL, is equal to the coefficient of linear expansion, α, multiplied by the original length, L_0, and the change in temperature, ΔT:$$\Delta L = \alpha L_0 \Delta T$$$$= (60 \times 10^{-6} \text{ K}^{-1})(0.20 \text{ m})(10 \text{ K})$$$$= 1.2 \times 10^{-4} \text{ m}$$

69. **(D)**	Use the ideal gas law. The change in both pressure and volume must offset the increase by a factor of 6 in temperature. $$PV = nRT$$ $$(\_\_\_P)(\_\_\_V) = nR(6T)$$ Find the combination that does not work. $$(15P)\left(\frac{1}{3}V\right) \neq nR(6T)$$ Increasing the pressure by a factor of 15 and decreasing the volume to $\frac{1}{3}$ will only raise the temperature to 5 times its initial value.
70. **(A)**	When the threshold frequency is reached, photoelectrons will be emitted with a kinetic energy above 0. According to the graph, that occurs when the frequency is equal to 1.0×10^{15} Hz. That narrows the choices to A and C. The work function is the amount of energy that must be added to reach the threshold frequency. The y-intercept of the graph, -4 eV, indicates the energy of the electrons occupying the lowest energy level inside the atom. To reach the threshold frequency, $+4$ eV of energy must be added to these electrons. Therefore, the work function is $+4$ eV, which is also consistent with answers A and C. Planck's constant is the slope of the function. $$h = \text{slope} = \frac{4 \text{ eV} - 0 \text{ eV}}{(2 \times 10^{15} \text{ Hz}) - (1 \times 10^{15} \text{ Hz})} = 4 \times 10^{-15} \text{ eV} \cdot \text{s}$$ This narrows the answer to A.
71. **(E)**	The slope of the line is Planck's constant. Constants do not change, so the slope of the line cannot change.
72. **(D)**	Beta decay occurs when a neutron releases an electron and becomes a proton. The result will be an increase in the atomic number by 1 but no change to the atomic mass because the mass of a neutron is essentially the same as that of a proton. The loss of mass from the release of an electron is insignificant.
73. **(C)**	In a balanced nuclear equation, the sum of the mass numbers on the left side of the equation must equal the sum of the mass numbers on the right side of the equation. Similarly, the sum of the atomic numbers on both sides of the equation must be equal. The sum of the mass numbers on the left side is 236. A liberation of 3 neutrons would result in a total mass number of 236 on the right side of the equation.
74. **(C)**	The result of an object's reaching the speed of light is that its mass will increase and its length will shorten in the direction of travel.
75. **(B)**	The motion of stars and galaxies is not consistent with the mass that can be seen with telescopes. Mathematically, a large portion of the mass of the galaxies is missing. Since the missing mass cannot be observed directly, it has been called dark matter. A search is underway to identify the validity, composition, and properties of dark matter.

ANSWER SHEET
Practice Test 3

1. Ⓐ Ⓑ Ⓒ Ⓓ Ⓔ
2. Ⓐ Ⓑ Ⓒ Ⓓ Ⓔ
3. Ⓐ Ⓑ Ⓒ Ⓓ Ⓔ
4. Ⓐ Ⓑ Ⓒ Ⓓ Ⓔ
5. Ⓐ Ⓑ Ⓒ Ⓓ Ⓔ
6. Ⓐ Ⓑ Ⓒ Ⓓ Ⓔ
7. Ⓐ Ⓑ Ⓒ Ⓓ Ⓔ
8. Ⓐ Ⓑ Ⓒ Ⓓ Ⓔ
9. Ⓐ Ⓑ Ⓒ Ⓓ Ⓔ
10. Ⓐ Ⓑ Ⓒ Ⓓ Ⓔ
11. Ⓐ Ⓑ Ⓒ Ⓓ Ⓔ
12. Ⓐ Ⓑ Ⓒ Ⓓ Ⓔ
13. Ⓐ Ⓑ Ⓒ Ⓓ Ⓔ
14. Ⓐ Ⓑ Ⓒ Ⓓ Ⓔ
15. Ⓐ Ⓑ Ⓒ Ⓓ Ⓔ
16. Ⓐ Ⓑ Ⓒ Ⓓ Ⓔ
17. Ⓐ Ⓑ Ⓒ Ⓓ Ⓔ
18. Ⓐ Ⓑ Ⓒ Ⓓ Ⓔ
19. Ⓐ Ⓑ Ⓒ Ⓓ Ⓔ
20. Ⓐ Ⓑ Ⓒ Ⓓ Ⓔ

21. Ⓐ Ⓑ Ⓒ Ⓓ Ⓔ
22. Ⓐ Ⓑ Ⓒ Ⓓ Ⓔ
23. Ⓐ Ⓑ Ⓒ Ⓓ Ⓔ
24. Ⓐ Ⓑ Ⓒ Ⓓ Ⓔ
25. Ⓐ Ⓑ Ⓒ Ⓓ Ⓔ
26. Ⓐ Ⓑ Ⓒ Ⓓ Ⓔ
27. Ⓐ Ⓑ Ⓒ Ⓓ Ⓔ
28. Ⓐ Ⓑ Ⓒ Ⓓ Ⓔ
29. Ⓐ Ⓑ Ⓒ Ⓓ Ⓔ
30. Ⓐ Ⓑ Ⓒ Ⓓ Ⓔ
31. Ⓐ Ⓑ Ⓒ Ⓓ Ⓔ
32. Ⓐ Ⓑ Ⓒ Ⓓ Ⓔ
33. Ⓐ Ⓑ Ⓒ Ⓓ Ⓔ
34. Ⓐ Ⓑ Ⓒ Ⓓ Ⓔ
35. Ⓐ Ⓑ Ⓒ Ⓓ Ⓔ
36. Ⓐ Ⓑ Ⓒ Ⓓ Ⓔ
37. Ⓐ Ⓑ Ⓒ Ⓓ Ⓔ
38. Ⓐ Ⓑ Ⓒ Ⓓ Ⓔ
39. Ⓐ Ⓑ Ⓒ Ⓓ Ⓔ
40. Ⓐ Ⓑ Ⓒ Ⓓ Ⓔ

41. Ⓐ Ⓑ Ⓒ Ⓓ Ⓔ
42. Ⓐ Ⓑ Ⓒ Ⓓ Ⓔ
43. Ⓐ Ⓑ Ⓒ Ⓓ Ⓔ
44. Ⓐ Ⓑ Ⓒ Ⓓ Ⓔ
45. Ⓐ Ⓑ Ⓒ Ⓓ Ⓔ
46. Ⓐ Ⓑ Ⓒ Ⓓ Ⓔ
47. Ⓐ Ⓑ Ⓒ Ⓓ Ⓔ
48. Ⓐ Ⓑ Ⓒ Ⓓ Ⓔ
49. Ⓐ Ⓑ Ⓒ Ⓓ Ⓔ
50. Ⓐ Ⓑ Ⓒ Ⓓ Ⓔ
51. Ⓐ Ⓑ Ⓒ Ⓓ Ⓔ
52. Ⓐ Ⓑ Ⓒ Ⓓ Ⓔ
53. Ⓐ Ⓑ Ⓒ Ⓓ Ⓔ
54. Ⓐ Ⓑ Ⓒ Ⓓ Ⓔ
55. Ⓐ Ⓑ Ⓒ Ⓓ Ⓔ
56. Ⓐ Ⓑ Ⓒ Ⓓ Ⓔ
57. Ⓐ Ⓑ Ⓒ Ⓓ Ⓔ
58. Ⓐ Ⓑ Ⓒ Ⓓ Ⓔ
59. Ⓐ Ⓑ Ⓒ Ⓓ Ⓔ
60. Ⓐ Ⓑ Ⓒ Ⓓ Ⓔ

61. Ⓐ Ⓑ Ⓒ Ⓓ Ⓔ
62. Ⓐ Ⓑ Ⓒ Ⓓ Ⓔ
63. Ⓐ Ⓑ Ⓒ Ⓓ Ⓔ
64. Ⓐ Ⓑ Ⓒ Ⓓ Ⓔ
65. Ⓐ Ⓑ Ⓒ Ⓓ Ⓔ
66. Ⓐ Ⓑ Ⓒ Ⓓ Ⓔ
67. Ⓐ Ⓑ Ⓒ Ⓓ Ⓔ
68. Ⓐ Ⓑ Ⓒ Ⓓ Ⓔ
69. Ⓐ Ⓑ Ⓒ Ⓓ Ⓔ
70. Ⓐ Ⓑ Ⓒ Ⓓ Ⓔ
71. Ⓐ Ⓑ Ⓒ Ⓓ Ⓔ
72. Ⓐ Ⓑ Ⓒ Ⓓ Ⓔ
73. Ⓐ Ⓑ Ⓒ Ⓓ Ⓔ
74. Ⓐ Ⓑ Ⓒ Ⓓ Ⓔ
75. Ⓐ Ⓑ Ⓒ Ⓓ Ⓔ

Practice Test 3

Do not use a calculator. To simplify numerical calculations, use $g = 10 \text{ m/s}^2$.

PART A

> **Directions:** In this section of the exam, the same lettered choices are used to answer several questions. Each group of questions is preceded by five lettered choices. When answering questions in each group, select the best answer from the available choices and fill in the corresponding bubble on the answer sheet. Each possible answer may be used once, more than once, or not at all.

Questions 1–4

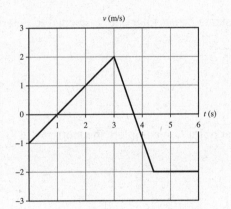

The above velocity-time graph depicts the motion of an object during a 6-second interval. Specific characteristics of this motion are encountered at the following times.

    (A) 1 s
    (B) 2 s
    (C) 3 s
    (D) 4 s
    (E) 5 s

1. At which of the above times is the object instantaneously at rest?

2. At which of the above times is the direction of the object changing?

3. At which of the above times does the object experience the greatest constant acceleration?

4. Which of the above times is the first time that the object returns to the initial starting position that it occupied when $t = 0$?

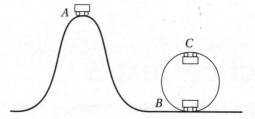

A roller coaster is initially at rest on the top of a hill at point *A*. It then rolls down the hill, passing through point *B* at the surface of Earth. Next, the roller coaster enters a circular loop and passes through point *C*, where it is upside down at the very top of the loop. Consider the speed at point *C* to be greater than the minimum speed to complete the loop, unless the question specifies minimum speed at point *C*. Choices A through E below list data that may be needed to determine the answers to questions 5 to 7.

(A) The mass, m, of the roller coaster.
(B) The height, h, of the initial hill at point *A*.
(C) The radius, R, of the loop at point *C*.
(D) Both the height, h, at point *A* and the radius, R, of the loop at point *C*.
(E) The normal force, N, exerted by the track on the roller coaster.

In addition to known constants, what other data is needed to determine

5. the speed at point *B*

6. the speed at point *C*

7. the minimum speed at point *C* to complete the loop

Questions 8–9

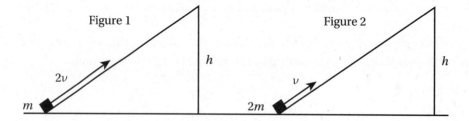

Two different masses are pushed upward along identical frictionless inclines. In Figure 1 a mass m is pushed along the incline with a speed $2v$, reaching a height h. In Figure 2 a mass $2m$ is pushed along the incline with a speed v reaching the same final height h. Air resistance is negligible. Use the responses below to answer the questions that follow.

(A) The applied force, F, needed to move each mass.
(B) The work of the applied force, W_F, done on each mass.
(C) The change in gravitational potential energy, ΔU_g, of each mass.
(D) The power, P, required to move each mass.
(E) None of these

8. Which of the above quantities is greater for the mass m moving up the incline in Figure 1?

9. Which of the above quantities is equal for both masses moving up their respective inclines?

Questions 10–12

Substance 1 is a liquid at room temperature. Substance 2 is a solid that is initially at a higher temperature than substance 1. At time $t = 0$, solid substance 2 is submerged in liquid substance 1, and the system is insulated from the environment. When time equals t the system reaches thermal equilibrium. Each substance has unique values for specific heat, c, and each substance has a different mass, m. The quantity of heat transferred is denoted as Q, the rate of heat transferred as H, and the change in temperature as ΔT. Use the following expressions to answer questions 10 to 12.

(A) ΔT_2

(B) $H_1 t$

(C) $m_2 c_2 t$

(D) $\dfrac{Q_2}{m_1 \Delta T_1}$

(E) $\dfrac{m_2 c_2 \Delta T_2}{m_1 c_1}$

10. What is the change in temperature of substance 1, ΔT_1, from $t = 0$ to time t?

11. What is the specific heat of substance 1?

12. How much heat energy is transferred during this process?

Questions 13–15

(A) Electric field
(B) Electric force
(C) Electric potential
(D) Electric potential energy
(E) Capacitance

13. A scalar quantity that reflects the ability or tendency of a charge or group of charges to create electric energy on another charge that occupies a particular point in space.

14. The ratio of stored electric charge to the potential difference between the objects that hold the charges.

15. A vector quantity representing the alteration of space surrounding point charges and charged objects that is associated with electric force.

(A)

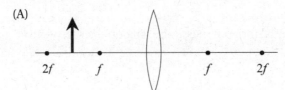

(B)

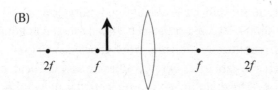

(C)

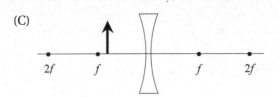

(D)

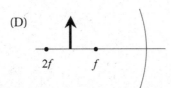

(E)

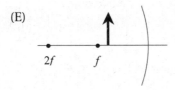

16. Which lens/mirror is a divergent optical instrument?

17. Which lens/mirror will produce a real image on the far side of the optical instrument?

18. Which lens/mirror will produce a large upright image on the near side of the optical instrument?

Which of the products of nuclear reactions listed below is released during the following nuclear reactions?

(A) $_{-1}^{0}e$

(B) $_{1}^{1}p$

(C) $_{0}^{1}n$

(D) $_{1}^{1}H$

(E) $_{2}^{4}He$

19. Alpha decay

20. Beta decay

21. Fission of uranium-235, resulting in a chain reaction.

> **Directions:** This section of the exam consists of questions or incomplete statements followed by five possible answers or completions. Select the best answer or completion, and fill in the corresponding bubble on the answer sheet.

22. A mass initially at rest undergoes a uniform acceleration a during a time interval t. The mass reaches a speed v when it has been displaced a distance x. The mass continues moving with acceleration a. What are the speed and displacement of the mass when the time has doubled, $2t$?

	Velocity	Displacement
(A)	$2v$	$\sqrt{2}\,x$
(B)	$\sqrt{2}\,v$	$2x$
(C)	$2v$	$2x$
(D)	$2v$	$4x$
(E)	$4v$	$4x$

23. Which is NOT true regarding acceleration?

(A) Some accelerating objects have constant speed.
(B) Acceleration is the rate of change in velocity.
(C) Acceleration is the slope of a position-time graph.
(D) Acceleration can be positive, negative, or zero.
(E) Acceleration can involve a change in either speed or direction.

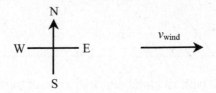

24. The pilot of an airplane needs to travel directly north. However, a strong crosswind is blowing out of the west, as shown above. Which diagram below indicates the direction in which the pilot must steer the plane to compensate for the crosswind so that the plane moves north?

(A)

(B)

(C)

(D)

(E)

25. A ball thrown horizontally with a speed v drops a distance y and moves horizontally a distance x. If the ball is thrown with the same horizontal speed v from a height of $2y$, what is its new horizontal displacement?

(A) x
(B) $\sqrt{2}\,x$
(C) $2x$
(D) $4x$
(E) $16x$

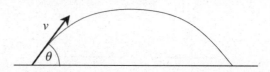

26. A projectile is launched with an initial velocity v at an upward angle θ, as shown above. During the entire motion, which variable(s) remain(s) constant?

   I. Horizontal component of velocity
  II. Vertical component of velocity
 III. Acceleration

(A) I only
(B) II only
(C) III only
(D) I and II only
(E) I and III only

Questions 27–28

The diagram below depicts a mass, m, being pulled along a rough horizontal surface by a force, F, acting at an angle θ above the horizontal. The force of friction acting on the block is f, and the coefficient of friction is μ.

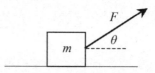

27. Which of the following describes the frictional force, f, if the resulting motion is constant velocity?

(A) 0
(B) $\mu m g$
(C) $\mu m g \cos \theta$
(D) $F \cos \theta$
(E) $F \sin \theta$

28. Which of the following is true regarding the normal force?

   (A) $N = mg$
   (B) $N = mg \cos \theta$
   (C) $N = mg \sin \theta$
   (D) $N = mg - F \cos \theta$
   (E) $N = mg - F \sin \theta$

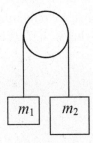

29. Two masses, $m_1 = 3.0$ kilograms and $m_2 = 5.0$ kilograms, are connected by a string, which is draped over a pulley. The masses are released from rest. Determine the magnitude of the acceleration of mass m_1.

   (A) 2.5 m/s^2
   (B) 3.0 m/s^2
   (C) 4.0 m/s^2
   (D) 4.5 m/s^2
   (E) 5.0 m/s^2

30. When an object is thrown upward in the absence of air resistance and reaches the top of its trajectory, it has an instantaneous velocity of zero. At this point, the net force acting on the object is

   (A) 0
   (B) equal to the weight of the object
   (C) equal to the mass of the object
   (D) g
   (E) changing direction

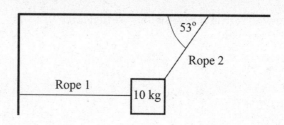

31. A 10-kilogram mass is suspended by two ropes, as shown above. Rope 1 is attached to a vertical wall. Rope 2 is attached to the horizontal ceiling at an angle $\theta = 53°$ as measured from the ceiling. The tensions in the ropes are T_1 and T_2, and the weight of the object is w. Which of the following is NOT true?

(A) $T_2 > T_1$
(B) $w > T_1$
(C) $w > T_2$
(D) $T_2 \cos \theta = T_1$
(E) $w = T_2 \sin \theta$

32. Two masses, connected by a string, are pulled along a frictionless surface by a 6-newton force, as shown above. Determine the tension in the string between the masses.

(A) 1 N
(B) 2 N
(C) 3 N
(D) 4 N
(E) 5 N

33. An object moving in a circular path completes one circumference. Which statement(s) below is (are) true?

 I. The displacement is zero at the conclusion of its motion.
 II. The average speed is zero at the conclusion of its motion.
 III. Velocity changes during its motion.
 IV. The acceleration is zero during its motion.

(A) I only
(B) I and II only
(C) I and III only
(D) IV only
(E) II and IV only

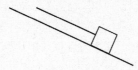

34. A mass is pulled up a rough incline at constant speed, as shown above. The coefficient of kinetic friction is μ. Which diagram is correct?

(A)

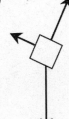

(B)

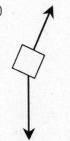

(C)

(D)

(E)

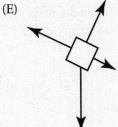

35. A mass is moving in a circle as shown above. The motion is uniform, having constant speed. Which set of vectors correctly depicts the velocity and acceleration at point P, shown in the diagram?

(A)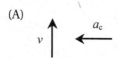

(B)

(C)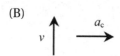

(D)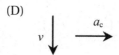

(E)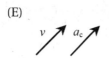

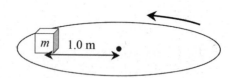

36. Mass m is positioned 1.0 meter from the center of a circular disk that is rotating with increasing speed, as shown above. The coefficient of friction between the mass and the disk is 0.40. What maximum speed can the mass obtain before it slips off the rotating disk?

(A) 1.0 m/s
(B) 1.4 m/s
(C) 2.0 m/s
(D) 2.8 m/s
(E) 4.0 m/s

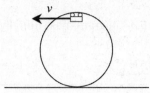

37. A roller-coaster car passes through a frictionless vertical circular loop in the track, as shown in the diagram above. When the roller-coaster car is at the very top of the loop, its instantaneous velocity is directed in the $-x$-direction. Which force diagram is correct at the instant that the roller-coaster car is at the top of the loop?

(A)

(B)

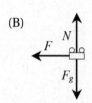

(C)

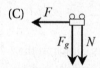

(D)

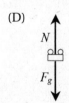

(E)

Questions 38–39

A 25-newton force is applied to a 5.0-kilogram mass initially at rest on a frictionless surface.

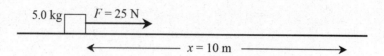

38. Determine the work done on the mass to move it 10 meters.

    (A)   2.5 J
    (B)   50 J
    (C)   125 J
    (D)   250 J
    (E) 1,250 J

39. Determine the speed of the mass when it has been displaced 10 meters.

    (A)   2 m/s
    (B)   4 m/s
    (C)   5 m/s
    (D) 10 m/s
    (E) 20 m/s

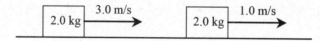

40. Two identical 2.0-kilogram masses are moving in the +x-direction as shown in the diagram above. The left mass has a speed of 3.0 meters per second, and the right mass has a speed of 1.0 meter per second. When the masses collide, they stick to one another. What is the final speed of the combined masses after the collision?

    (A) 1.0 m/s
    (B) 1.5 m/s
    (C) 2.0 m/s
    (D) 4.0 m/s
    (E) 8.0 m/s

41. The magnitude of impulse is equal to:

    I. $F\Delta d$
    II. $F\Delta t$
    III. mv
    IV. $m\Delta v$

    (A) I only
    (B) II only
    (C) II and III only
    (D) II and IV only
    (E) II, III, and IV only

42. Several planets orbit a large central star that has a mass M. The mass of each planet is given in terms of m, and their orbital radii are given in terms of r. Which planet experiences the greatest gravitational force pulling it toward the central star?

	Mass	Orbital radius
(A)	m	r
(B)	$2m$	r
(C)	m	$2r$
(D)	$4m$	$2r$
(E)	$9m$	$3r$

Questions 43–44

A large sphere located at the origin contains a +20-coulomb charge. A small sphere located 4 meters to the right of the large sphere has a charge of –1 coulomb.

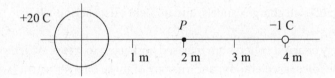

43. The force that the –1-coulomb charge pulls on the +20-coulomb charge is

    (A) $\frac{1}{20}$ the force that the +20 C charge pulls on the –1 C charge

    (B) $\frac{1}{5}$ the force that the +20 C charge pulls on the –1 C charge

    (C) $\frac{5}{4}$ the force that the +20 C charge pulls on the –1 C charge

    (D) $\frac{1}{16}$ the force that the +20 C charge pulls on the –1 C charge

    (E) the same as the force that the +20 C charge pulls on the –1 C charge

44. The force acting on the –1-coulomb charge at its present position is F_E. What will the force acting on the –1-coulomb charge become if the –1-coulomb charge is moved to point P?

(A) $\frac{1}{4} F_E$

(B) $\frac{1}{2} F_E$

(C) F_E

(D) $2F_E$

(E) $4F_E$

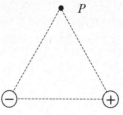 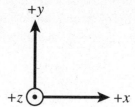

45. Two charges and point P form an equilateral triangle as shown above. One charge is negative, and the other is positive. However, the magnitude of the charges is the same. What is the direction of the electric field at point P?

(A) $+x$
(B) $-x$
(C) $+y$
(D) $-y$
(E) The field of each charge cancels, and no field exists at point P.

46. Two positively charged spheres are released and begin to move. Which statement regarding the force, acceleration, and velocity of these charges is true?

(A) Force increases, acceleration increases, and velocity decreases.
(B) Force increases, acceleration increases, and velocity increases.
(C) Force decreases, acceleration decreases, and velocity decreases.
(D) Force decreases, acceleration decreases, and velocity increases.
(E) Force decreases, acceleration increases, and velocity decreases.

47. A proton and an electron are placed between two plates of equal and opposite charge. Assume the attraction of each particle to the other is insignificant compared with their attraction to the plates. Which of these is true?

(A) The proton and electron move in the same direction, and the proton has a higher acceleration.
(B) The proton and electron move in the same direction, and the electron has a higher acceleration.
(C) The proton and electron move in opposite directions, and the proton has a higher acceleration.
(D) The proton and electron move in opposite directions, and the electron has a higher acceleration.
(E) The proton and electron move in opposite directions and have the same accelerations.

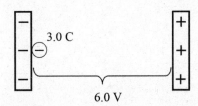

48. What is the velocity of a 1.0-kilogram sphere with a –3.0-coulomb charge that is accelerated from rest through a 6.0 V potential difference, as shown above?

(A) 2.0 m/s
(B) 3.0 m/s
(C) 6.0 m/s
(D) 12 m/s
(E) 18 m/s

49. How is the capacitance of a capacitor affected when the area of the plates is doubled and the distance between the plates is also doubled?

(A) $\frac{1}{4}$ its original value

(B) $\frac{1}{2}$ its original value

(C) remains the same
(D) 2 times greater
(E) 4 times greater

Questions 50–52

The following diagram depicts a 120-volt household circuit containing three electrical components with resistances $R_1 = 120$ ohms, $R_2 = 240$ ohms, and $R_3 = 240$ ohms. Switches X, Y, and Z are all originally in the open position.

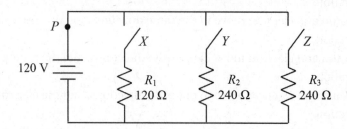

50. What current is following through point P when switches X and Y are closed?

(A) 1.0 A
(B) 1.5 A
(C) 2.0 A
(D) 3.0 A
(E) 6.0 A

51. What current is flowing through R_1 when switches X and Y are closed?

(A) 1.0 A
(B) 1.5 A
(C) 2.0 A
(D) 3.0 A
(E) 6.0 A

52. Switch Z is closed. How much power is dissipated in R_3?

(A) 15 W
(B) 20 W
(C) 30 W
(D) 40 W
(E) 60 W

53. Which answer is consistent with a replacement lightbulb that will increase the intensity (brightness) of the light in a room?

(A) Use a lightbulb with less resistance, which restricts the current flow and results in a lower rate of energy dissipation.

(B) Use a lightbulb with less resistance, which allows more current to flow and results in a higher rate of energy dissipation.

(C) Use a lightbulb with less resistance, which allows more current to flow and results in a lower rate of energy dissipation.

(D) Use a lightbulb with more resistance, which restricts the current flow and results in a lower rate of energy dissipation.

(E) Use a lightbulb with more resistance, which allows more current to flow and results in a higher rate of energy dissipation.

Questions 54–56

A horseshoe magnet is positioned as shown in the diagram below.

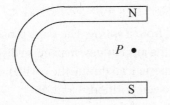 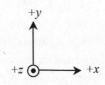

54. What is the direction of magnetic force acting on an electron if it is moving in the $-x$-direction at the instant that it is located at point P?

(A) The magnetic force is 0 and has no direction.
(B) $+y$
(C) $-y$
(D) $+z$
(E) $-z$

55. What is the direction of magnetic force acting on a proton if it is moving in the $+y$-direction at the instant that it is located at point P?

(A) The magnetic force is 0 and has no direction.
(B) $+y$
(C) $-y$
(D) $+z$
(E) $-z$

56. What is the direction of magnetic force acting on a wire passing through point P that is carrying a current into the page in the $-z$-direction?

(A) The magnetic force is 0 and has no direction.
(B) $+x$
(C) $-x$
(D) $+y$
(E) $-y$

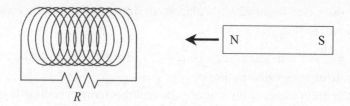

57. A magnet is moved into a coil of wire that contains a resistor with resistance R. The magnet enters the coil at constant velocity. Which statement is NOT correct?

(A) As the magnet enters the coil, an emf, $\mathcal{E}$, is induced in the wire.
(B) As the magnet enters the coil, a current, $\mathcal{E}/R$, is induced in the wire.
(C) Doubling the speed of the magnet doubles the induced current.
(D) Reversing the direction of the magnet has no effect on either the magnitude or the direction of current as long as the speed remains the same.
(E) Doubling the magnetic field doubles the induced emf.

Questions 58–60

The diagram below shows mass m suspended from a spring. The mass is oscillating between positions I and III. These positions represent the amplitudes, maximum displacements above and below the equilibrium position, experienced during the oscillation. Position II is the equilibrium position and lies midway between positions I and III.

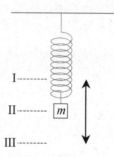

58. At which position(s) is the useful potential energy of the oscillation system zero?

(A) I only
(B) II only
(C) III only
(D) I and II only
(E) I and III only

59. At which position(s) is the net force on the mass the greatest?

(A) I only
(B) II only
(C) III only
(D) I and II
(E) I and III

60. What is the effect on the period, T, of the oscillation if mass m is replaced with mass $2m$?

   (A) $\frac{1}{4}T$

   (B) $\frac{1}{2}T$

   (C) $\frac{\sqrt{2}}{2}T$

   (D) $\sqrt{2}T$

   (E) $4T$

---

61. In a standing wave, a node is the position where

   (A) constructive interference occurs
   (B) destructive interference occurs
   (C) amplitude is maximum
   (D) Both A and C
   (E) Both B and C

62. Which of the following must occur in order to increase the intensity of light?

   (A) Increase the wavelength
   (B) Increase the frequency
   (C) Increase the amplitude
   (D) Decrease the wavelength
   (E) Decrease the frequency

63. Two waveforms interfere with each other in a manner that creates a beat frequency. One wave has a frequency of 12 Hz, and the other has a frequency of 8.0 Hz. Determine the beat frequency.

   (A) 0.67 Hz
   (B) 1.5 Hz
   (C) 4.0 Hz
   (D) 10 Hz
   (E) 20 Hz

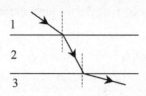

64. In the diagram above, a ray of light refracts as it travels through three mediums: 1, 2, and 3. Rank the indexes of refraction from greatest to least.

(A) $n_1 > n_2 > n_3$
(B) $n_2 > n_1 > n_3$
(C) $n_3 > n_1 > n_2$
(D) $n_1 > n_3 > n_2$
(E) $n_2 > n_3 > n_1$

Questions 65–66

Light with wavelength λ is incident on two narrow slits spaced d meters apart. A distinct pattern of alternating bright and dark regions is visible on a screen located L meters behind the slits. The second bright region (maximum) is observed to be x meters from the central bright maximum. The paths of the rays of light, extending from each slit to the second bright maximum, are shown converging at the second maximum.

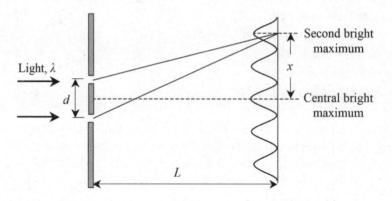

65. The path difference is

(A) 2λ
(B) $2x\lambda$
(C) $\dfrac{x\lambda}{L}$
(D) $\dfrac{2x}{L}$
(E) $2x$

66. Which of the following is true regarding the above experiment?

(A) The experiment provides evidence that light has a wave characteristic.
(B) Light passing through each slit diffracts into circular wave fronts.
(C) The light from each slit interferes to create the resulting bright maximums and dark minimums seen on the screen.
(D) The path differences, measured from each slit to the bright maximums, are integer multiples of the wavelength.
(E) All of these are true.

---

67. Which statement is NOT true when two systems with different temperatures come into contact with each other?

(A) Probability will drive the system toward a more disorganized equilibrium state.
(B) The natural direction of heat flow will be from the system with the higher temperature to the system with the lower temperature.
(C) Heat flow continues until thermal equilibrium is reached.
(D) Once thermal equilibrium is reached, no net heat flow occurs.
(E) Entropy is reduced in the process.

68. Two metal blocks are identical in every respect except for their thickness. Block 2 is twice as thick as block 1. A heat source is applied equally to both blocks. During time t, heat Q is transferred through block 1. During the same amount of time, how much heat is transferred through block 2?

(A) $\frac{1}{4}Q$
(B) $\frac{1}{2}Q$
(C) Q
(D) $2Q$
(E) $4Q$

69. During a thermodynamic process, 2,400 joules of heat are removed from a gas while 600 joules of work are done by the gas. What is the change in internal energy of the gas?

(A) −3,000 J
(B) −1,800 J
(C) 0 J
(D) +1,800 J
(E) +3,000 J

Questions 70–71

In the photoelectric-effect experiment, photons of light are shined on a metal, resulting in the ejection of electrons that then flow as a current through a circuit. The amount of current can be measured with an ammeter placed in series in the circuit. An adjustable battery is reversed in the circuit. It can be adjusted so that the current is canceled.

When the flow of current is stopped, the voltage of the battery is equal to the voltage of the photocell. It may be assumed that the threshold frequency has been reached.

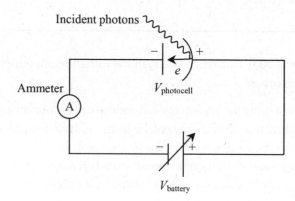

70. What is the effect of increasing the intensity of the light shined on the photocell?

   (A) The current flowing in the circuit is increased.
   (B) The threshold frequency is increased.
   (C) The kinetic energy of the emitted electrons is increased.
   (D) The voltage of the photocell is increased.
   (E) Both C and D

71. What is the effect of increasing the frequency of the light shined on the photocell?

   (A) The current flowing in the circuit is increased.
   (B) The threshold frequency is increased.
   (C) The kinetic energy of the emitted electrons is increased.
   (D) The voltage of the photocell is increased.
   (E) Both C and D

$$4\,^{1}_{1}\mathrm{H} \rightarrow\, ^{4}_{2}\mathrm{He} + 2\left(\,^{0}_{+1}\mathrm{e}\right) + 2\nu_e + 2\gamma$$

72. The nuclear reaction shown above is an example of

   (A) Alpha decay
   (B) Beta decay
   (C) Gamma decay
   (D) Fission
   (E) Fusion

73. Which word best describes uranium-238 and uranium-235?

(A) Molecules
(B) Ions
(C) Isotopes
(D) Nucleons
(E) Nuclides

74. Which statement regarding the forces in the nucleus is NOT correct?

(A) The strong force attracts protons to protons.
(B) The strong force attracts neutrons to both protons and other neutrons.
(C) At short distances, the strong force is stronger than the electric force.
(D) Adding neutrons to a nucleus adds to both the strong force and the electrostatic force.
(E) When the ratio of neutrons to protons falls outside of an optimal range, the nucleus becomes unstable.

75. Which of the following is an electrical component that can amplify a signal and act as a switch?

(A) Capacitor
(B) Power supply
(C) Resistor
(D) Superconductor
(E) Transistor

ANSWER KEY
Practice Test 3

1.	A	26.	E	51.	A
2.	A	27.	D	52.	E
3.	D	28.	E	53.	B
4.	B	29.	A	54.	E
5.	B	30.	B	55.	A
6.	D	31.	C	56.	C
7.	C	32.	B	57.	D
8.	E	33.	C	58.	B
9.	D	34.	E	59.	E
10.	E	35.	B	60.	D
11.	D	36.	C	61.	B
12.	B	37.	E	62.	C
13.	C	38.	D	63.	C
14.	E	39.	D	64.	B
15.	A	40.	C	65.	A
16.	C	41.	D	66.	E
17.	A	42.	B	67.	E
18.	B	43.	E	68.	B
19.	E	44.	E	69.	A
20.	A	45.	B	70.	A
21.	C	46.	D	71.	E
22.	D	47.	D	72.	E
23.	C	48.	C	73.	C
24.	B	49.	C	74.	D
25.	B	50.	B	75.	E

DIAGNOSTIC CHART

Subject Area	Question Numbers	Questions Incorrect	Chapter(s) to Study
Mechanics	1, 2, 3, 4, 5, 6, 7, 8, 9, 22, 23, 24, 25, 26, 27, 28, 29, 30, 31, 32, 33, 34, 35, 36, 37, 38, 39, 40, 41, 42, 58, 59, 60		1–9
Electricity and Magnetism	13, 14, 15, 43, 44, 45, 46, 47, 48, 49, 50, 51, 52, 53, 54, 55, 56, 57		10–13
Waves and Optics	16, 17, 18, 61, 62, 63, 64, 65, 66		14–17
Heat and Thermodynamics	10, 11, 12, 67, 68, 69		18–19
Modern Physics	19, 20, 21, 70, 71, 72, 73, 74		20–22
Miscellaneous	75		23

SCORING YOUR TEST

How to Determine Your Raw Score

Your raw score is the number of correctly answered questions minus the incorrectly answered questions multiplied by ¼. An incorrectly answered question is one that you bubbled in but was incorrect. If you leave the answer blank, it does not count as an incorrect answer.

Number of correctly answered questions: _____

Number of incorrectly answered questions: _____ $\times \dfrac{1}{4}$ = _____

_____ – _____ = _____

Number Correct   Number Incorrect $\times \dfrac{1}{4}$ Raw Score

ANSWERS EXPLAINED

1. **(A)**	The object is instantaneously at rest when its velocity equals zero, which occurs at 1 second.
2. **(A)**	Prior to 1 second, the object had a negative velocity and was moving in the negative direction. After 1 second, the object had a positive velocity and was moving in the positive direction. Be careful, as the turning point of the graph at 3 seconds is not the turning point of the object's motion.
3. **(D)**	Acceleration is the slope of a velocity versus time graph. The steepest slope corresponds to the greatest acceleration.
4. **(B)**	To return to the starting location, an object's displacement must be zero. On a velocity versus time graph, displacement is the area between the function and the time axis. The negative area from 0 to 1 second is equal to the positive area from 1 second to 2 seconds. These equal and opposite areas add to zero at 2 seconds.
5. **(B)**	Use conservation of energy. Set the total mechanical energy at point A equal to the total mechanical energy at point B. $$mgh_A = \frac{1}{2}mv_B{}^2$$ $$v_B = \sqrt{2gh_A}$$ Other than the known constant g, only the height at point A is needed to determine the speed at point B.
6. **(D)**	Use conservation of energy. Set the total mechanical energy at point A equal to the total mechanical energy at point C. The height at point C is the diameter of the circular loop, which equals $2R$. $$mgh_A = \frac{1}{2}mv_C{}^2 + mg(2R)$$ $$v_C = \sqrt{2g(h_A - 2R)}$$ Other than the known constant g, the height of the hill at point A, h_A, and the radius of the loop, R, are needed to determine speed at point C.
7. **(C)**	When the roller coaster completes the loop at point C with minimum speed the normal force will be zero. The only force acting on the roller coaster at this instant is the force of gravity, and it is pointing to the center of the circle. $$F_c = F_g$$ $$m\frac{v_{min}{}^2}{R} = mg$$ $$v_{min} = \sqrt{Rg}$$ Other than the known constant g, only the radius at point C is needed to determine the minimum speed at point C.

8. **(E)**	The applied force is equal and opposite the component of gravity along the incline, $mg\sin\theta$. The work of the applied force to push the object upward to a height h is equal to mgh. The change in gravitational potential energy is equal to the work, mgh. The smaller mass, m, will require less force, experience less change in energy, and do less work. Power is the product of force and velocity, $mgv\sin\theta$. The power needed to move the smaller mass, m, at a speed $2v$ is $2mgv\sin\theta$. The power needed to move the larger mass, $2m$, at a speed v is also $2mgv\sin\theta$. As a result, the smaller mass, m, is not greater in any category.
9. **(D)**	As seen in the previous answer, the power needed to move either mass is equal.
10. **(E)**	When two substances come into contact in a closed system, heat, Q, will flow from the hotter object to the colder object until they reach the same final temperature at thermal equilibrium. The total amount of heat lost by the hotter object will be gained by the colder object. $$Q_1 = Q_2$$ $$m_1 c_1 \Delta T_1 = m_2 c_2 \Delta T_2$$ $$\Delta T_1 = \frac{m_2 c_2 \Delta T_2}{m_1 c_1}$$
11. **(D)**	$$Q_1 = Q_2$$ $$m_1 c_1 \Delta T_1 = Q_2$$ $$\frac{Q_2}{m_1 \Delta T_1}$$
12. **(B)**	The rate of heat transfer, H, is the amount of heat, Q, divided by time, t. $$H = \frac{Q}{t} \qquad Q = Ht$$ Since the amount of heat transferred from substance 2 to substance 1 is the same during the same time interval, then the rate of heat transfer is the same. As a result, this can be solved for either substance 1 or substance 2.
13. **(C)**	Also known as voltage, electric potential is the potential ability to generate electric energy should a charged particle occupy a particular point in space. The charged objects surrounding the point in space generate this potential. When a new charged object is positioned at the point in space, it acquires potential energy due to the surrounding charges.
14. **(E)**	Capacitance is the ratio of charged stored on the plates of a capacitor to the potential difference between the plates, $C = Q/V$.

15. **(A)**	Charges are surrounded by an electric field. This is similar to the electric potential in question 13. The electric field is the capability of the charge to create an electric force on any other charge that occupies a particular location within the electric field.
16. **(C)**	The concave lens is a diverging lens.
17. **(A)**	A converging lens will form a real and inverted image on the far side of the lens from the object if the object is placed outside of the focal length.
18. **(B)**	A converging lens will form a virtual and upright image on the near side of the lens if the object is placed inside of the focal length. This is the effect produced by a magnifying lens.
19. **(E)**	An alpha particle is the nucleus of a helium atom. It has a charge of +2.
20. **(A)**	A beta particle is an electron ejected from a neutron. As a result, the neutron becomes a positive proton. Beta particles have a charge of –1.
21. **(C)**	Neutrons are generally released as a by-product of fission. Neutrons have a neutral charge.
22. **(D)**	The final speed can be determined with the following kinematic equation: $$v_f = v_i + at$$ When the initial speed equals zero, the equation simplifies to $v_f = at$, and the final speed is directly proportional to time. Doubling the time will double the final speed. $$(2v_f) = a(2t)$$ Displacement can be described with the following kinematic equation: $$x = v_i t + \frac{1}{2}at^2$$ When the initial speed equals zero, the equation simplifies to $x = \frac{1}{2}at^2$, and displacement is proportional to the square of time. Doubling the time will quadruple the displacement. $$(4x) = \frac{1}{2}a(2t)^2$$
23. **(C)**	Acceleration is the slope of a velocity-time graph. It is *not* the slope of a position-time graph. Choice A is good distracter. However, this scenario can be true for objects that are changing direction at constant speed, such as objects in uniform circular motion.
24. **(B)**	To travel north, the plane must have a component of velocity directed north. To overcome the eastward wind, the plane must also have a component of velocity that opposes the wind and is therefore directed to the west. Resolving the components results in a true velocity directed northwest.

| 25. **(B)** | Since the ball is thrown horizontally, the time, t, for the ball to hit the ground is the same as if it were simply dropped from a height of y.

$$y = \frac{1}{2}gt^2$$

$$t = \sqrt{\frac{2y}{g}}$$

Time is proportional to the square root of vertical displacement. Doubling the vertical displacement to $2y$ increases the time to $\sqrt{2}t$.

$$\sqrt{2}t = \sqrt{\frac{2(2y)}{g}}$$

The horizontal motion solves independently, and it has constant velocity.

$$x = v_x t$$

Horizontal displacement is directly proportional to time. If the time is $\sqrt{2}t$, then the horizontal displacement is $\sqrt{2}x$.

$$\sqrt{2}x = v_x\left(\sqrt{2}t\right)$$ |
|---|---|
| 26. **(E)** | Acceleration due to gravity acts on only the motion in the vertical direction. The acceleration remains a constant 10 meters per second squared downward. This action continually changes the vertical component of velocity. Horizontal velocity is unaffected by acceleration due to gravity and continues unabated in the horizontal direction. |
| 27. **(D)** | At first glance, answer B may appear to be the obvious choice. However, the normal force is *not* always equal to an object's weight. The y-component of force F must be included.

$$N = mg - F\sin\theta$$
$$f = \mu(mg - F\sin\theta)$$

This is not one of the available answers. The key to this problem is the motion experienced by the object. In order for the velocity to be constant, the frictional force must be opposite in direction but equal in magnitude to the component of force acting in the direction of motion.

$$f = F_x = F\cos\theta$$ |
| 28. **(E)** | The component of force F pulling in the vertical direction can be described as $F\sin\theta$. The normal force, N, is reduced by the amount of force pulling in the vertical direction. |
| 29. **(A)** | The masses are connected by a string and therefore act as if they were a single system with a mass of 8.0 kg. Mass m_1 is being pulled in one direction by the force of gravity. Mass m_2 is being pulled in the other direction by the force of gravity.

$$\Sigma F_{sys} = F_{g2} - F_{g1}$$
$$(m_1 + m_2)a = m_2 g - m_1 g$$
$$(3.0 \text{ kg} + 5.0 \text{ kg})a = (5.0 \text{ kg})(10 \text{ m/s}^2) - (3.0 \text{ kg})(10 \text{ m/s}^2)$$
$$a = 2.5 \text{ m/s}^2$$ |

30. **(B)**	The force acting on the object remains mg, its weight, throughout its flight. As the object rises, the force acts to slow it down. Upon reaching its highest point, the force acts to change the object's direction and bring it back to the ground. However, the force acting on the object never changes.
31. **(C)**	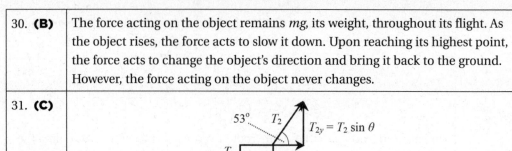 In order for the mass to remain stationary, the vertical component of T_2 ($T_2 \sin \theta$) must be equal to w. Since T_2 is the hypotenuse of a right triangle, it is greater than $T_2 \sin \theta$. Therefore, T_2 must be greater than w.
32. **(B)**	Determine the acceleration of the entire system. $$\Sigma F_{sys} = F$$ $$(m_1 + m_2)a = F$$ $$(1 \text{ kg} + 2 \text{ kg})a = 6$$ $$a = 2 \text{ m/s}^2$$ The string is pulling the 1 kg mass. Sum the forces for this mass only. $$\Sigma F_1 = T$$ $$m_1 a = T$$ $$(1 \text{ kg})(2 \text{ m/s}^2) = T$$ $$T = 2 \text{ N}$$
33. **(C)**	Since the object returns to its original position, displacement is zero. Direction is continuously changing, and this means that velocity is changing. Therefore, answers I and III are true. Speed is a measure of distance divided by time. Even though displacement is zero, the distance traveled is not zero. This eliminates answer II. Acceleration changes direction during the motion but is never zero, eliminating answer IV.
34. **(E)**	Four forces are present: the force of gravity acting downward, the normal force acting perpendicular to the incline, tension pulling the object up the incline, and friction acting to oppose the motion.
35. **(B)**	Velocity is tangent to the path in the direction of motion. Centripetal acceleration is always directed toward the center of rotation and is perpendicular to the velocity.

36. **(C)**	The friction force creates the centripetal force needed to keep the object moving in a circle. $$F_c = f$$ $$m\frac{v^2}{r} = \mu mg$$ $$v = \sqrt{\mu gr}$$ $$= \sqrt{(0.40)(10 \text{ m/s}^2)(1.0 \text{ m})}$$ $$= 2.0 \text{ m/s}$$	
37. **(E)**	Only two forces are acting on the car as it is in the loop: the force of gravity, F_g, and the normal force, N. The force of gravity always acts downward. The normal force always acts perpendicularly to the plane of the surface upon which the mass rests. In this case, that would be downward. The velocity is directed in the $-x$-direction, but that is not a force. Instead, it is the result of the inertia of the car.	
38. **(D)**	$$W = F \Delta d$$ $$W = (25 \text{ N})(10 \text{ m}) = 250 \text{ J}$$	
39. **(D)**	The work–kinetic energy theorem states that work is equal to a change in kinetic energy. Use the work found in the previous answer. $$W = \Delta K$$ $$W = \frac{1}{2}mv_f^2 - \frac{1}{2}mv_i^2$$ The object is initially at rest, $v_i = 0$. $$(250 \text{ J}) = \frac{1}{2}(5.0 \text{ kg})v_f^2 - 0$$ $$v_f = 10 \text{ m/s}$$	
40. **(C)**	Momentum is conserved before and after the collision. $$m_1 v_1 + m_2 v_2 = (m_1 + m_2)v$$ $$(2.0 \text{ kg})(3.0 \text{ m/s}) + (2.0 \text{ kg})(1.0 \text{ m/s}) = (2.0 \text{ kg} + 2.0 \text{ kg})v$$ $$v = 2.0 \text{ m/s}$$	
41. **(D)**	Impulse, J, is equal to the force multiplied by the change in time. It is also equal to the mass multiplied by the change in velocity, which is known as the change in momentum. It is *not* equal to the momentum itself but, rather, the *change* in momentum.	
42. **(B)**	Try each possibility in Newton's law of gravitation. $$F_g = G\frac{mM}{r^2}$$ $$(2F_g) = G\frac{(2m)M}{(r)^2}$$ Choice B results in the largest possible gravity.	

43. **(E)**	According to Newton's third law, the two charges pull on each other with the same force.
44. **(E)**	$$F_E = K\frac{q_1 q_2}{r^2}$$ Coulomb's law is very similar to Newton's law of universal gravitation. When the –1 C charge is at point P, the radial distance will be $\frac{1}{2}$ of what it was originally. The result is consistent with the inverse-square law. $$4F_E = K\frac{q_1 q_2}{\left(\frac{1}{2}r\right)^2}$$
45. **(B)**	Identify the charges as charge 1, q_1, and charge 2, q_2. The electric field points toward the negative charge, so the electric field of charge 1, E_1, points toward charge 1. The electric field points away from the positive charge, so the electric field of charge 2, E_2, points away from charge 2. These two vectors can be added using vector addition to find the total electric field due to both charges. The resulting electric field points in the $-x$ direction.
46. **(D)**	Since the spheres are positively charged, they will repel one another. As they begin to move farther away from one another, the force acting on them will decrease according to Coulomb's law. A reduction in force leads to a reduction in acceleration. Velocity, however, will continue to increase because even though acceleration is decreasing, it continues to act in the direction of motion and continues to increase the speed of the charged spheres.
47. **(D)**	Protons and electrons have the same magnitude of charge, e, but opposite signs. As a result, the charged electric plates apply an equal electric force on the similarly charged proton and electron, but in opposite directions. The acceleration of each particle is dependent upon its mass. Electrons have a much smaller mass than protons. As a result, the same force applied to an electron will cause the electron to have a greater acceleration than the more massive proton.

48. **(C)**	This problem can be solved using conservation of energy. The electric potential energy is converted into kinetic energy. $$U_E = K$$ $$qV = \frac{1}{2}mv^2$$ $$(3.0 \text{ C})(6.0 \text{ V}) = \frac{1}{2}(1 \text{ kg})v^2$$ $$v = 6.0 \text{ m/s}$$	
49. **(C)**	Capacitance is proportional to the area of the plates divided by the distance between them. Doubling both area and distance will result in the capacitance's, remaining the same.	
50. **(B)**	The resistors can be added in parallel to determine their total resistance. $$\frac{1}{R_p} = \frac{1}{R_1} + \frac{1}{R_2} = \frac{1}{120 \ \Omega} + \frac{1}{240 \ \Omega} = \frac{3}{240 \ \Omega}$$ $$R_p = 80 \ \Omega$$ Apply Ohm's law, $V = IR$, to find the current. $$I = \frac{V}{R} = \frac{120 \text{ V}}{80 \ \Omega} = 1.5 \text{ A}$$	
51. **(A)**	Resistor 1 is wired in parallel, and the voltage drop across it will be equal to the voltage of the battery. Use Ohm's law, $V = IR$, to solve for the current flowing through R_1. $$I = \frac{V}{R} = \frac{120 \text{ V}}{120 \ \Omega} = 1.0 \text{ A}$$	
52. **(E)**	Power can be determined two ways. The current flowing through R_3 can be found in the same manner as in the previous problem. $$I = \frac{V}{R} = \frac{120 \text{ V}}{240 \ \Omega} = 0.5 \text{ A}$$ Then, the power can be determined as follows: $$P = IV = (0.5 \text{ A})(120 \text{ V}) = 60 \text{ W}$$ Instead, it could have been determined directly using: $$P = \frac{V^2}{R} = \frac{(120 \text{ V})^2}{(240 \ \Omega)} = 60 \text{ W}$$	
53. **(B)**	A bulb with less resistance will allow more current to flow. In households, the voltage is constant. So, increasing the current will increase the power according to $P = IV$. Power is the rate of energy dissipation. Increasing the power increases a bulb's brightness.	
54. **(E)**	The right-hand rule states that the thumb of the right hand points in the direction of motion of a charged particle (the $-x$-direction in this case), the extended fingers point in the direction of the magnetic field (down the page in this case), and the palm of the hand points in the direction of the force (out of the page in this case). However, the particle is an electron. Electrons experience the complete opposite force. So, the direction of force is into the page, $-z$-direction. As an alternative, you can use the left hand to determine the direction of magnetic force on negative charges.	

55. **(A)**	In this case, the charged proton is moving parallel to the magnetic field. No magnetic force acts on the proton if it is moving completely parallel to the field.
56. **(C)**	The right-hand rule states that the thumb of the right hand points in the direction of either the motion of a charged particle or the current in a wire (the $-z$-direction in this case), the extended fingers point in the direction of the magnetic field (down the page in this case), and the palm of the hand points in the direction of the force (to the left of the page in this case, the $-x$-direction).
57. **(D)**	Reversing the magnet would reverse the direction of the current.
58. **(B)**	At position II, the total energy is in the form of kinetic energy, K, and potential energy is zero.
59. **(E)**	At position I, the constant force of gravity is greater than the variable force of the spring. At position III, the variable force of the spring is greater than the constant force of gravity. At position II, the constant force of gravity is equal to but in the opposite direction of the variable force of the spring. Therefore, there is no net force at position II.
60. **(D)**	The period of a spring depends on the mass and the spring constant. $$T_s = 2\pi\sqrt{\frac{m}{k}}$$ The period of an oscillating spring-mass system is proportional to the square root of the suspended mass. Doubling mass m would result in increasing the period to $\sqrt{2}\,T$. $$\sqrt{2}T_s = 2\pi\sqrt{\frac{(2m)}{k}}$$
61. **(B)**	At the nodes, waves add destructively, and there is zero amplitude.
62. **(C)**	Light intensity is directly proportional to the amplitude of a wave. Frequency and wavelength have no effect on the intensity of light.
63. **(C)**	Beat frequency is the difference between two interfering waveform frequencies. $$f_{\text{beat}} = f_1 - f_2 = 12 \text{ Hz} - 8.0 \text{ Hz} = 4.0 \text{ Hz}$$
64. **(B)**	As the index of refraction increases, the angle measured between the light ray and a normal line drawn perpendicular to the surface of the medium decreases. Medium 2 has the smallest angle and therefore the greatest index of refraction. Medium 1 has a slightly larger angle and a slightly smaller index of refraction. Medium 3 has the largest angle and the smallest index of refraction.
65. **(A)**	Interference is constructive when the path-length difference is a whole number of wavelengths. The second bright maximum occurs at a path length difference of 2λ.
66. **(E)**	Each of these statements is true.
67. **(E)**	Entropy always increases for an isolated system that is reaching equilibrium. This is the second law of thermodynamics.

68. **(B)**	The amount of heat, Q, transferred through an object is inversely proportional to the length, L, that that heat must transverse while moving through the object, as shown in the equation below. $$\frac{Q}{\Delta t} = \frac{kA\Delta T}{L}$$ Doubling the distance that heat must travel cuts the amount of heat transferred in half. $$\left(\frac{1}{2}Q\right)\frac{1}{\Delta t} = \frac{kA\Delta T}{(2L)}$$
69. **(A)**	Adding heat to a system of gas is positive and removing heat is negative. Doing work on a system is positive while work done by the system is negative. Use the first law of thermodynamics. $$\Delta U = Q + W$$ $$\Delta U = (-2{,}400 \text{ J}) + (-600 \text{ J}) = -3{,}000 \text{ J}$$
70. **(A)**	The current is directly proportional to the intensity of the light.
71. **(E)**	When the frequency of the light is increased, the energy of each photon is increased, $E = hf$. Photons with higher energies will emit electrons with higher kinetic energies, $K = hf - \phi$. The ejected electrons arrive at the opposite plate of the photocell, creating a potential energy and a proportional potential difference between the plates of the photocell. $$qV = K$$ Therefore, increasing the frequency increases both the kinetic energy of the ejected electrons and the resulting voltage of the photocell.
72. **(E)**	Fusion results when there is an increase in the atomic number. Hydrogen has an atomic number of 1, and helium has an atomic number of 2. The fusing together of hydrogen atoms produces helium.
73. **(C)**	Isotopes have the same elemental symbol but a different number of neutrons. Both of these forms of uranium have the same atomic number, 92, but their masses vary depending upon the number of neutrons. Uranium-238 has 146 neutrons, while uranium-235 has 143 neutrons.
74. **(D)**	Adding neutrons adds to only the strong force. Adding neutrons cannot add to the electrostatic force, because neutrons have a neutral charge.
75. **(E)**	This is the definition of a transistor.

Visit *online.barronsbooks.com* for access to an
additional full-length online practice test.

Appendix I:
Key Equations

MECHANICS
Kinematics

Constant velocity

$$v = \frac{\Delta x}{t}$$

Kinematic equations

$$v_f = v_i + at$$

$$v_f^2 = v_i^2 + 2a\Delta x$$

$$\Delta x = v_i t + \frac{1}{2}at^2$$

Dynamics

Net force

$$\vec{F}_{net} = \Sigma \vec{F} = m\vec{a}$$

Friction force

$$\vec{f} = -\mu \vec{N}$$

Restoring force (Hooke's law)

$$\vec{F}_s = |k\vec{x}|$$

Force of gravity (weight)

$$\vec{w} = \vec{F}_g = m\vec{g}$$

Uniform Circular Motion

Period

$$T = \frac{t}{\text{cycles}} = \frac{1}{f}$$

Frequency

$$f = \frac{\text{cycles}}{t} = \frac{1}{T}$$

Speed in uniform circular motion

$$v = \frac{2\pi r}{T}$$

Centripetal acceleration

$$a_c = \frac{v^2}{r}$$

Centripetal force

$$F_c = ma_c$$

Energy, Work, and Power

Potential energy of gravity

$$U_g = mgh = -G\frac{m_1 m_2}{r}$$

Elastic potential energy

$$U_s = \frac{1}{2}kx^2$$

Kinetic energy

$$K = \frac{1}{2}mv^2$$

Conservation of energy

$$K_i + U_i = K_f + U_f$$

Work

$$W = \Delta E = Fd \cos \theta$$

Power

$$P = \frac{\Delta E}{t} = \frac{W}{t} = Fv \cos \theta$$

Momentum

$$\vec{p} = m\vec{v}$$

Impulse

$$\vec{J} = \Delta \vec{p} = \vec{F} \Delta t$$

Conservation of momentum
(elastic and inelastic)

$$m_1 \vec{v}_{1i} + m_2 \vec{v}_{2i} = m_1 \vec{v}_{1f} + m_2 \vec{v}_{2f}$$

Conservation of momentum
(perfectly inelastic)

$$m_1 \vec{v}_{1i} + m_2 \vec{v}_{2i} = (m_1 + m_2) \vec{v}_f$$

Conservation of momentum
(explosions)

$$(m_1 + m_2) \vec{v}_i = m_1 \vec{v}_{1f} + m_2 \vec{v}_{2f}$$

Explosion, initially at rest

$$|m_1 \vec{v}_{1f}| + |m_2 \vec{v}_{2f}|$$

Newton's law of gravity

$$F_g = G \frac{m_1 m_2}{r^2}$$

Gravity field

$$g = G \frac{m}{r^2}$$

ELECTRICITY AND MAGNETISM
Electric Fields and Force

Electric force

$$\vec{F}_E = q\vec{E}$$

Electric field (charged plates)

$$|\vec{E}| = \frac{\Delta V}{d}$$

Electric field (point charges)

$$|\vec{E}| = k \frac{q}{r^2}$$

Coulomb's law

$$|\vec{F}_E| = k \frac{q_1 q_2}{r^2}$$

Electric Potential and Energy

Potential difference (charged plates)

$$\Delta V = |\vec{E}\, d|$$

Potential (point charges)

$$V = k \frac{q}{r}$$

Electric potential energy

$$U_E = qV = k \frac{q_1 q_2}{r}$$

Work in electric fields

$$W = -\Delta U_E = -q \, \Delta V$$

Capacitors

$$C = \frac{\mathcal{E}_0 A}{d}$$

Charge stored on a capacitor

$$Q = CV$$

Energy of a capacitor

$$U_c = \frac{1}{2} QV = \frac{1}{2} CV^2$$

Circuits

Ohm's law

$$V = IR$$

Equivalent resistance in series

$$R_T = R_1 + R_2 + R_3 + \cdots$$

Equivalent resistance in parallel

$$1/R_T = 1/R_1 + 1/R_2 + 1/R_3 + \cdots$$

Joule's law

$$Q = I^2 Rt$$

Power

$$P = IV = V^2/R = I^2 R$$

Magnetism

Magnetic field of a long, straight, current-carrying wire

$$B = \frac{\mu_0 I}{2\pi r}$$

Force of magnetism on a point charge

$$\vec{F}_B = q\vec{v}\vec{B}$$

Force of magnetism on a current-carrying wire

$$\vec{F}_B = I\vec{L}\vec{B}$$

Magnetic flux

$$\phi = BA$$

Faraday's law (induced emf)

$$\mathcal{E} = \frac{\Delta\phi}{t} = BLv$$

SIMPLE HARMONIC MOTION

Period

$$T = \frac{t}{\text{cycles}} = \frac{1}{f}$$

Frequency

$$f = \frac{\text{cycles}}{t} = \frac{1}{T}$$

Period of a spring

$$T_s = 2\pi\sqrt{\frac{m}{k}}$$

Period of a pendulum

$$T_p = 2\pi\sqrt{\frac{L}{g}}$$

Conservation of energy (oscillating spring)

$$\frac{1}{2}kx_1^2 + \frac{1}{2}mv_1^2 = \frac{1}{2}kx_2^2 + \frac{1}{2}mv_2^2$$

Conservation of energy (oscillating pendulum)

$$mgh_1 + \frac{1}{2}mv_1^2 = mgh_2 + \frac{1}{2}mv_2^2$$

WAVES AND OPTICS

Wave speed

$$v = f\lambda = \frac{d}{t}$$

Law of reflection

$$\theta_i = \theta_r$$

Index of refraction

$$n = \frac{c}{v}$$

Snell's law

$$n_1 \sin\theta_1 = n_2 \sin\theta_2$$

Critical angle

$$n_1 \sin\theta_c = n_2 \sin 90°$$

Magnification

$$M = \frac{h_i}{h_o} = -\frac{d_i}{d_o}$$

Focal length

$$f = \frac{R}{2}$$

Focal length, object distance,
and image distance

$$\frac{1}{f} = \frac{1}{d_o} + \frac{1}{d_i}$$

Young's double-slit experiment

$$x_m \approx \frac{m\lambda L}{d}$$

$$d \sin \theta = m\lambda$$

THERMAL PHYSICS/ THERMODYNAMICS

Average kinetic energy of gas particles

$$K_{\text{avg}} = \frac{3}{2} k_B T$$

Ideal gas law

$$PV = nRT$$

Rate of heat transfer

$$\frac{Q}{\Delta t} = \frac{kA\Delta T}{L}$$

Specific heat capacity

$$Q = mc\,\Delta T$$

Latent heat

$$Q = mL$$

First law of thermodynamics

$$\Delta U = Q + W$$

$W_{\text{on a gas}}$ is positive

$W_{\text{by a gas}}$ is negative

Work of a heat engine

$$W = Q_H - Q_C$$

Efficiency of heat engines

$$e = \left| \frac{W_{net}}{Q_H} \right|$$

$$e = \frac{T_H - T_C}{T_H}$$

ATOMIC AND MODERN PHYSICS

Energy of a photon

$$E = hf = \frac{hc}{\lambda}$$

Photoelectric effect

$$K_{\text{max}} = hf - \Phi$$

$$K_{\text{max}} = \frac{hc}{\lambda} - \Phi$$

Mass-energy equivalence

$$E = (\Delta m)c^2$$

Special relativity

$$t = \frac{t_0}{\sqrt{1 - v^2/c^2}}$$

$$L = L_0 \sqrt{1 - v^2/c^2}$$

$$m = \frac{m_0}{\sqrt{1 - v^2/c^2}}$$

Appendix II: Physical Constants

Acceleration of gravity at the surface of Earth $g = 10 \text{ m/s}^2$

Universal gravity constant $G = 6.67 \times 10^{-11} \text{ m}^3/\text{kg} \cdot \text{s}^2$

Charge of an electron and a proton $e = 1.6 \times 10^{-19} \text{ C}$

Mass of an electron $m_e = 9.11 \times 10^{-31} \text{ kg}$

Mass of a proton $m_p = 1.67 \times 10^{-27} \text{ kg}$

Mass of a neutron $m_n = 1.67 \times 10^{-27} \text{ kg}$

Coulomb's law constant $k = 9 \times 10^9 \text{ N} \cdot \text{m}^2/\text{C}^2$

Vacuum permeability (magnetism) $\mu_0 = 4\pi \times 10^{-7} \text{ T} \cdot \text{m/A}$

Avogadro's number $n = 6.02 \times 10^{23} \text{ mol}^{-1}$

Universal gas constant $R = 8.31 \text{ J/mol} \cdot \text{K}$

Planck's constant $h = 6.63 \times 10^{-34} \text{ J} \cdot \text{s}$

$h = 4.14 \times 10^{-15} \text{ eV} \cdot \text{s}$

Speed of light in a vacuum $c = 3 \times 10^8 \text{ m/s}$

Appendix III: Conversion Factors

![striped bar divider]

METRIC CONVERSION FACTORS

tera-	$T = 10^{12}$
giga-	$G = 10^9$
mega-	$M = 10^6$
kilo-	$k = 10^3$
centi-	$c = 10^{-2}$
milli-	$m = 10^{-3}$
micro-	$\mu = 10^{-6}$
nano-	$n = 10^{-9}$
pico-	$p = 10^{-12}$

OTHER CONVERSION FACTORS

Atmospheres and pascals	$1 \text{ atm} = 1.01 \times 10^5 \text{ Pa}$
Liters and meters cubed	$1 \text{ L} = 1 \times 10^{-3} \text{ m}^3$
Electron volts and joules	$1 \text{ eV} = 1.6 \times 10^{-19} \text{ J}$
Atomic mass units and kilograms	$1 \text{ u} = 1.66 \times 10^{-27} \text{ kg}$

Glossary

A

Absolute temperature: A measure of the average kinetic energy of molecules in a solid, liquid, or gas. Measured in kelvins. To convert degrees Celsius to kelvins, add 273.

Absolute zero: The theoretical lowest temperature, in kelvins. All molecular motion would cease. 0 kelvin is equal to –273° Celsius.

Absorption spectrum: A continuous color spectrum, emitted by a star, that has dark lines that match the emission spectrum of atoms found in the atmosphere of the star. The atoms in the atmosphere of the star absorb radiation in the form of specific frequencies, and the dark lines indicate those frequencies.

Acceleration: A vector quantity. A change in magnitude and/or direction indicates a change in acceleration. Acceleration is the rate of change of velocity.

Adiabatic: A thermodynamic process of rapid expansion or compression of a system of gas that changes internal energy and temperature but has no heat exchange between the system and its surroundings. Heat exchange during an adiabatic process is equal to zero.

Alpha decay: The spontaneous emission of a helium nucleus—2 protons, 2 neutrons, but no electrons—from a large, radioactive element such as uranium, thorium, radium, and other large elements. During alpha decay, the emitting element reduces its mass number by 4 and its atomic number by 2, causing the atom to become a new element.

Alpha particle: A helium nucleus emitted from a radioactive element. Has a charge of +2. Can be stopped by a sheet of paper or by skin. Dangerous if ingested.

Ammeter: A device used to measure the amount of current flowing in a circuit. Ammeters are always located in series in the circuit.

Ampere: The SI unit of current. It is the number of coulombs flowing per second.

Amplitude: The maximum displacement of an oscillating particle or wave relative to its rest position. It is a measure of intensity of the oscillation and is directly proportional to the energy imparted into the oscillating system. The amplitude does not affect the period or the frequency.

Angle of incidence: The angle between a normal line (drawn perpendicular to the surface) and a ray of light striking that surface.

Angle of reflection: The angle between a normal line to the surface and a ray of light reflecting off of that surface.

Angle of refraction: The angle between a normal line to an optically transparent medium and a ray of light emerging into that medium.

Antinode: A point along a standing wave that has maximum amplitude or displacement. Antinodes are midway between the nodes.

Aphelion: The farthest distance a planet is away from the Sun during its orbit. The opposite of perihelion.

Apogee: The farthest approach in an orbit around any celestial body. The opposite of perigee.

Atmosphere: The unit of pressure at sea level is 1 atmosphere. It is equivalent to 101.3 kPa and to 14 psi.

Atomic mass unit (u): The mass of 1/12 of a carbon-12 atom. Protons and neutrons each have a mass of 1 atomic mass unit.

Atomic number: The number of protons in the nucleus of an atom.

Avogadro's number: The number of particles (atoms, molecules, ions, etc.) contained in one mole of that particle.

B

Battery: A chemical storage of charge that converts chemical energy into electrical energy. Charge flows from the positive terminal to the negative terminal through a completed circuit.

Beat frequency: The frequency at which destructive interference occurs between two sound waves of similar but not the same frequency. Can be determined by subtracting the two dissimilar frequencies.

Beta decay: The spontaneous emission of an electron from the nucleus of a radioactive atom. When beta decay occurs, the atomic number is affected but the mass number remains unchanged.

Beta particle: High-energy and high-speed electrons emitted from the nucleus of a radioactive atom. Can be stopped by aluminum foil.

C

Capacitance: The amount of charge Q that a capacitor can hold per volt V. Units of farads. Capacitance depends upon the area of the plates, the distance between the plates, and the dielectric constant of the material between the plates.

Capacitor: A pair of conducting plates used in circuits to store charge and electrical energy.

Carnot cycle: A thermodynamics term indicating a specific four-step process of an ideal gas in a cylinder's being alternately expanded and then compressed by a piston. It includes an isothermal expansion followed by an adiabatic expansion, an isothermal compression, and finally an adiabatic compression.

Cathode ray tube (CRT): A vacuum tube containing a source of electrons and a fluorescent screen. Electrostatic deflection plates and/or electromagnets are used to deflect the electron source to strike a particular section of the fluorescent screen.

Center of curvature: A point that is at a distance equal to the radius of curvature of a spherical mirror. The point is also equal to twice the focal length of a spherical mirror.

Centripetal acceleration: Acceleration acting on a mass that is moving along a curved path. This vector points in the direction the object is turning and is always perpendicular to the instantaneous tangential velocity of the turning object. As the object follows its curved path, the direction of the centripetal acceleration vector continually changes. For objects in uniform circular motion, the centripetal acceleration always points to the center of the circular path. In addition, uniform circular motion is a special case where centripetal acceleration has constant magnitude.

Coherent light: Light waves having the same frequency, wavelength, and phase.

Concave lens: See "Diverging lens."

Concave mirror: See "Converging mirror."

Conductor: A material that allows for the movement of electrical charges. Typically, conductors are metallic.

Conservation of energy: For an isolated system, the total energy may change forms but remains constant during the interactions within the system.

Conservation of mass-energy: During the conversion of mass into energy or energy into mass, the total amount of mass-energy must remain the same according to the equation $E = mc^2$.

Conservation of momentum: The total momentum of an isolated system will remain constant as long as no external forces act on the system.

Conservative force: A force in which the work done in moving a particle between two points is independent of the path taken.

Constructive interference: Two or more waves undergoing superposition such that their displacements add constructively.

Continuous spectrum: A continuous band of colors formed by the diffraction of white light into its component colors. The acronym ROYGBIV can be used to remember the order of the colors produced.

Converging lens: A lens that causes parallel rays of light to refract and converge at a focal point on the other side of the lens. Also known as a convex lens.

Converging mirror: A mirror that reflects light rays that arrive parallel to the optical axis so the waves converge at the focal point of the mirror. Concave mirrors are convergent mirrors.

Convex lens: See "Converging lens."

Convex mirror: See "Diverging mirror."

Coordinate system: Used to describe points in space known as a frame of reference. In two-dimensional space, two coordinate points are necessary to assign a specific location.

Coulomb: The SI unit of charge. An electron or a proton has a value of 1.60×10^{-19} coulombs.

Coulomb's law: The electrostatic force between two point charges is directly proportional to the product of their charges and inversely proportional to the square of the distance that separates them. The force can be attractive or repulsive depending upon the charges involved. Like charges repel; opposite charges attract.

Critical angle of incidence: The angle of incidence within a transparent medium such that the angle of refraction is 90° relative to a normal line drawn perpendicular to the surface. Can occur only when a light beam is moving from a more optically dense transparent medium to a less optically dense transparent medium. Any angle greater than the critical angle of incidence will result in total internal reflection.

Crystallography: See "X-ray diffraction."

Current: The amount of charge passing a given point in an electric circuit per second. Measured in amperes and is a scalar quantity.

D

Destructive interference: Two or more waves undergoing superposition such that their displacements add destructively.

Deuterium: An isotope of hydrogen consisting of 1 proton and 1 neutron in its nucleus.

Diffraction: The spreading out of a wave as it encounters the edge of a barrier or the edges of an opening (slit). The effect is most pronounced in narrow openings where the size of the opening approaches the wavelength of the wave passing through the opening. Diffraction is the result of the complex interference of the individual wavelets generated by every individual oscillating particle that makes up the larger aggregate wave.

Diffraction grating: A transparent or reflective surface with many thousands of closely spaced lines such that light is diffracted and creates a color spectrum.

Direct current: Current in an electrical circuit that flows in one direction continuously.

Dispersion: Light separated into its component colors when passed through a medium in which the index of refraction is slightly different for different wavelengths of light. An example of a dispersive medium is a glass prism.

Displacement: The change in position of an object from a beginning point to an ending point. A vector quantity. Displacement is independent of the path taken and is a straight-line distance from the beginning point to the ending point. Displacement can be zero if an object moves from point *A* and returns to point *A*.

Distance: A measure of the total length of a path taken by an object from a beginning point to an ending point. A scalar quantity. Distance is not independent of the path taken. Distance is not zero if an object moves from point *A* and returns to point *A*.

Diverging lens: A lens that causes parallel rays of light to diverge away from a focal point behind the lens. Also known as a concave lens.

Diverging mirror: A mirror that reflects light rays so the rays diverge from the focal point of the mirror. Convex mirrors are convergent mirrors.

Doppler effect: The perceived change in frequency by an observer of a moving source relative to the observer. The source frequency does not change. Only the frequency perceived by the observer changes. The perceived frequency increases as source and observer move toward one another. The perceived frequency decreases if they move away from one another. The perception of wavelength will also change, but the wavespeed will remain constant. The red-shift noticed in the spectrum of stars receding away from Earth is a result of the Doppler effect.

Dynamic friction: See "Kinetic friction."

E

Efficiency: The ratio of work output to energy input. For all machines, efficiency cannot reach or exceed 1 because some of the energy input is lost to the surroundings.

Elastic collision: A collision in which two objects rebound without sticking together in any way or generating any heat. Kinetic energy is conserved before and after an elastic collision.

Elastic potential energy: The stored energy of a spring. Work must first be done to stretch or compress the spring.

Electric field: A region in space that exerts a force on a charged particle. The force can be attractive or repulsive based upon the charge of the particle relative to the charge of the electric field.

Electric force: The force produced on a charged particle by another charged particle or an electric field.

Electrical potential energy: The energy of a point charge due to its location within an electric field.

Electromagnetic induction: The process of inducing (creating) an emf (induced potential difference) in a conducting loop. Electromagnetic induction occurs when there is a change in magnetic flux within a closed loop of conducting material.

Electromagnetic spectrum: The range of electromagnetic waves that all travel at the speed of light. From lowest frequency to highest, the range includes radio waves, microwaves, infrared waves, visible light, ultraviolet light, X-rays, and gamma rays. Waves with the lowest frequency transmit the least amount of energy.

Electromagnetic wave: Generated by oscillating electric charges, electromagnetic waves produce an interacting electric and magnetic field that carries energy proportional to its frequency and travels at the speed of light in a vacuum. A medium is not required for the transmission of an electromagnetic wave. However, upon encountering a medium, the velocity and wavelength of an electromagnetic wave are affected.

Electromotive force (emf): The potential difference (voltage) generated by either a battery or a changing magnetic field. The word "force" is misleading. It is not an actual force. It is a potential difference and is measured in volts.

Electron: A subatomic particle with a negative elementary electric charge, e, of -1.6×10^{-19} C. Orbits the nucleus of an atom.

Electron volt (eV): The amount of energy necessary to move 1 unit of elementary electric charge, e, through a potential difference of 1 volt. One electron volt equals 1.6×10^{-19} J.

Electroscope: A simple instrument used in detecting the presence of an electrostatic charge.

Elementary charge: The fundamental unit of charge, e, ascribed to a proton or to an electron. It has a value of 1.6×10^{-19} C. It is positive in the case of a proton and negative in the case of an electron.

Emission spectrum: A discrete series of colored lines, at particular frequencies, observed through a diffraction grating or spectroscope. The spectrum is produced when an atom is excited by heat, photon absorption, electric current, or atomic collision and then releases photons at discrete frequencies specific for that atom. Can be used as an atomic fingerprint to identify an atom. Atoms absorb radiation energy at specific frequencies, as is noticed in the absorption spectrum of stars.

Energy: The property of an object, or a system, that allows it to do work. A scalar quantity. Energy comes in several interchangeable forms, such as mechanical, electrical, chemical, thermal, and nuclear.

Energy level: Each atom has discrete levels of energy at which an electron may be found outside of the nucleus. Electrons may move from one level to another through excitation by heat, photon absorption, electric current, or atomic collision. Electrons may return to a lower level, or to their original energy level, by emitting a photon of a discrete frequency determined by the difference in the change in energy levels. Such photon frequencies make up the emission spectrum for that atom.

Entropy: The amount of disorder and randomness of a system. The second law of thermodynamics defines the role of entropy.

Equilibrium: Occurs when the sum of all external force vectors acting on an object is zero. Forces can occur, but their vector sum must be zero. An object in equilibrium will move at a constant velocity that can be either zero or a nonzero value. There is no net force and, therefore, no net acceleration for an object that is in equilibrium.

Escape velocity: The minimum initial velocity necessary so that an object will not return to the planet from which it was launched.

Excitation: The absorption of energy by an atom that causes the atom's electrons to rise to a higher energy level. The process can include but is not limited to adding heat, photon absorption, electric current, or atomic collision.

Excited state: The state of an atom in which its electrons are occupying a higher energy level.

F

Farad: The unit for capacitance. One farad is equal to 1 coulomb per volt.

Field: A region in space that is created by a mass, a magnet, or an electric charge. Another mass, magnet, or electric charge placed into this region will experience a force based on its relationship to the field.

First law of thermodynamics: A version of the law of conservation of energy but specifically for thermodynamic systems. It states that the change in internal energy for a system is equal to the sum of the heat and work supplied and/or removed from a system. The first law is expressed mathematically as $\Delta U = Q + W$.

Fission: The splitting of large atomic nuclei into smaller atomic nuclei. Fission is typically induced by bombarding large, unstable nuclei with neutrons. An example is U-235's being bombarded by a neutron and then splitting into Kr-92 and Ba-141 and releasing 3 free neutrons to continue the reaction with other U-235 atoms. Tremendous amounts of heat and some gamma radiation are also released.

Flux: The amount of magnetic field lines passing through a loop of wire at a given instant. The amount can be changed based on varying the strength of the magnetic field, varying the orientation of the loop of wire to the field, or varying the diameter of the wire in the magnetic field. Changes in flux induce a current in the loop of wire and produce an electromotive force (emf). This is known as electromagnetic induction. Static flux does not induce a current. Measured in units of T/m^2.

Focal length: The distance between the center of a lens and its focal point along the principal axis.

Focus: For a converging lens or mirror, the point where parallel rays converge. For an ellipse (such as planetary orbit), one of the two points within the ellipse for which the sum of the distances to any point on the ellipse is a constant.

Force: A push or a pull caused by an interaction that accelerates an object. A vector quantity. Measured in units of newtons.

Frequency: The number of oscillations per second. Can be found by taking the reciprocal of the period of oscillation. Measured in Hertz or in s^{-1}.

Friction: A force that acts in the opposite direction relative to the motion of an object sliding along a surface.

Fusion: The combining of smaller nuclei, such as hydrogen, into larger atoms, such as helium. The amount of activation energy required to induce a fusion reaction is greater than the amount of activation energy required to induce a fission reaction. The amount of energy released due to the mass defect in a fusion reaction is also greater than the amount of mass-defect energy released in a fission reaction.

G

Gamma radiation (gamma rays): High-energy photons emitted by naturally decaying radioisotopes and by fission reactions. Gamma rays are high-frequency, high-energy, electromagnetic waves and travel at the speed of light. They can penetrate most substances. Lead is a dense substance that can stop gamma radiation.

Gamma rays: See "Gamma radiation."

Gravitational field: A field that surrounds all objects having mass. The gravity field of one mass will create a force of gravity on all other masses.

Ground state: The lowest energy level for an electron in orbit around the nucleus of an atom.

H

Half-life: The amount of time required for half of the radioactive isotopes of an element to decay.

Heat: The quantity of thermal energy transferred from one system to another system. A scalar quantity. Measured in units of joules.

Heat capacity: The amount of heat, in joules, needed to change the temperature of a fixed quantity of a particular substance by 1 kelvin.

Heat engine: A device that converts thermal energy into other forms of energy.

Heat of fusion (latent heat of fusion): The heat energy needed to convert 1 kilogram of a substance from its solid form to its liquid form.

Heat of vaporization (latent heat of vaporization): The heat energy needed to convert 1 kilogram of a substance from its liquid form to its gaseous form. The heat of vaporization is always significantly larger than the heat of fusion. More energy is required for the phase change from liquid to gas than from solid to liquid.

Heat pump: A device that transfers heat opposite the natural direction. It moves heat from a region of low temperature to one of high temperature.

Heat reservoir (heat sink): Also known as a thermal reservoir. It is a source or recipient of heat from a thermodynamic system of gas. The heat capacity of a heat reservoir is very high such that its temperature will not change when in thermal contact with a system. Heat reservoirs are also known as heat sinks for their ability to absorb or deliver heat without affecting their own temperature. Pouring a cup of hot coffee into the ocean is an example of the ocean's acting as a heat reservoir (heat sink). The ocean's temperature will not change as a result.

Heat sink: See "Heat resevoir."

Hertz: The unit for frequency; equivalent to s^{-1}.

Hooke's law: The distance of extension or compression of a spring is directly proportional to the applied force.

I

Ideal gas: A collection of gas particles that behaves according to the kinetic-molecular theory for gas particles.

Impulse: The change in momentum. Impulse can also be found as the product of force and the time interval of the force acting on a mass or as the area underneath a force vs. time graph. A vector quantity. The units for impulse are newton-seconds or kg • m/s.

Index of refraction: A ratio of the speed of light in a vacuum to the speed of light in an optically transparent medium. Denser mediums cause light to move slower and the index of refraction to increase. The index of refraction for a medium is specific to different wavelengths of light and causes the dispersion of light in a prism.

Inelastic collision: A collision in which the objects undergoing the collision stick together. Some kinetic energy is transformed into heat and/or light during the collision and is therefore not conserved.

Inertia: Associated with an object's mass. It is the tendency of an object to resist the action of an applied force.

Instantaneous velocity: The velocity of an object at a specific instant of time. Can be determined by taking the slope of a tangent line on a displacement vs. time graph.

Insulator: A poor conductor of heat or electricity.

Interference: Two or more waves superimposing to form a resultant wave of greater or lesser displacement.

Internal energy: The total energy of a system. The change in internal energy is important in the study of thermodynamic systems of gases. In thermodynamic systems, all energies are held constant except thermal energy. The change in thermal energy is therefore equal to the change in internal energy for a thermodynamic system. A change in internal energy always includes a change in the temperature of the system.

Isobaric: A thermodynamic process in which the pressure of a gas remains the same.

Isochoric (isometric): A thermodynamic process in which the volume of a gas remains the same. No work is done on the gas or by the gas during an isochoric process.

Isometric: See "Isochoric."

Isotherm: A curve on a pressure vs. volume graph in which each point on the curve represents the same temperature.

Isothermal: A thermodynamic process in which the temperature of a gas remains the same and there is no change in internal energy. During an isothermal process, the amount of heat, Q, added to or removed from a gas is equal in quantity but opposite in sign to the amount of work, W, done to or by the gas ($Q = -W$). Following the equation of the first law of thermodynamics ($U = Q + W$), there will be no resulting change in the internal energy of the gas when the temperature remains constant.

Isotope: An atom with the same atomic number (number of protons) but a different atomic mass number (protons plus neutrons). The atom has the same number of protons but a different number of neutrons.

J

Joule: The SI unit of work and of all forms of energy. Can also be expressed as a newton-meter or as kg • m^2/s^2.

K

Kelvin: The SI unit of temperature. It is the absolute number of degrees above absolute zero where all molecular motion stops. Temperatures measured in degrees Celsius can be converted to kelvins by adding 273.

Kepler's first law: The orbital paths of all planets are elliptical with the Sun at one focus.

Kepler's second law: An imaginary line from the Sun to a planet will sweep out an equal area in an equal amount of time during the orbit of the planet about the Sun.

Kepler's third law: The square of the period of an orbiting body is proportional to the cube of the average orbital radius.

Kinetic energy: The amount of energy based on the mass of an object and its velocity. A scalar quantity. Measured in units of joules.

Kinetic friction (dynamic friction): The friction of one surface sliding over another.

Kinetic-molecular theory: A theory that assumes the following concepts for a system of gas particles are true. (1) The particles are much smaller than the distances between them. (2) The particles collide elastically. (3) The particles are in constant random motion. (4) The particles do not lose kinetic energy when colliding with each other or with the walls of a container. (5) The collection of gas particles has the same average kinetic energy at a given temperature.

L

Laser: An acronym standing for *L*ight *A*mplification by the *S*timulated *E*mission of *R*adiation. A beam of intense, coherent (all in the same phase), monochromatic (all having the same frequency) light.

Latent heat of fusion: See "Heat of fusion."

Latent heat of vaporization: See "Heat of vaporization."

Law of refraction: See "Snell's law."

Lenz's law: A current induced in a closed conducting loop, due to a changing flux within the loop, flows in a direction so that the newly created magnetic field associated with the induced current opposes the change.

Longitudinal wave: A wave in which the vibration of oscillating particles is in a direction parallel to the direction of wave propagation.

M

Magnetic field: A region surrounding a permanent magnet, a moving charged particle, or an electric current that induces a force in magnetic materials or moving charged particles.

Mass: The property of matter that quantifies the amount of inertia inherent in that matter. Can be determined by the ratio of applied force to its subsequent acceleration. Measured in units of kilograms.

Mass defect: Following nuclear reactions, the total amount of mass at the end of the reaction is slightly different than the total amount of mass before the reaction. The change in mass is known as the mass defect, Δm. It equals the energy either required as a reactant or released as a product during the reaction according to the formula $E = (\Delta m)c^2$. In radioactive decay, fission, and fusion, the mass is less at the end of the reaction, and the mass defect is a product.

Mass number: The total number of protons plus neutrons for an atom. It is also the mass of a single atom measured in atomic mass units. It is the molar mass, measured in kilograms per mole in physics.

Mole: See "Avogadro's number."

Momentum: The product of the mass and velocity of an object. A vector quantity. The change in momentum equals impulse. Momentum is measured in units of newton-seconds or kg • m/s.

N

Natural frequency: A frequency at which an object will vibrate. Natural frequency is distinct to the object. A forced vibration with a frequency equal to the natural frequency of an object will cause the object to resonate (vibrate).

Net force: The vector sum of all forces acting on a mass. If the net force is zero, the mass is in equilibrium and moves at a constant velocity with no acceleration.

Neutron: An electrically neutral nucleon that has a mass of 1 atomic mass unit (amu) and works with protons to produce the strong nuclear force to keep the nucleus of an atom intact. The atomic mass number of an atom is the sum of the number of protons and neutrons in its nucleus.

Newton: The SI unit for force. Can also be expressed as a kg • m/s^2.

Newton's first law of motion: A statement of inertia. An object in motion will stay in motion unless acted upon by an applied force. An object at rest will remain at rest unless acted upon by an applied force.

Newton's second law of motion: The acceleration, a, produced by a net force, F, on an object is directly proportional to the magnitude and direction of that net force and is inversely proportional to the mass, m, of the object. It is summarized by the equation $a = F/m$, most often expressed as $F = ma$.

Newton's third law of motion: For every action of a force by one object upon another, there is an equal force in the opposite direction.

Nonconservative force: A force, such as friction or air resistance, that reduces the amount of kinetic energy inherent in a mass after the nonconservative force has done work to the mass.

Normal force: The force on an object due to contact with a surface. According to Newton's third law, the normal force is equal in magnitude but opposite in direction to the force with which the object pushes against the surface.

Nuclear force: See "Strong nuclear force."

Nucleon: The name given to either a proton or a neutron when it resides in the nucleus of an atom.

O

Ohm: The SI unit of electric resistance. It is represented by the Greek letter Ω and is equal to 1 volt per amp.

Ohm's law: The resistance, R, in a circuit is directly proportional to the potential difference (voltage), V, applied across the circuit and is inversely proportional to the current, I, flowing through the circuit. It is summarized by the equation: $V = IR$.

P

Pascal: The SI unit of pressure. It is the amount of force per area and has units of newtons per meter squared.

Perigee: The closest approach in an orbit around any celestial body. The opposite of apogee.

Perihelion: The closest distance a planet is away from the Sun during its orbit. The opposite of aphelion.

Period: The time of one complete cycle (revolution, rotation, orbit, etc.), measured in seconds, for any type of event that repeats itself regularly, such as circular motion and simple harmonic motion.

Phase change: A physical change of matter from one phase (solid, liquid, or gas) to another phase. During a phase change, the temperature remains constant.

Photoelectric effect: The process by which electrons in the surface of a piece of metal can be ejected above a certain minimum (threshold) frequency of electromagnetic radiation (light) that is incident upon the surface of the metal. Increasing the frequency above the threshold will cause the electrons to leave with greater kinetic energy. Increasing the intensity (brightness) of the electromagnetic radiation (light) will induce more electrons to be ejected but will not affect their kinetic energy.

Photon: A quantum particle of electromagnetic energy that travels at the speed of light and has a discrete frequency. It is created by an electron's returning to its ground-state energy level.

Pitch: The frequency of a sound wave.

Planck's constant: A fundamental constant with a value of 6.63×10^{-34} kg • m^2/s. When Planck's constant is multiplied by the frequency of an electromagnetic wave, the amount of energy propagated by the wave can be found.

Polarization: The alignment of vibrations of a transverse wave of light so they lie in a single plane. Polarization of light is evidence of the transverse-wave behavior of light.

Polarized light: Light that has been polarized by passing it through a polarizing filter that allows only one alignment of light waves to pass. Partially polarized light can also be caused by the reflection of light off of a reflective surface such as glass, metal, or water.

Positron: The antimatter version of an electron. It has the same mass as an electron but has a positive charge.

Potential difference: The difference in electric potential between two points. Synonymous with voltage.

Potential energy: The energy associated with an object's position. Gravitational potential energy is related to an object's height in a gravity field. Elastic potential energy is related to the stretch of an elastic object, such as a spring. Electric potential energy is related to the position of a charge in an electric field.

Power: The rate at which work is done or energy is transformed. A scalar quantity. It is equal to the ratio of work done to the time needed to complete the work. Measured in units of watts or in joules per second.

Pressure: The force per unit area. A scalar quantity. Measured in units of pascals or in newtons per meter squared.

Prism: An optically transparent and triangularly shaped medium that takes advantage of the differing indexes of refraction for wavelengths of light and of Snell's law to disperse the light into a continuous spectrum.

Proton: A positively charged nucleon that has a mass of 1 atomic mass unit (amu). It works with neutrons to produce the strong nuclear force to keep the nucleus of an atom intact. The atomic number of an atom is equal to the number of protons in its nucleus.

Pulse: A sudden fluctuation in a medium that causes a wave of energy to propagate.

R

Radioactive decay: The spontaneous change of a radioactive nucleus into other atomic particles including alpha particles, beta particles, and gamma radiation, and/or the transmutation to other atoms.

Ray: A thin beam of light. A straight line used to illustrate the direction of travel for a light wave.

Real image: An image formed by the convergence of light rays by a converging lens or mirror. Real images can be projected onto a screen and are in focus at only one point.

Red-shift: Light observed from an object that is moving away will display an increased wavelength and its spectrum will shift toward the red end of the spectrum. Red-shift is a result of the Doppler effect. Red-shift is evidence of an expanding universe because light from all distant stars visible from Earth has so far shown a red-shift.

Refraction: The bending of light due to a change in optical density and index of refraction as light changes mediums. Caused by the change in the velocity of light as it enters the medium. Refraction can occur only if a light ray strikes the surface at a nonzero angle of incidence relative to a normal line to the surface of the medium. A zero angle of incidence will result in a reduction in velocity but no refraction (bending).

Resistance: The ratio of potential difference (voltage) across a circuit to the current flowing in the circuit. A scalar quantity. Measured in units of ohms.

Resonance: A dramatic increase in the amplitude of a vibrating object when the frequency of forced vibrations matches the natural frequency of the object.

Revolution: The motion of an object turning around another object. An example is the yearly revolution of Earth about the Sun.

Rotation: The spinning motion of an object about its own axis. An example is the daily motion of Earth.

S

Scalar: A physical quantity that has magnitude but no specific direction. Can be positive or negative in value. Examples include but are not limited to speed, distance, energy, work, heat, mass, pressure, voltage, current, resistance, and power.

Second law of thermodynamics: Heat cannot flow spontaneously from a colder source to a hotter source. Heat can flow spontaneously from a hotter source to a colder source. Also the entropy of the universe is always increasing. It is therefore not possible to convert a given quantity of heat energy completely into useful work energy—some energy will be lost to the surroundings. The second law of thermodynamics dictates that perpetual-motion machines are not possible.

Snell's law: Also known as the law of refraction. The ratio of the sine of the angle of refraction, θ_2, to the sine of the angle of incidence, θ_1, is equal to the ratio between the indexes of refraction of the medium of incidence, n_1, and the medium of refraction, n_2. $n_1 \sin \theta_1 = n_2 \sin \theta_2$.

Solenoid: A coil of wire in which a current may be run to create a magnetic field. Conversely, a current may be created in the coil of wire by changing the flux of a magnetic field inside of the solenoid.

Specific-heat capacity: The amount of energy needed to raise the temperature of 1 kilogram of a specific substance 1 degree Celsius.

Spectroscope: A device used to disperse an incoming light source into its light spectrum. Often used to analyze light from distant stars or heated elements.

Spectrum: The range of wavelengths and frequencies that comprise all types of electromagnetic radiation. The spectrum of white light always follows the same pattern and can be remembered with the acronym ROYGBIV.

Speed: The distance an object moves in a specific amount of time. A scalar quantity. Measured in units of meters per second.

Spring constant: The force needed to stretch or compress a spring by a specific length. It is always constant for a given spring but may vary from spring to spring. Measured in units of newtons per meter.

Standard pressure: Air pressure at sea level. Equal to 1 atmosphere or to 101.3 kilopascals.

Standing wave: A stationary wave pattern that appears to be moving. The medium of the wave remains stationary while the energy of the wave moves through the medium and reflects back through the medium. It is the result of the interference between an incident and a reflected wave.

Static friction: A force of friction that prevents two objects from sliding past one another. Once static friction is overcome by adding a force sufficient to start an object moving, the object will experience kinetic (dynamic) friction.

Strong nuclear force: The force responsible for holding neutrons and protons together in the nucleus of an atom and overcoming the electrostatic repulsive force among protons. It is stronger than either gravity or electromagnetic forces. However, it works only at very close distances, such as those found among nucleons in the nucleus of an atom.

Superposition: The sum of the amplitudes of two or more waves overlapping.

T

Temperature: The measure of the relative feeling of hot or cold of an object. Also a measure of the average kinetic energy of molecules in a gas. A scalar quantity. Heat energy will spontaneously flow from an object with a greater temperature to one with a lesser temperature.

The units for temperature are in kelvins. To convert to kelvins from degrees Celsius, add 273 to the Celsius value.

Tesla: The SI unit of strength of a magnetic field. Can also be expressed as a $kg/C \cdot s$.

Test charge: A small, positively charged particle with negligible mass that is used to test for the presence of an electric field.

Thermal equilibrium: When two objects that are in contact with each other reach the same temperature and have no net heat flow between them.

Threshold frequency: The minimum frequency of an electromagnetic wave necessary to strike the surface of a piece of metal and cause the ejection of electrons as per the photoelectric effect.

Total internal reflection: The complete reflection of light at a boundary. When light strikes the interface from a more optically dense medium to a less optically dense medium at or above the critical angle of incidence, the angle of refraction is then 90 degrees or greater. So, the light is reflected back into the more optically dense medium.

Total mechanical energy: The sum of kinetic and potential energies in a mechanical system. Total mechanical energy is conserved unless a nonconservative force, such as heat, light, or sound loss, acts upon the system to reduce it.

Transmutation: The process by which one atom can become another atom by the addition or loss of protons in its nucleus. Alpha decay, beta decay, fission, and fusion are all examples of transmutation.

Transverse wave: A wave with oscillations that are at right angles to the direction of propagation. Waves on a string that has been shaken are an example of transverse waves. Electromagnetic waves are transverse waves.

U

Uniform circular motion: An object in circular motion that has a constant speed. The magnitude of the velocity of the object also remains constant. However, since the object is continually changing direction, the velocity itself is not constant. Velocity is a vector quantity. So, a change in velocity can come from a change in magnitude, a change in direction, or both.

Uniform motion: Straight-line motion at a constant speed.

Universal law of gravitation: The gravitational force between any two masses is directly proportional to the product of their masses and is inversely proportional to the square of the distance that separates them.

V

Vector: A physical quantity that has magnitude and a specific direction. It can be represented by an arrow on a Cartesian coordinate system. The magnitude of a vector is always positive. The direction of a vector along an axis can be indicated as negative. Examples include but are not limited to velocity, acceleration, displacement, force, weight, impulse, and momentum.

Velocity: The speed of an object in a specific direction. A vector quantity. Measured in units of meters per second.

Virtual image: An image formed by either refraction or reflection that can be seen by an observer but cannot be projected onto a screen. Virtual images can be formed from either converging or diverging lenses and mirrors.

Visible light: The range of the electromagnetic spectrum that can be perceived by the human eye as colors. The range of the colors can be remembered with the acronym ROYGBIV. The wavelength range of colors is from about 700 nm for red light to about 400 nm for blue light. Blue light has the shortest wavelength and highest frequency of visible light.

Volt (potential difference): The SI unit of potential difference. Can also be expressed as a J/C.

Voltage: Electric potential measured in volts.

Voltmeter: A device used to measure the voltage across a circuit or portions of a circuit. It is always placed in parallel with the circuit.

W

Watt: The SI unit of power. It is equivalent to the amount of energy in joules that can be expended in a second. Can also be expressed as a J/s.

Wavelength: The distance between two successive points in phase along a wave. Usually measured in meters. For electromagnetic waves, the wavelength may be measured in nanometers.

Weak nuclear force: Along with gravity, electromagnetism, and the strong nuclear force, it is one of the fundamental forces in nature. The weak nuclear force is responsible for the radioactive decay of subatomic particles in processes such as beta decay.

Weight: The force of gravity on an object in a gravitational field. A vector quantity. Measured in units of newtons.

Work: A measure of the amount of change in mechanical energy. Can be determined by multiplying the force on an object by the distance traveled while the force is being applied. Can also be determined by the area under a force vs. distance graph. A scalar quantity. Measured in units of joules.

Work function: The minimum amount of energy needed to eject an electron from the surface of a piece of metal as per the photoelectric effect. The work function divided by Planck's constant will reveal the threshold frequency needed to eject an electron. Measured in units of joules.

X

X-ray diffraction (crystallography): A method of determining the arrangement of atoms within a crystal by passing X-rays through the crystal and exposing a photosensitive plate.

Y

Young's double-slit experiment: Demonstrated the wavelike nature of light by passing coherent light through two small openings and revealing a pattern of constructive and destructive interference. These patterns could occur only if light were a wave as opposed to being a particle.

Index

A

Absolute zero, 365, 443

Absorption, 356, 400

Acceleration
    centripetal, 143–143
    in kinematics, 75–76
    radial, 144
    simple harmonic motion, 297
    time interval charting, 52
    uniform, 142

Action force, 116–117

Adiabatic process, 384

Agent, 110

Alpha decay, 421

Alpha particle, 417

Amplitude, 290, 306

Angle of reflection, 324

Angular displacement, 145, 146

Angular velocity, 146

Antinodes, 315

Applied force, 111

Area, 50–51

Astrophysics, 444

Atom, planetary model, 444

atomic and quantum phenomena,
    development of, 395–399

Atomic mass units, 416

Atomic number, 416

Atomic theory
    Bohr's contribution, 399
    development of, 395–399
    Einstein's contribution, 398
    electron volt, 398
    emissions of light, 400–401
    energy and work, 403
    energy levels, 402
    energy-level transition, 399–400
    gold-foil experiment, 396–397
    photoelectric effect, 403–406
    Planck's contribution, 397–398
    plum-pudding model, 395–396
    Rutherford's contribution, 396–397
    spectroscopy, 397
    Thompson's contribution, 395–396
    variables used in, 395

Atwood, George, 126

Atwood machine, 126–129

B

Balmer series, emission of light, 401

Battery, 247–248

Beats, 316–317

Beta decay, 422

Beta particle, 417–418

Bohr, Niels, 399, 444

Boltzmann's constant, 367

Brahe, Tycho, 201

C

Capacitance, 236

Capacitors, 236–238

Carbon-12, 416, 418

Carbon-14, 418, 423

Celsius scale, 365

Centrifugal force, 147

Centripetal acceleration, 143–144

Centripetal force, 145

Chain reaction, 424

Chaos theory, 444

Charged objects, 210

Charged plates versus gravity
    field, 212

Charges
    defined, 210
    in electric fields, 212
    motion of, 215
    point charges, 216–220
    uniform electric fields, 212–215
Charging methods, 210–211
Circuits
    components of a, 247–249
    DC, 247–249
    heat and power dissipation, 257
    parallel circuits, 252–254
    power dissipation in, 256–257
    series circuits, 249–252
    series-parallel circuits, 255–257
    summary of, 221
    variables used in, 247
Circular motion
    angular displacement in, 145–146
    angular velocity, 146
    centrifugal force, 147
    dynamics in, 145
    in elliptical orbits, 201–202
    horizontal, 148
    orbital, 198–202
    summary of, 202
    uniform, 141–144
    variables used in, 141
    vertical, 149–150
Circular orbits, 198–201
Coefficient of friction, 113
Coefficient of kinetic friction, 113
Coefficient of static friction, 113
Coils, 277
Collisions, 187–189
Color, in physical optics, 356
Compound-body problems, 123–125
Concave lens, 334–335
Concave mirrors, 335
Conduction, 211, 369
Conductor, 210–211
Conservation of charge, 210
Conservation of energy
    with conservative forces, 170–171
    conservative forces, 169
    nonconservative forces, 169, 174
    oscillations in, 297

overview, 168–169
    and pendulums, 172
    and springs, 173
Conservation of momentum, 187–189
Conservative forces, 169, 170–171
Constant force, 162
Constructive interference, 314, 348–349
Contemporary physics, 444–445
Convection, 369, 370
Converging lenses, 330–332
Converging mirrors, 335–337
Convex lens, 331–333
Convex mirrors, 335
Cooling curve, and heating, 372–373
Coordinate system, 59–60, 267–268
Coordinate-axis system, 68
Coulomb, Charles-Augustin De, 442
Coulomb's law, 220, 442
Current, 278
Current-carrying wires, 267–268, 273–275
Cycle, 142

D

Dark energy, 444
Dark matter, 444–445
DC circuit, 247–249
Decay rate, 423
Dependent variable, 49
Destructive interference, 314, 349
Deuterium, 425
Diffraction, 347–348
Diffuse reflection, 324
Dispersion, in physical optics, 356
Displacement
    area, 52
    and distance, 73–74
    versus force, 164–165
    horizontal, 62
    true velocity and, 93
    vectors, 160
Distance, and displacement, 73–74
Diverging lenses, 334–335
Diverging mirrors, 337–338
Domains, 266
Doppler, Christian, 442
Doppler effect, 311–313, 442
Double-slit experiment (Young), 349–352

Dynamic equilibrium, 116
Dynamics
    in circular motion, 145
    compound-body problems, 123–125
    force, 110–112, 117–119
    force diagrams, 114–115
    friction, 112–113
    inclines, 120–121
    inertia, 110
    kinematics and, 122
    Newton's Laws of Motion, 115–117
    one-dimension variables, 109
    pulley problems, 126–129
    springs, 114
    tension, 114
    See also Force

E

Efficiency of heat engines, 386–387
Einstein, Albert, 398, 443
Elastic collisions, 187–188
Elastic potential energy, 158
Electric fields
    charge, 210
    imaginary positive test charge, 217
    magnitude of point charges of, 216–218
    of point charges, 216–220
    summary of, 221
    superposition, 217–219
    variables used in, 209
    See also Uniform electric fields
Electric force, in uniform electric fields, 213
Electric potential
    capacitance, 236
    energy in, 231–235
    equipotential lines, 228
    motion of charges, 233
    of uniform fields, 228–229
    variables used in, 227
    work of electricity, 233–234
Electricity, work, 233–234
Electricity and magnetism, 442
Electromagnetic fields, 442
Electromagnetic induction
    contributions for, 442
    description of, 275
    emf, 277–278

    induced currents, 278
    Lenz's law, 279
    magnetic flux, 275–277
    motional emf, 275–277
    transformers, 279–280
Electromagnetic spectrum, 310
Electromagnetic waves, 308–309, 353, 442
Electromagnets, 269
Electron volt, 398
Electrostatic constant, 209, 217, 220
Elliptical orbit, 201
emf, 277–278
Emission of light, 400–401
Energy
    in collisions, 187
    conservation of, 168–173, 234–235
    elastic potential, 158
    explosions, 189
    gravitational potential, 158
    ionization, 403
    kinetic, 158
    levels, 399, 402
    mechanical, 157
    model, 382–384
    nonconservative forces, 174
    in oscillations, 296–297
    pendulums and, 172
    of photons, 398
    potential of, 158
    springs and, 173
    summary of, 175
    total mechanical, 159–160
    transfers in thermodynamics, 382–383
    variables used in, 157
Energy method, for work, 163
Energy-level transitions, 399–400
Engine, efficiency of, 386–387
Entropy, heat, 387–388
Equilibrium, 115–116
Equilibrium position, 291
Equipotential lines, 228–229
Equivalent resistance, 249
Exam
    content breakdown, 2–3
    diagnostic test, 11–35
    format, 3–4
    overview, 1–2

scoring, 37–38

strategies, 4–5

Explosions, 189

External forces, 123

F

Fahrenheit scale, 365

False centrifugal force, 147

Faraday, Michael, 277, 442

Faraday's law, 277–278

First harmonic, 315

First law of motion, 441

First law of planetary motion, 442

First law of thermodynamics, 384–385

First postulate of special relativity, 433

Fission, 424

Fixed magnets, 266, 274

Force

applied, 111

caused by uniform fields, 270–273

centripetal, 145

commonly, 111–113

components of, 119

constant, 162

on current-carrying wires, 274

diagrams, 114–115

and direction, 110–111

versus displacement, 164–165

friction, 112

of gravitational attraction, 196

of gravity, 111

on an incline, 120–121

net, 110, 122

normal, 112

one-dimension variables, 109

simple harmonic motion, 297

solving for, 117

superposition of, 220

tension, 114

in work, 160–161

See also Dynamics

Force-time graph, 185–186

Free-body diagram, 112, 114–115

Frequency, 142, 289–290, 315–316

Friction, 112–114, 211

Fundamental frequency, 315

Fusion, 425

G

Galilee, Galileo, 441

Gamma rays, 308, 418, 422–423

Gases, ideal, 366–367

Geometric optics

concave lens, 334–335

converging lenses, 330–333

converging mirrors, 335–337

convex lens, 331–332

diverging lenses, 334–335

diverging mirrors, 337–338

index of refraction, 325–326

pinhole camera, 328–329

plane mirror, 324–325

ray model of light, 324

reflection, 324

refraction, 325–327

Snell's law, 326–327

thin lenses, 329–330

total internal reflection, 327–328

variables used in, 323

See also Physical optics

Gold-foil experiment, 396–397, 444

Graph(s)

force versus displacement, 164–165

identifying, 52–53

interpreting, 51–52

for physics equation illustration, 52–53

simple harmonic motion, 295

variables, 49–51

velocity vs time, 51

Gravitational field, 111, 197

Gravitational potential energy, 158

Gravity

acceleration of, 76

circular orbits, 198–201

force of, 111

gravitational field, 197

initial motion, 110–111

inverse-square law, 196

Kepler's laws, 201–202

Newton's third law, 196

orbital velocities, 199–200

surface, 197–198

universal gravity, 196
variables used in, 195
waves, 445
work through, 162–163

H
Half-life, 423
Harmonics, 315–316
Heat
    and cooling curve, 372–373
    engines, 386–387
    of fusion, 372
    latent, 371–372
    overview, 369
    and power dissipation, 257
    pump, 388
    reservoir, 383
    in thermodynamics, 383–384
    of transformation, 371–372
    of vaporization, 372
Heat engines, efficiency of, 386–387
Heat transfer, 369
Heating and cooling, 370–371
Hooke's law, 114, 158, 291
Horizontal circular motion, 148
Horizontal displacement, 62
Hubble, Edwin, 444
Huygen, Christian, 348
Huygens' principle, 348

I
Ideal gas law, 368
Ideal gases, 366–367
Images, 328–329
Imaginary positive test charge, 217
Impulse, 184–186
Impulse-momentum theorem, 184–185
Incline(s), solving for, 120–121
Index of refraction, 325–326
Induced currents, 278
Induction, charging method, 211
Inelastic collisions, 188–189
Inertia, 110
Inertial reference frames, 433
Infrared light waves, 308
Insulator, 210–211
Interference of light, 348–349

Interferometer, 443
Internal forces, 123
Inverse-square law, 196, 217
Ion, 210
Ionization energy, 403
Isobaric process, 383
Isochoric process, 383
Isolated system, 388
Isometric process, 383
Isothermal process, 382, 385
Isotopes, 418

J
Joule, James, 443
Joule's law, 257

K
Kelvin, Lord, 443
Kepler, Johannes, 201, 442
Kepler's laws, 201–202
Kinematics
    dynamics, 122
    equations, 92–93
    equations with gravity, 96
    in one-dimensional, 73–84
    in two-dimensional, 91–99
    in uniform electric fields, 213–214
    See also One-dimensional kinematics;
        Two-dimensional kinematics
Kinetic energy, 158, 169, 187
Kinetic friction, 113

L
Laser Interferometer Gravitation Wave
        Observatory (LIGO) project, 445
Latent heat, 371–372
Law of conservation of energy, 364
Law of gravity, 441
Law of inertia, 115–116
Law of motion, 115–117
Law of reflection, 324
Law of thermodynamics, 384–386, 388–389
Length contraction, 435
Lenses, 329–333
Lenz, Heinrich, 442
Lenz's law, 279, 442
Light, 348–353

Light waves, 308–309, 313, 324
Lightbulb(s), 249, 258
Linear expansion, 365–366
Linear momentum, 183–184
Longitudinal waves, 306
Lyman series, emission of light, 401

M

Magnetic field
    current-carrying wires, 267–269
    versus electric field, 266
    magnitude of the, 268–269
    right-hand rule, 267–268
    two current-carrying wires, 275
    uniform, 266, 270
    visualizing the field, 266–267
Magnetic flux, 275–276
Magnetism
    electromagnetic induction, 275–278
    fixed magnets, 274
    permanent or fixed magnets, 266–267
    solenoid and electromagnets, 269
    variables used in, 265
    work done by, 273
Magnets, 266–267
Magnification, 329
Magnitude, 62–63, 213
Mass, 110
Mass defect, 419
Mass-energy equivalence, 419–420, 443
Maxwell, James Clerk, 442
Mechanical energy, 157, 159–160, 364–365, 383
Mechanical wave, 308
Medium, 307–308
Michelson, Albert, 443
Michelson-Morley experiment, 443
Microprocessor, 445
Microwaves, 308
Millikan, Robert, 215
Millikan Oil Drop, 214
Mirrors, 324–325, 335–338
Modified Atwood machine, 128–129
Momentum, 183–184
Momentum effects, 435–436
Monochromatic light, 349
Morley, Edward, 443

Motion
    analyzing, 52
    caused by uniform fields, 270–273
    horizontal, 59
    independence of, 92–93
    in relation to force, 110–111
Motional emf, 278
Moving charges, current-carrying wires, 273

N

Net force, 110, 122
Neutrino, 417
Newton, Sir Isaac, 109, 196, 441
Newtonian mechanics, 441
Newton's law of universal gravitation, 196–198, 220
Newton's laws of motion
    first law, 115–116
    second law, 116
    third law, 111, 116–117, 196
Nodes, 315
Nonconservative forces, 169, 174
Normal force, 112
Nuclear reactions
    alpha decay, 421
    beta decay, 421
    fission, 424, 425
    gamma rays, 422–423
    identifying, 425
    mass-energy equivalence, 419–420
    quarks, 415
    radioactive decay, 420
    strong force, 419
    subatomic particles, 416–418
Nucleons, 415

O

Ohm, Georg Simon, 442
Ohmic resistors, 248
Ohm's law, 248, 442
One-dimensional forces, 109
One-dimensional kinematics
    acceleration, 75–76
    displacement and distance, 73–74
    equations, 79–81
    graphs, 82–83
    sign conventions, 77–79

variables, 73
velocity and speed, 74–75
See also Kinematics
Optics
See Geometric optics; Physical optics
Orbital velocity, 198–201
Orbiting satellite, 201
Ordinary collisions, 188
Oscillations
of an electromagnetic field, 310
energy in, 296–297
of pendulums, 295
perpendicular, 353
of spring, 291–294
trends in, 296–297

P

Parallel circuits, 252–254
Parallel force, 160
Parallelogram method, 67
Pendulums, 172, 295
Perfectly inelastic collisions, 188–189
Period, 142, 289, 293–294
Permanent magnets, 266
Permittivity of free space, 236
Perpendicular oscillations, 353
Photocell, 404
Photoelectric effect, 403–406, 443
Photoelectron, 400
Physical optics
absorption and reflection, 356
color, 356
dispersion, 356
double-slit experiment, 349–352
interference of light, 348–349
polarizing filter, 354–355
scattering, 356–357
single-slit experiment, 352–353
variables used in, 347
See also Geometric optics
Pinhole camera, 328–329
Planck, Max, 397, 443
Planck's constant, 398
Plane mirror, 324–325
Plane of polarization, 353
Planetary model of the atom, 444
Plasma, 425

Plum-pudding model, 395–396
Point charges, 216, 230–231
Polarization of light, 353–355
Polarizing filter, 354–355
Potential
See Electric potential
Potential difference, 233
Potential energy, 158
Power, 167–168, 257, 441
Practice exercises
atomic and quantum phenomena, 410–411
circuit elements and DC circuits, 260–262
circular motion, 152–153
conventions and graphing, 55–56
dynamics, 131–135
electric fields, 222–224
electric potential, 241–243
energy, work, and power, 177–179
geometric optics, 341–343
gravity, 203–205
historical figures and contemporary physics, 446
kinematics in two dimensions, 102–104
magnetism, 283–286
momentum and impulse, 191–192
nuclear reactions, 428–430
one-dimension kinematics, 86–88
physical optics, 358–360
relativity, 437–438
simple harmonic motion, 300–302
thermal properties, 376–378
thermodynamics, 391–392
two-dimensional kinematics, 102–104
waves, 319–321
Pressure, 367–368
Principle of inertia, 441
Prism, 356
Projectile motion, 62–63, 97–98
Projectile velocity vectors, 92–93
Pulley problems, 126–129
Pythagorean theorem, 65

Q

Quantum theory, 443
Quarks, 415

R

Radial acceleration, 144
Radiation, 369
Radio waves, 308
Radioactive decay, 420
Radioactive particles, 422
Raisin cake model, 395–396
Rate of heat transfer, 370
Ray model, 324
Reaction force, 116–117
Real images, 329
Reflected ray, 324
Reflection, 324–325, 356
Refraction, 325–327
Relative-velocity, 93–96
Relativity
    inertial reference frames, 433
    momentum effects, 435–436
    postulates of special relativity, 433
    special theory of, 433–434
    time dilation, 434–435
Resistance, 248
Resistor(s), 248
Restorative force, 291
Restoring force, 114
Right triangles, 64–65
Right-hand rule, 267, 272
Rutherford, Ernest, 396–397, 444

S

Scalars, 60, 74–76
Scattering, in physical optics, 356–357
Scores, improving, 38
Second law of motion, 441
Second law of planetary motion, 442
Second law of thermodynamics, 388–389
Second postulate of special relativity, 433
Second-harmonic waveforms, 316
Semiconductor, 445
Series circuits, 249–252
Series-parallel circuits, 255–257
Sign conventions, 77–79
Simple harmonic motion
    graphical representation, 295
    oscillations in, 291–294
    period of a spring in, 293–294

spring oscillations, 291–294
    terms related to, 289–290
Single-slit interference, 352–353
Sinusoidal waveform, 324
Slit, 348
Slope(s), 49–50
Snell's law, 326–327
Solenoid(s), 269
Solving
    force problems, 117
    incline problems, 120–121
    x- and y-force problems, 118
Sonic boom, 313
Sound, 311–313
Sound wave, 308
Special theory of relativity, 433–434, 443
Specific heat, 370–371
Spectroscopy, 397
Specular reflection, 324
Speed
    in circling object, 142
    Doppler effect and, 313
    and velocity, 74–75
    of a wave, 307
    work-kinetic energy, 166
Spherical mirrors, 335
Spherical point charge, 217
Spring(s)
    conservation of energy, 13
    constant, 114, 291–292
    force, 114
    oscillations, 291–294
    period determination of a, 293–294
    work through a, 163–164
Standard model, 415
Standing waves, 314–315
Static equilibrium, 116
Static friction, 113
Stationary object, 313
String theory, 445
Strong force, 419
Subatomic particles, 416–418
Subsonic object, 313
Superconductivity, 445
Superposition, 314
Supersonic, 313

Surface gravity, 197–198
Switch, 249

T

Tangential orbital velocities, 199–200
Tangential velocity, 143
Temperature, 365
Tension, 114
Thermal conductivity, 370
Thermal energy, 364–365
Thermal energy transfer, 383
Thermal expansion, 365–366
Thermal properties
    conduction, 369
    heat and heat transfer, 369–370
    heating and cooling, 370–372
    ideal gases, 366–367
    linear expansion, 365–366
    pressure, 367–368
    radiation, 370
    temperature, 365
    thermal conductivity, 370
    thermal energy, 364–365
    thermal expansion, 365–366
    thermal systems, 363
    variables used in, 364
Thermal systems, 363
Thermodynamics
    energy model, 384
    energy transfers in, 382–383
    entropy, 387–388
    first law of, 384–385
    heat, 383–384
    internal energy, 381–382
    second law of, 388–389
    variables used in, 381
    work through, 382–383
Third law of planetary motion, 441, 442
Thomson, J. J., 395, 443
Time dilation, 434–435
Tip-to-tail method, 65–66
Total internal reflection, 327–328
Total mechanical energy, 159–160
Total momentum, 184
Transformers, 279–280
Transistor, 445
Transmutation, 415

Transverse waves, 306
Triangles, right, 64–65
True displacement, 93
True velocity, 93–96
Two-dimensional kinematics
    independence of motion, 92
    projectile motion, 96–97
    relative velocity equations, 94
    true velocity and displacement, 93–95
    variables used in, 91
    velocity vectors, 92
    See also Kinematics

U

Ultraviolet light, 308
Uniform acceleration, 75–76
Uniform circular motion, 141–144, 167
Uniform electric fields
    electric force in, 213
    kinematics in, 213–214
    magnitude, 213
    Millikan Oil Drop, 214
    motion of charge in, 215
    potential of, 228–229
    visualizing, 212
    See also Electric fields
Uniform magnetic fields, 266–267, 270–273
Units, 47–48, 52
Universal gravity, 196

V

Variable(s)
    dependent, 49
    graphing, 49–51
    identifying, 77
    one-dimension forces, 109
    one-dimension kinematics, 73
    sign conventions, 77–79
    two-dimensional kinematics, 91
Vector method, for work, 163
Vectors
    adding, 65–66
    commonly used, 61–63
    components, 64–65
    coordinate system, 59–60
    direction, 61–62
    displacement and distance, 74

mathematics, 63
missing, 67–68
parallelogram method, 67
in projectile motion, 62–63
projectile motion, 99
right triangles, 64–65
scalars, 60
tip-to-tail method, 65–66
velocity, 74–75, 92
Velocity
angular, 146
orbital, 200
in relation to force, 110–111
relative, 93–96
and speed, 74–75
tangential, 143, 199
true, 93
Vertical circular motion, 149–150
Vertical compound bodies,
123–124
Virtual images, 329
Visible wave, 308

W
Watt, James, 441
Wave front model, 311
Wavelength, 306
Wave(s)
amplitude on a, 308
beats, 316–317
components of a, 306
doppler effect, 311–313
electromagnetic, 308–309
electromagnetic spectrum, 310
equation, 307
first harmonic, 315
fronts, 347–348
harmonic, 316

light, 308–309
mechanical, 308
second-harmonic, 316
sound, 308
speed, 307
standing, 314–315
superposition of, 314
traveling, 305–306
variables used in, 305
Weight, force of gravity, 111
Work
defined, 160
of electricity, 233–234
force and displacement, 160–161
function, 403
by gravity, 162–163
by magnetic force, 273
by a spring, 163–164
in thermodynamics, 382–383
in uniform circular motion, 167
using energy method, 163
using vector method, 163
Work-kinetic energy theorem, 166, 232

X
x-force problems, 118
x-rays, 308–310

Y
y-force problems, 118
Young, Thomas, 349, 442
Young's double-slit experiment, 349–352,
357, 442

Z
Zero value, 77
Zero velocity, 115
Zero-point reference, 169